Landscapes of Minnesota

Landscapes of MINNESOTA

A GEOGRAPHY

JOHN FRASER HART AND SUSY SVATEK ZIEGLER

With the cartographic collaboration of Mark B. Lindberg

MINNESOTA HISTORICAL SOCIETY PRESS

www.mhspress.org

The Minnesota Historical Society Press is a member of the Association of American University Presses.

Manufactured in China
10 9 8 7 6 5 4 3 2 1

♾ The paper used in this publication meets the minimum requirements of the American National Standard for Information Sciences—Permanence for Printed Library Materials, ANSI Z39.48-1984.

Front photograph credits: ice fishing at Mille Lacs, page ii, © Explore Minnesota Tourism; wind turbines near Hendricks, page vi, *Minneapolis Star Tribune,* May 31, 1998 (David Brewster), by permission of the *Star Tribune*; Hibbing Taconite Mine, page viii, courtesy Cleveland-Cliffs; Minneapolis milling district and St. Anthony Falls, page x, by the authors.

Book and cover design: Percolator Graphic Design, Minneapolis

International Standard Book Number
ISBN 13: 978–0-87351-591-7 (paper);
978-0-87351-611-2 (cloth)
ISBN 10: 0–87351-591-9 (paper);
0-87351-611-7 (cloth)

Library of Congress
Cataloging-in-Publication Data
Hart, John Fraser.
Landscapes of Minnesota : a geography /
John Fraser Hart and Susy Svatek Ziegler.
p. cm.
Includes bibliographical references and index.
ISBN-13: 978-0-87351-611-2 (cloth : alk. paper)
ISBN-10: 0-87351-611-7 (cloth : alk. paper)
ISBN-13: 978-0-87351-591-7 (pbk. : alk. paper)
ISBN-10: 0-87351-591-9 (pbk. : alk. paper)
1. Minnesota—Geography. 2. Minnesota—Environmental conditions. 3. Human geography—Minnesota. I. Ziegler, Susy S. II. Title.
F606.8.H37 2008
917.76'02—dc22
2007024116

Dedicated to the memory of our illustrious predecessors

Ralph H. Brown
Richard Hartshorne
John R. Borchert

CONTENTS

Acknowledgments ... ix
Introduction ... 1
1 Battleground of the Elements ... 4
2 The Shape of the Land ... 22
3 Evolving Ecosystems ... 42
4 Indians, Voyageurs, and Croupiers ... 56
5 Dividing the Land ... 76
6 The First Great Industry ... 88
7 Settlers Move In ... 104
8 From Wheat to Dairy Farming ... 120
9 The Dairy Belt and the Valley ... 148
10 The Corn Belt ... 166
11 Small Towns ... 188
12 The Iron Ranges ... 218
13 A Demographic Mini-Atlas ... 238
14 A Twin Cities Primer ... 250
15 The Lakeshore Resort and Retirement Belt ... 274
16 What Next? ... 292
Notes ... 298
Glossary ... 303
Index ... 306
Credits ... 315

ACKNOWLEDGMENTS

We are deeply grateful for the information, encouragement, and assistance we have received from John Adams, Arnie Alanen, Heather Anderson, Scott Anfinson, Dave Biesboer, Dwight Brown, Paul Christ, Bob Christensen, Ron Durst, Juan Estrada, Phil Gersmehl, John Hudson, Kathy Klink, Jodi Larson, Gordon Levine, Doug Magnus, John Maki, Becky Marty, Bob McMaster, Martha McMurry, Paulette Molin, George Raab, Clayton Rollins, Laura Smith, Bob Sparboe, Bob Spizzo, Rod Squires, Ron Tobkin, and Dennis Uittenbogaard.

Susy Ziegler thanks the College of Liberal Arts at the University of Minnesota for granting her a single-semester leave to focus on this project and thereby contribute to the research, teaching, creative activity, and service to society that are the university's mission.

All photographs, except those otherwise acknowledged, were taken by the authors.

All maps and graphs, except those otherwise acknowledged, were drafted in the University of Minnesota Cartography Laboratory with the assistance of Elizabeth Fairley, Julia Rauchfuss, Xuejin Ruan, and Jonathan Schroeder under the direction of Mark B. Lindberg. Jay Whitmore scanned the authors' slides in the Digital Content Laboratory at the University of Minnesota. Galen Schroeder of Dakota Indexing prepared the index. Percolator designed the book and cover.

We appreciate the support of the Minnesota Historical Society Press, including Director Greg Britton, Editor Marilyn Ziebarth, and Design and Production Manager Will Powers.

INTRODUCTION

Geography is about places, places as large as the entire globe, as small as the chair on which you are sitting, and all dimensions in between. Every place is a unique confection of the natural environment and the things people have done to that environment. Some places remain almost entirely natural, but others have been completely transformed by people and are dominated by human works. This book is a geography of a delightful place known as Minnesota, which is actually a complex mosaic of many smaller places.

Each piece of this mosaic has the solid earth beneath it and the atmosphere above it, but every place has its own distinctive mix of natural and human features, some closely related and interdependent, others coexisting by pure happenstance. All places are changing constantly, some quite rapidly, others almost imperceptibly slowly, so geography is about how places are becoming rather than about their mere being. The present is deeply rooted in the past, and it strongly influences the future.

The natural environment changes both geographically and historically. It varies from place to place, and it changes from time to time. The weather changes from day to day, hour to hour, even minute to minute. Plant life changes more slowly but equally surely. Even the surface of the land is not nearly as permanent as it might seem, because the elements of the atmosphere are constantly attacking and inexorably altering it, often over a span of time far greater than mere human experience.

The natural environment provides the raw material that people have to work with. In some places the natural environment may promote or preclude particular human activities, but in most places it is fairly permissive, and the ways in which people use and abuse the rocks, waters, soils, plants, animals, and even the wind are influenced by their numbers, their cultural values, and their technical competencies. The built environment, which is the product of past human activities, may influence contemporary human activities just as powerfully as the natural environment, both today and in the future.

People have built homes and businesses and many other kinds of structures that are clustered in settlements ranging from isolated farmsteads through hamlets, villages, towns, and cities to great sprawling metropolitan centers. They have built streets and roads and railroads to facilitate the circulation of people, goods, and ideas between and within these clusters.

Geography explores the myriad ways in which human beings have interacted with their environments to produce complex landscacapes of places and the ways in which these places have evolved, with the certain knowledge that they will continue to evolve, albeit in ways about which we can only speculate but have not the power to predict. Maps are essential tools to show where places are and to help us understand why they are where they are, why they have become what they are, and how they are changing. Even though this book has many maps, we simply cannot show all of the important places in Minnesota on a single page-size map, and you will need to keep a good state highway map handy while you are reading it.

The most immediately obvious aspect of any place is what we see when we visit it, so geography begins with the visible landscape but quickly transcends it in search of explanations and answers. Photographs are a useful substitute for actually visiting places, but the best way to learn about places, the best way to understand and appreciate them, is to see them for yourself. We hope that this book will incite you to explore Minnesota.

Minnesota Counties

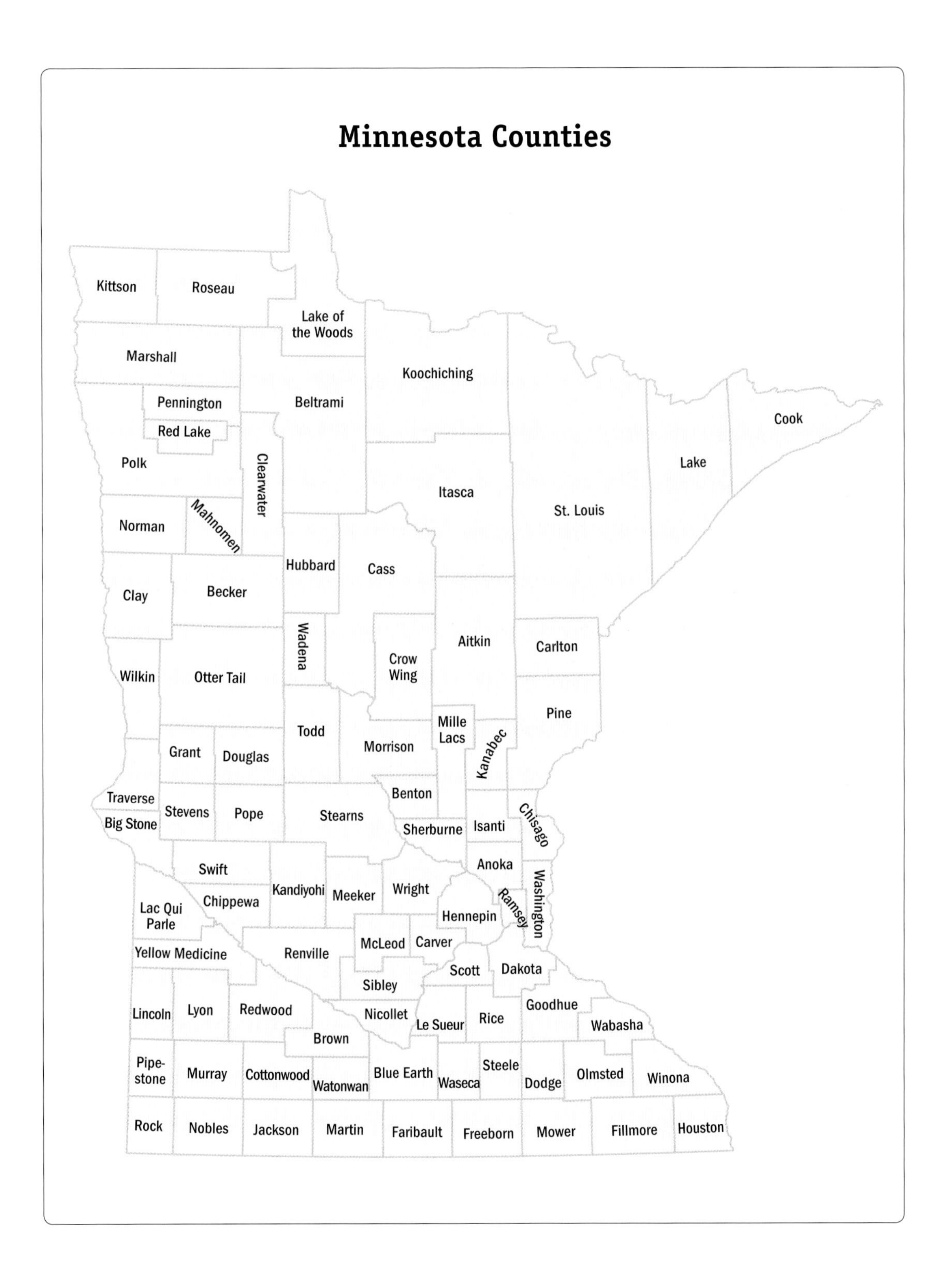

1. Lake Superior
2. Lake Mille Lacs
3. Tower
4. Twin Cities
5. Beardsley
6. Duluth
Kittson
Roseau
Lake of the Woods
Marshall
Beltrami
Koochiching
Pennington
Red Lake
Polk
Clearwater
Itasca
Norman
Mahnomen
St. Louis
Hubbard
Cass
Clay
Becker
Wadena
Aitkin
Carlton
Crow Wing
Wilkin
Otter Tail
Pine
Todd
Grant
Douglas
Morrison
Mille Lacs
Kanabec
Traverse
Stevens
Pope
Stearns
Benton
Big Stone
Sherburne
Isanti
Chisago
Swift
Anoka
Kandiyohi
Meeker
Wright
Washington
Lac Qui Parle
Chippewa
Hennepin
Ramsey
McLeod
Carver
Yellow Medicine
Renville
Scott
Dakota
Sibley
Lyon
Redwood
Nicollet
Goodhue
Rice
Le Sueur
Wabasha
Brown
Murray
Cottonwood
Watonwan
Blue Earth
Waseca
Steele
Dodge
Olmsted
Winona
Rock
Nobles
Jackson
Martin
Faribault
Freeborn
Mower
Fillmore
Houston

1

Battleground of the Elements

For many Americans, Minnesota's chief claim to fame is its awful winters, but many of us revel in them. We boast that there is nothing between us and the North Pole but a barbed wire fence. In fact, the state is more than 3,000 miles from the North Pole and about the same distance from the equator, although the Pole often does seem closer.

Minnesota is at the very heart of the North American land mass, about as far from salt water as it is possible to be, and the state has an extreme continental climate, with cold winters and hot summers. It is a battleground of the atmosphere, where cold, dry air masses from Canada slug it out with warm, moist air masses from the Gulf of Mexico, and a mass of relatively mild, dry air from the Pacific slips in every now and then. The state has just barely enough precipitation to allow farmers to grow crops such as corn and soybeans, although it has enough winter precipitation, when combined with severe cold, to create challenging driving conditions.

Weather and climate have two basic elements: temperature and precipitation. Minnesota's latitude affects its temperatures. Its continental location affects its precipitation and is responsible for the extremes of temperature about which Minnesotans like to brag.

When we speak of temperature, we mean the temperature of the air. Planet Earth receives its heat energy from our sun in the form of solar radiation, which is the basic force that drives weather and climate. Air is heated by the surface of the earth, not by the sun. Some of our coldest winter days have brilliant sunshine, so obviously the sun is not heating the air directly. The energy radiated from the sun consists of short waves that pass through the air without heating it. Solar radiation heats the planet's surface, and the heated surface re-radiates long waves that heat the air.

Solar radiation varies with latitude. It is greatest when the sun is directly overhead and decreases as the sun sinks lower in the sky. Solar radiation is greatest near the equator and decreases toward the poles, but not as much as one might expect. Higher latitudes have longer days in summer, so they receive solar radiation for a longer time. This partially compensates for the less intense radiation received due to the lower angle of the sun in the sky.

Some surfaces heat and cool faster than others. For example, an asphalt driveway warms up and cools off much more quickly than a grass lawn. Water bodies gain and lose temperature at only one-fifth the rate of land, so water bodies are cooler than the land in summer and warmer in winter. The interior of the continent is warmer than coastal areas at the same latitude in summer and much colder in winter. Minnesota presents an extreme example of this "continental effect."

Land surfaces also differ in the rate at which they gain and lose temperature. Bare ground heats and cools faster than vegetated areas, and different types of vegetation heat and cool at different rates. Solar radiation striking a water body at a low angle literally bounces off the surface instead of heating it. Snow-covered ground, which reflects solar radiation back into the atmosphere, is a virtual recipe for cold weather.

The long waves re-radiated from the earth's surface cannot pass freely through the atmosphere. Water vapor and carbon dioxide in the air act like an insulating blanket that absorbs heat, similar to the way that heat is trapped in a greenhouse or sun porch. Automobile exhaust, power plants, factories, and anything else that increases the amount of carbon dioxide in the air can enhance a "greenhouse effect" that raises temperatures, although this popular journalistic analogy can be misleading.

Air expands and rises when it is heated, creating an area of low atmospheric pressure at the ground surface. Cooler air, which is denser, settles and creates an area of high pressure. In the Northern Hemisphere, winds try to equalize pressure by swirling clockwise from areas of high pressure to areas of low pressure. An example can be seen in the reaction of land and sea breezes on the shores of the oceans and some large water bodies, such as Lake Superior or Mille Lacs. During the daytime the land is hotter, the heated air rises, and breezes blow from the water toward the land. After dark the land cools off more rapidly than the water, the cooled air settles over the land, and a breeze blows toward the water.

Cold air is dense, and it drains downhill. On clear winter nights it flows slowly downward and collects in the lowest areas, where it loses what little heat energy it might still have. The coldest official temperature ever recorded in Minnesota was in a low-lying frost pocket near Tower, where the thermometer read -59½° F at 9:10 AM on February 2, 1996.[1] The low temperature in the Twin Cities that day was only -32°F. A week later the Twin Cities had a high of 39°F. Such extreme variability of temperature, both from day to day and from year to year, is characteristic of continental climates.[2]

All these effects influence the temperature in Minnesota, which generally decreases from south to north; the North Shore of Lake Superior has a distinct "lake effect," with cooler summers and milder winters. In July the average temperature of the state is 70° F, slightly higher in the south, slightly lower in the north. Most summers have a week or so of temperatures above 90° F. The state's all-time high temperature of 114° F was recorded at Beardsley in western Big Stone County on July 29, 1917, and was matched at Moorhead on July 6, 1936.

January has a slightly greater range of temperatures, with an average of 2°F to 6°F in the northern third of the state and 12° to 14°F in the southern third. In winter the temperature often does not climb as high as 32°F, even at the warmest time of the day. The average last spring frost varies from the first week of May in the south to the end of May in the north, and the average first fall frost ranges from the second week of September in the north to the first week of October in the south.

The average growing season is 170 to 180 days in the northern third of the state and 200 to 210 days in the southern third. The Twin Cities have a slightly longer growing season than adjacent areas because they form an urban heat island. The temperature in the central city is 2° to 4°F higher than in areas thirty miles out in the country. The tall buildings and paved streets trap heat, so the city is hotter in the daytime and cools off more slowly at night than the country.

City buildings reduce the speed of the wind, which would transfer heat away from the city. Because urban air is more polluted, clouds blanket the city more often than the countryside. The city has less vegetation, and plants cool the air by absorbing heat to transpire moisture. The ground surface is drier in the city than in the countryside, because city people dislike rain and snow, and dispose of them as rapidly as possible through sewer and drain systems. In short, the city is warmer because it absorbs more solar radiation and re-radiates less heat energy to space.

THE MOISTURE CONTENT OF THE AIR is called its humidity. The amount of water vapor that can exist in air increases rapidly with the air's temperature. When the air is saturated with moisture, the water molecules condense on dust particles, salt crystals, or other microscopic nuclei, leading to the formation of clouds or fog. When the condensed particles become heavy enough, they fall to the ground as

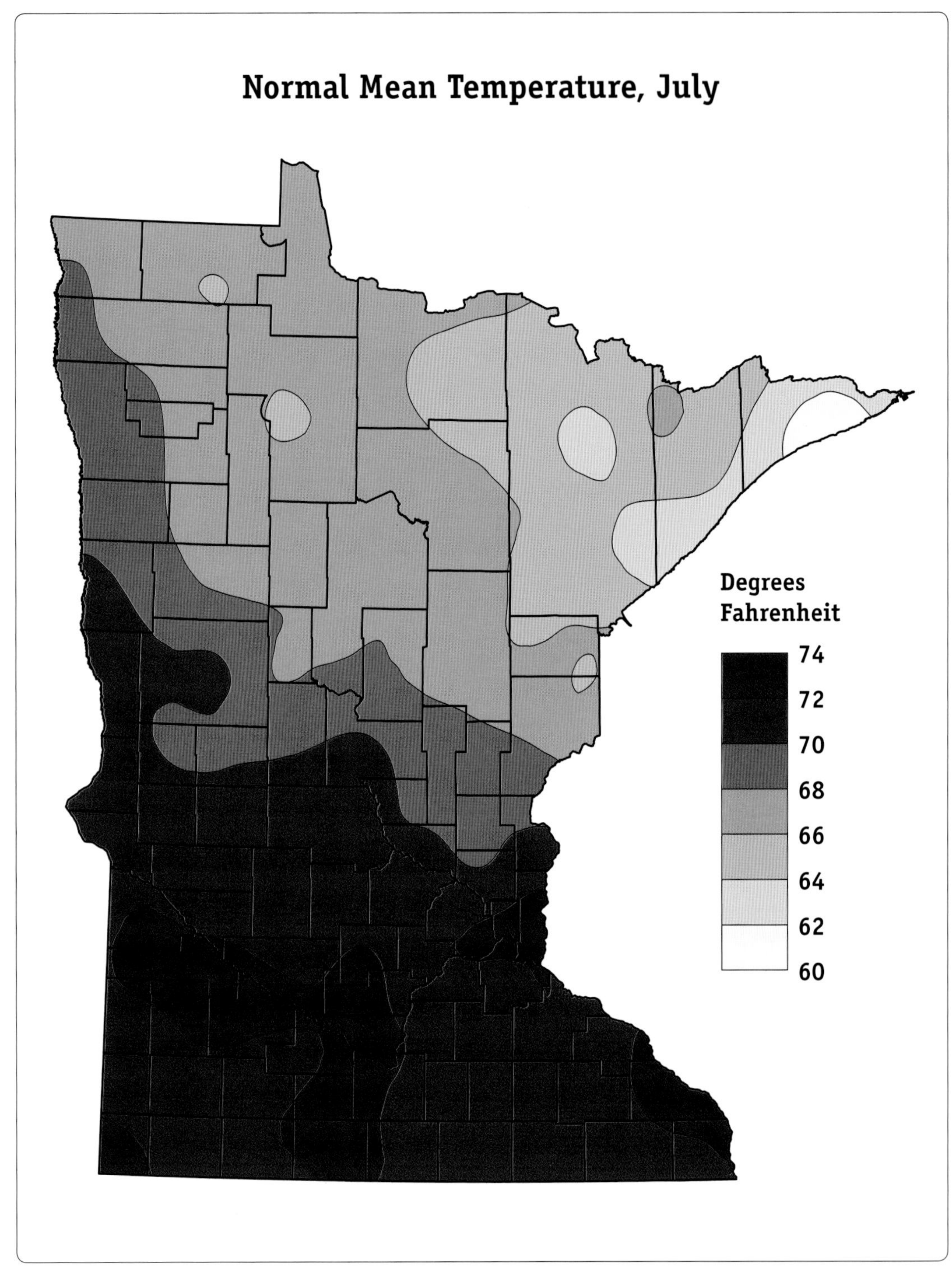

Southwestern Minnesota has the state's hottest summers.

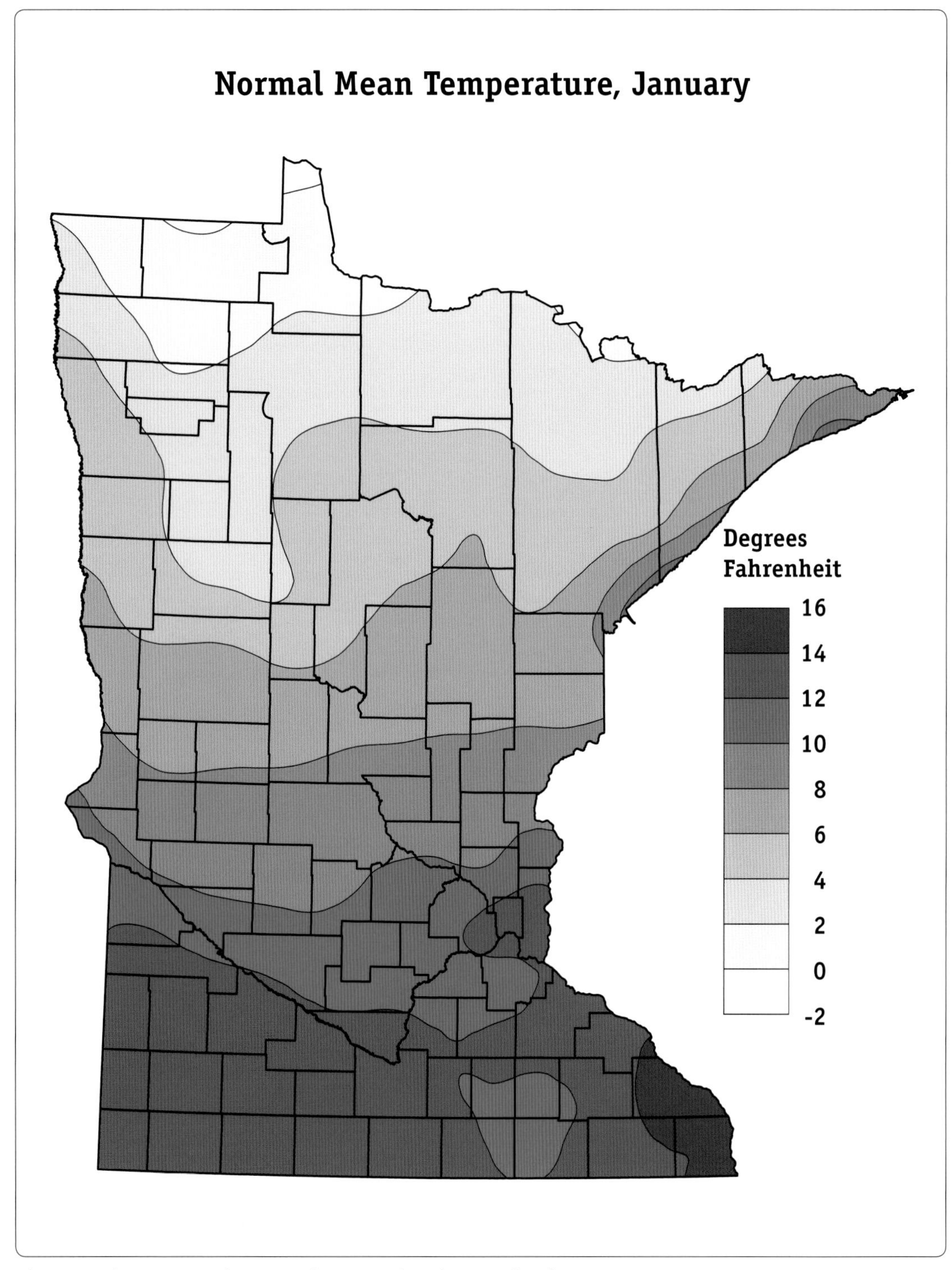

The range of temperature in January is greater than the range in July.

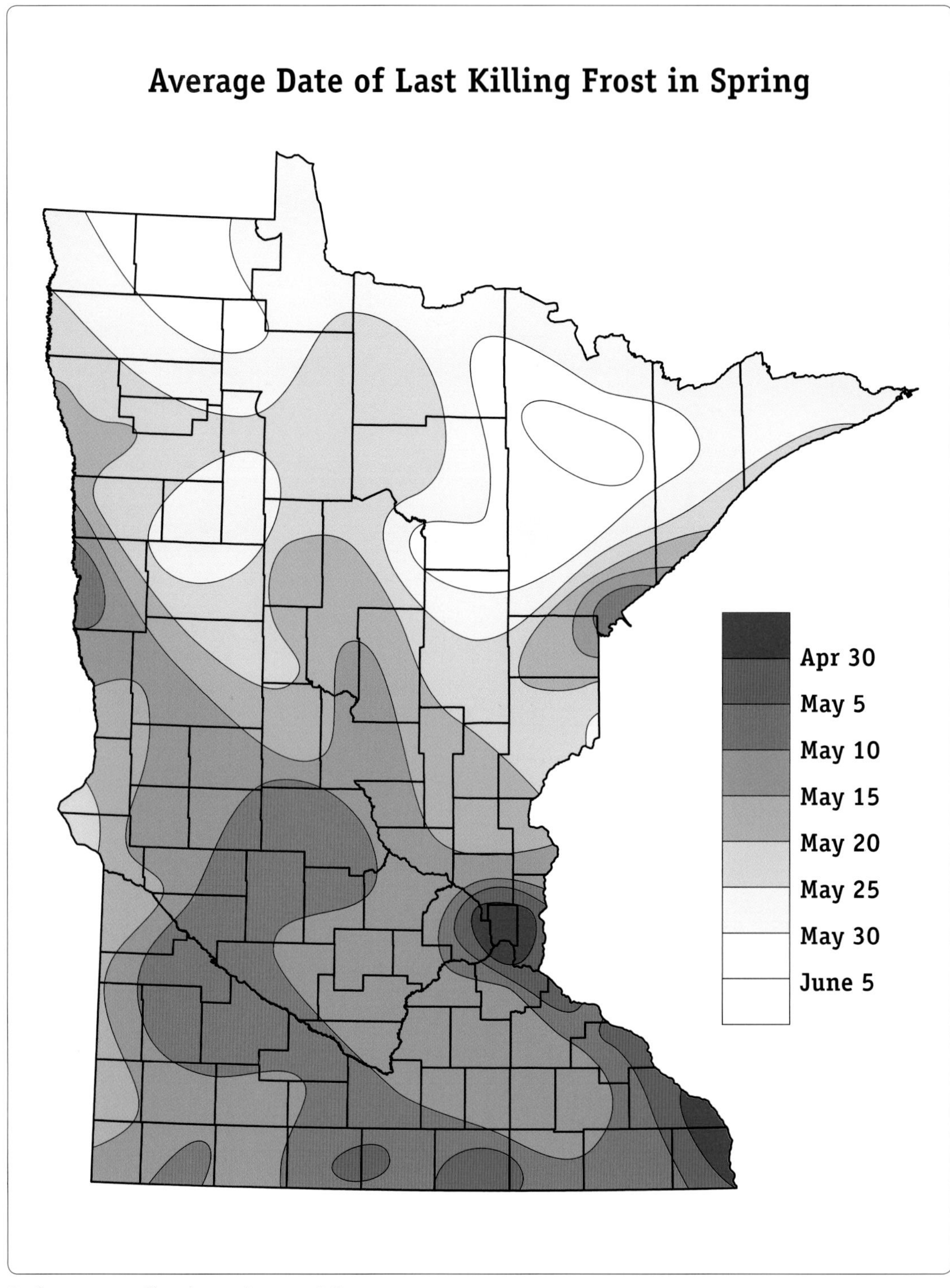

Spring comes earliest in southeastern Minnesota.

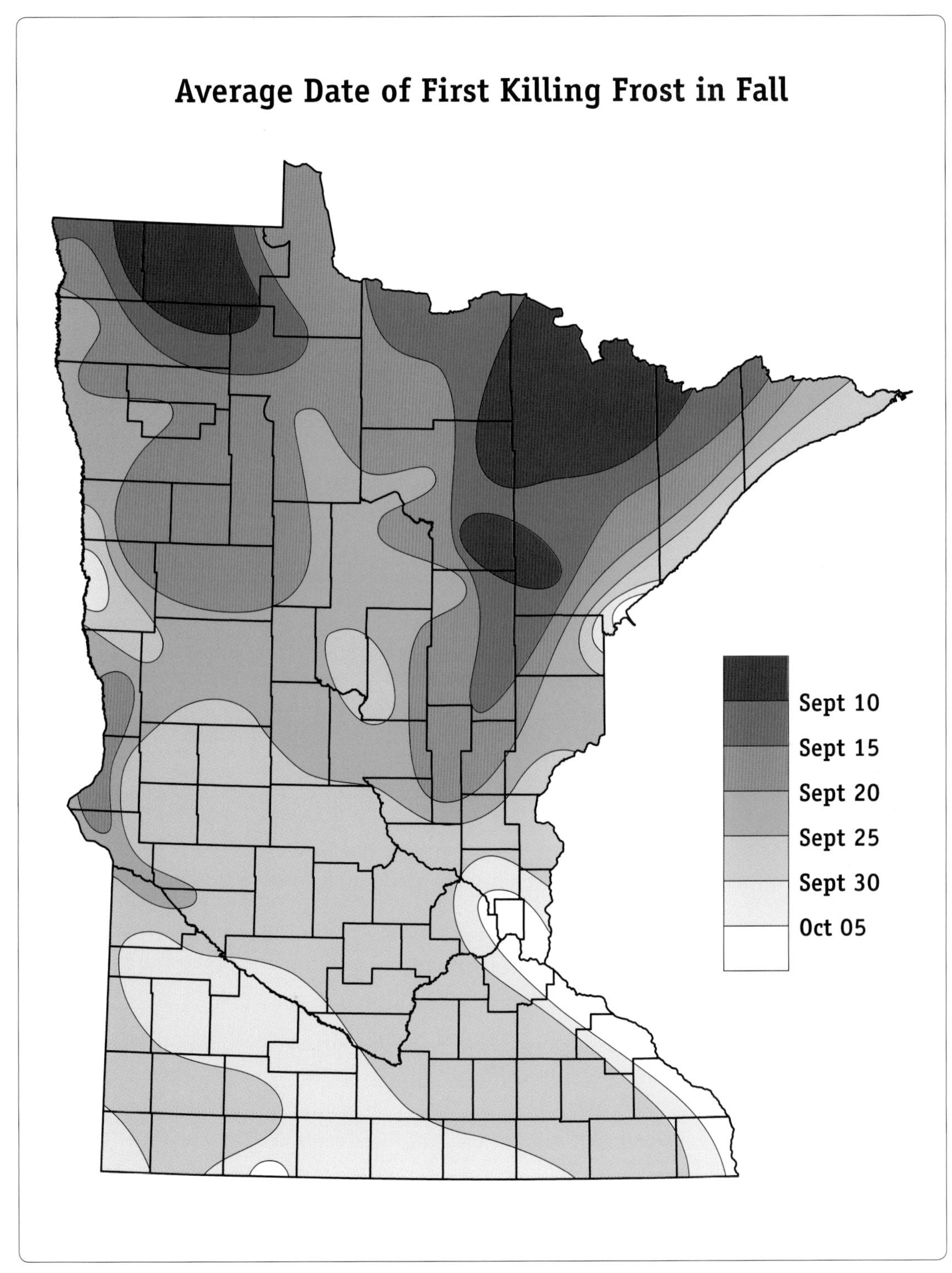

The first fall frost is nearly a month earlier in the north than in the south.

precipitation—rain, snow, sleet, or hail. Condensation releases heat, but evaporation consumes heat and thus is a cooling process. For example, the human body releases heat by perspiring and further cools itself when the perspiration evaporates. Perspiration cannot evaporate when the air is saturated with moisture, making humid, muggy days singularly unpleasant for many people.

Most of the moisture in Minnesota comes from the Gulf of Mexico. The air over Canada is too cold to hold significant moisture, and the continental land mass would not be a major source of moisture even if the air were warmer. Air is moist when it moves inland from the Pacific Ocean, but it loses most of its moisture when it crosses the Rocky Mountains, and it is fairly dry when it finally reaches Minnesota. Even air from the Gulf has lost much of its moisture by the time it reaches Minnesota, but still it is the principal source of moisture and precipitation in the state.

The precipitation map of Minnesota reflects the importance of the Gulf of Mexico as its moisture source. The northwestern corner of the state is driest, with just under twenty inches a year, and the southeastern corner is wettest, with slightly more than thirty inches. The North Shore is slightly wetter because of the lake effect from Lake Superior. Northeasterly winds pick up moisture when they blow across the open water. The air rises and cools when it hits the bluffs along the North Shore. The air's water vapor condenses when it cools and then falls as rain or snow, depending on the temperature of the air.

Two-thirds of Minnesota's precipitation falls between May and September, when plants need it to grow, but severe drought can occur about once every ten years in the western part of the state and about once every twenty-five years in the east.

From November through March much of the state's precipitation is snow. An inch of snow contains about as much moisture as one-tenth of an inch of rain. Minnesota has an annual snowfall of around forty-five inches, a bit less in the west and a bit more in the northeast, with more than sixty inches along the North Shore. November and March have the heaviest snowfalls, and midwinter has somewhat less. The snow that falls in November and March, which is damper and heavier than that in the other months, is sometimes called "heart-attack snow" because shoveling it requires such physical exertion.

Each winter Minnesota has an average of two blizzards, when heavy snow and winds stronger than thirty-five miles an hour are combined with bitterly cold temperatures. The blinding, wind-driven snow reduces visibility to less than a quarter of a mile and creates treacherous driving conditions. The infamous Halloween Blizzard of 1991 dumped twenty-eight inches of snow on the Twin Cities and thirty-seven inches on Duluth.

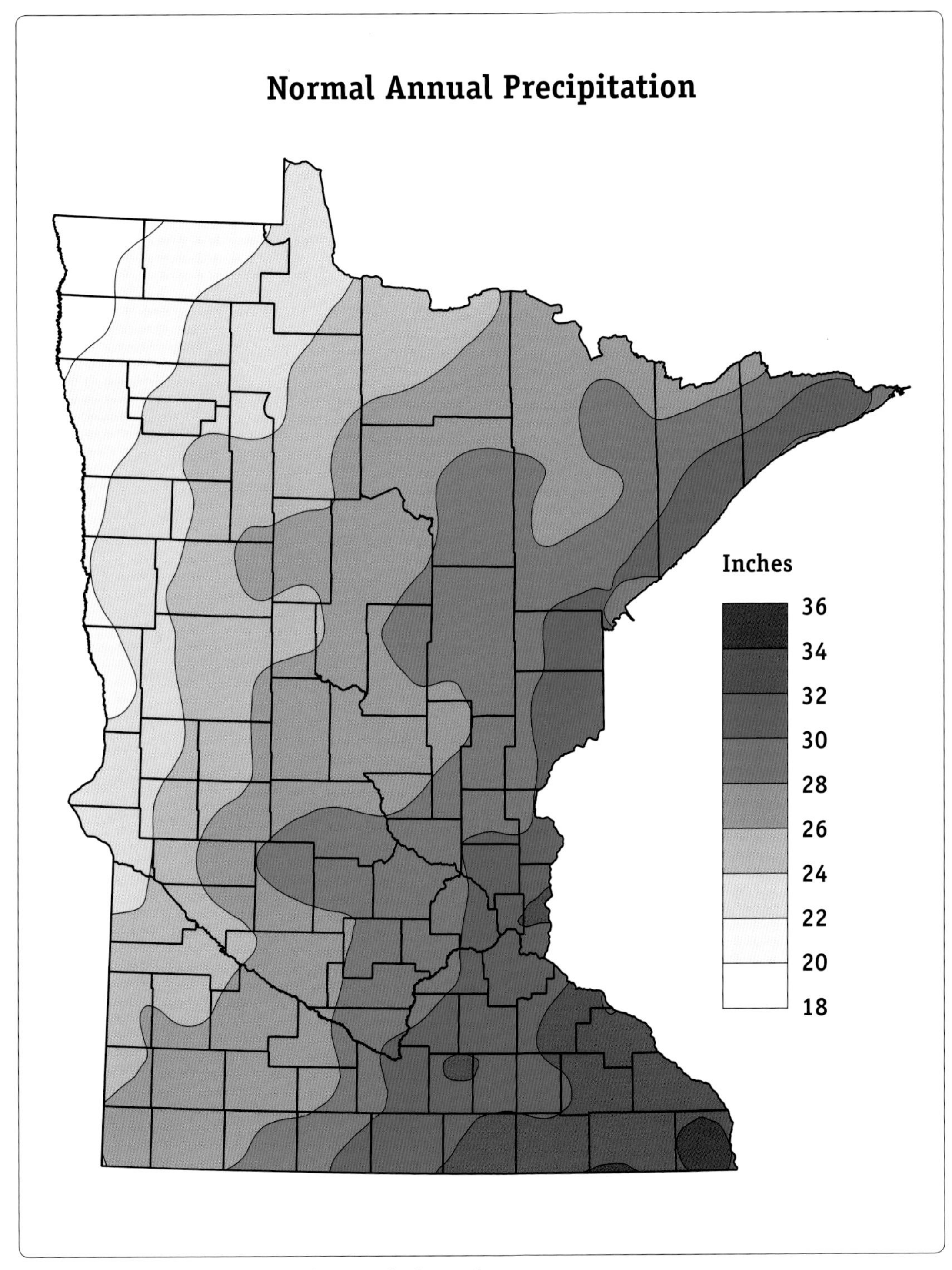

Rainfall is least in the northwest and greatest in the southeast.

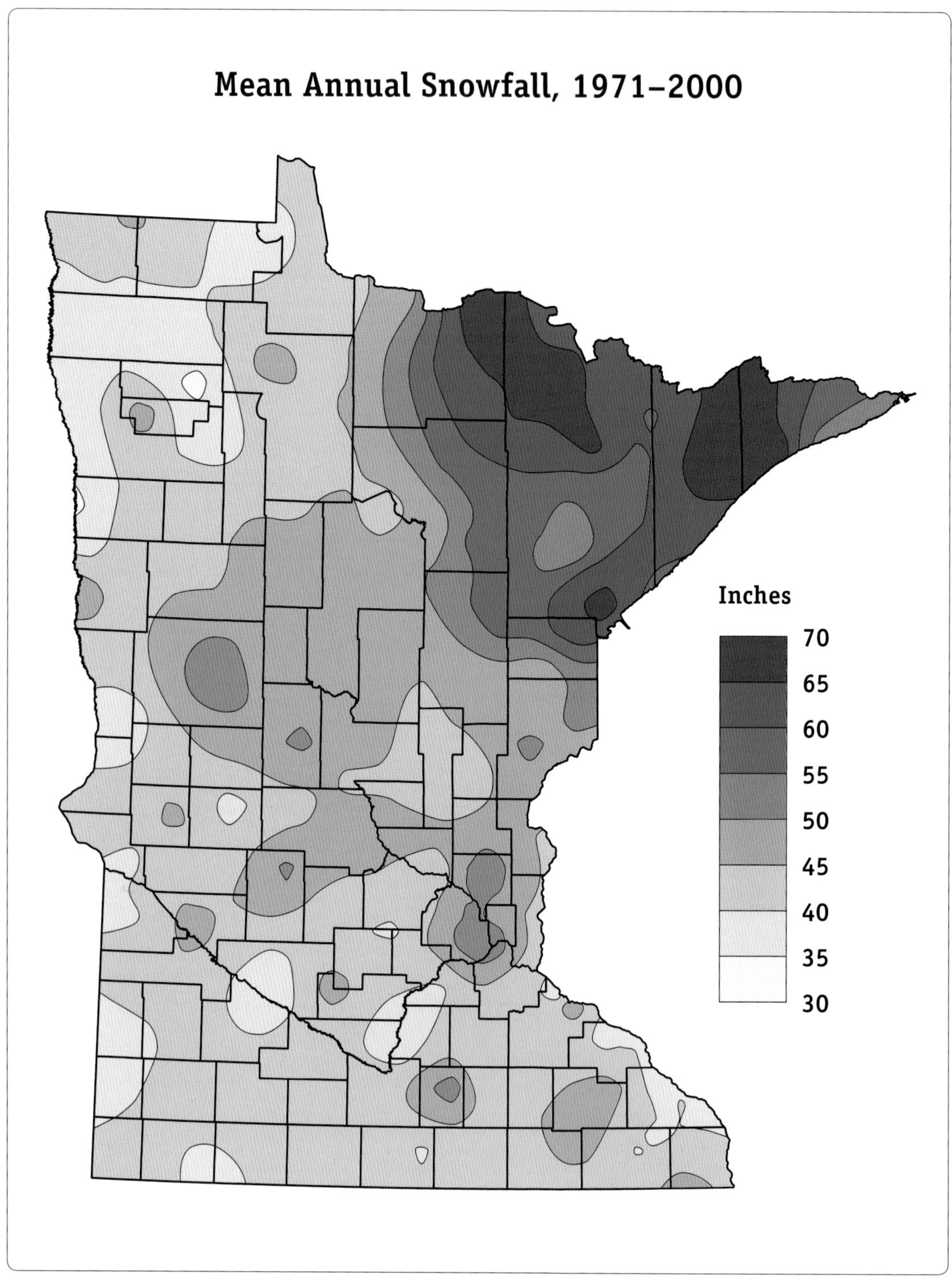

Northeastern Minnesota has the heaviest snowfall.

A winter "snow trench" downwind from Lake Calhoun in Minneapolis.

Wet snow sticks to branches like fine lace.

The snow that falls in October and early November usually melts, and Minnesota normally does not have a permanent winter snow cover of an inch or more until the end of November. The length of snow cover varies from eighty-five days in the south to more than 140 days in the northeast. The snow-covered ground keeps the air above it cold, postponing the arrival of spring forever, or so it seems.

The winter weather in Minnesota is dominated by cold, dry polar air from the northwest, with occasional incursions of bitterly cold arctic air. The combination of bone-chilling cold and strong winds can be life threatening. Two-thirds of the days in November and December are overcast with clouds. January is brighter, and many Minnesotans cherish winter's scintillating high barometric-pressure days, when the air is bitterly cold, and the sky is piercingly blue.

Winter has been the season to sled, skate, ski, snowmobile, and ice-fish. A mild winter can be disastrous for many businesses that are oriented toward outdoor activities. The winter of 2000 was so mild that the governor seriously considered declaring a mild-winter emergency and requesting federal relief funds. The winter days are short, and the winters sometimes seem interminable for those who do not enjoy winter sports. By February or March, some Minnesotans suffer "cabin fever" and, if they can afford it, take short vacations in the Sunbelt and other warmer places.

In the spring Minnesota warms up slowly because of its snow-covered ground, but by late spring, the days may turn hot almost overnight.[3] April and May are the months of thaw, when the days begin to warm up but the nights are still cold. Heavy rains on melting snow can overload streams and cause severe floods. Ice breakup on large lakes can be spectacular, as strong winds shove sheets of ice onshore, crushing buildings and blocking highways. Ice may be pushed as far as seventy-five feet inland, with jagged shards sticking twenty feet into the air.[4]

This farmstead has an L-shaped windbreak of trees on its northern and western sides.

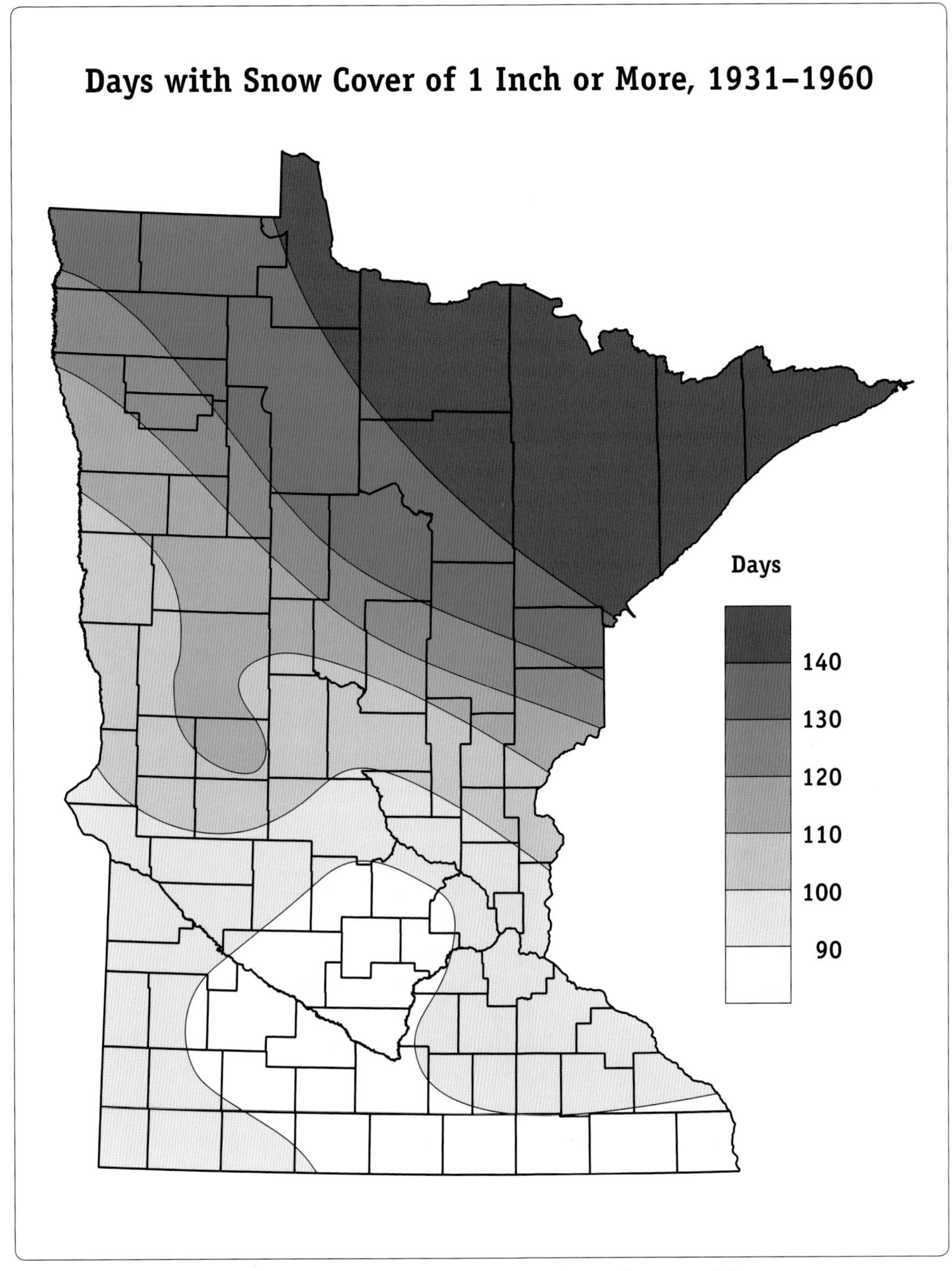

Northern Minnesota has snow cover almost twice as many days as the southern part of the state.

Warm days and cold nights of spring easily create road potholes. Ice and snow melt during the day, and the water seeps into cracks in the pavement. At night, when the water refreezes, it expands, widening the cracks and crumbling the road surface, which breaks up under the pounding of traffic. Dodging potholes is an annual rite of spring in Minnesota.

Late spring and early summer is a time for severe rainstorms. Warm air from the Gulf of Mexico is beginning to push northward, but cold polar air is still strong enough to push back. Their struggle generates extreme and highly variable weather. The boundary between two air masses is called a "front." The edge of an advancing mass of warm air is called a "warm front." Less dense than cold air, warm air skids up over the cold air. Warm fronts often are associated with several days of overcast weather and light, drizzly rain.

The front edge of an advancing mass of cold air is called a "cold front." Because the cold air is denser than the warm air, it undermines the warm air and forces it to rise. The rising air cools quickly, its moisture condenses, and condensation releases large amounts of energy, which is manifested in thunder and lightning. The passage of a cold front across Minnesota brings a drop in temperature and air pressure, heavy rain and thundershowers, strong winds, and occasionally hail.

A cold front can spawn tornadoes if the contrast between the two air masses is great enough. A tornado is a destructive wind greater than seventy miles an hour and ranging up to 318 miles an hour. These winds twist hot air into a dark funnel that can whirl up from the ground or down from the sky. May through August is peak tornado season in Minnesota, which has thirty-five in an average year. They can occur anywhere, even in cities, and every county in Minnesota has been hit by at least one. Most are relatively mild, although no one should be outside even in a mild one. Severe tornadoes can wreak incredible havoc along narrow paths.

The atmosphere's polar jet stream steers the movement of air masses and fronts across Minnesota. The jet stream is a great river of wind, six to eight miles above the ground, that blows steadily from west to east at average speeds of 110 miles per hour. It undulates north and south in great sweeping curves and is about twice as strong in winter as in summer. It directs cold, dry polar air into Minnesota when it swings southward and warm, moist Gulf of Mexico air into the state when it swings northward.

In summer the jet stream flows farther north than Minnesota, whose weather is dominated by warm, moist, sometimes sultry air from the Gulf of Mexico. Air-conditioning is welcome for most of the summer, and for a couple of weeks or so it seems essential. The first real heat wave of summer can spawn power outages when everyone turns on their air conditioners at the same time.

Summer days are long, bright, and sunny. The ground is so warm that the air above it rises to a level where it condenses and forms fleecy, flat-bottomed cumulus clouds that look like giant balls of popcorn. Cumulus clouds usually indicate fair weather, but on especially hot days these clouds can tower high enough to create thunderheads, with spectacular flashes of lightning.

Tornado damage at Comfrey.

Southern Minnesota has an average of forty-five thunderstorms a year, and northern Minnesota has thirty. Some storms are triggered by surface heating, others by the passage of a cold front. Thunderstorms, and the strong winds associated with them, do more total property damage in the state than tornadoes, which strike only small areas.

Summer gives way to winter slowly and reluctantly. Many people consider fall the best season of the year in Minnesota. Days are warm but not hot, the sky is sunny, and nuisance bugs have died by the truckload. Trees stop producing green chlorophyll as the days grow shorter, and leaves turn brilliant orange and yellow when the green color starts to fade and unmasks the other pigments that were latent in the leaves throughout the growing season.[5] Some botanists argue that the leaves start producing color only in autumn, either to warn away insects that might lay damaging eggs on them or as a kind of sunscreen to protect the leaves from damage by excessive sunlight when the chlorophyll starts to fade.[6] Most people do not really care why the leaves change color; they simply enjoy the spectacular autumn foliage, which attracts tourists to the state.

By fall many Minnesotans seem to have forgotten that ice and snow can be slippery. The

Pigments that were latent in the leaves appear in the fall when the days grow shorter and trees stop producing green chlorophyll.

In spring the leaves are brilliant new green in color.

Autumn foliage after the leaves' chlorophyll starts to fade.

first heavy snowstorm can be challenging, as many drivers continue to race along the highways at the speed limit, or even faster, with some cars caroming off guardrails or skidding into ditches.[7] Fender benders and multiple-car pileups are frequent. Repair shops stay busy, and hospitals attend to motorists with minor injuries. People who live in a battleground of the elements slowly learn to adjust to the changing seasons.

The climate of Minnesota is ever so slowly becoming milder and moister. Temperature and precipitation fluctuate greatly from year to year, but the long-term temperature trend is definitely upward. Since 1900 the average annual temperature of the state has risen nearly 2°F, the average annual precipitation is about five inches more than it used to be, and these rates of increase are up to twice as fast as in some other parts of the world.

A few people may continue to debate whether this climate change is natural fluctuation or the result of human activities, but Minnesotans know that lake ice-out dates are earlier, lilacs are blooming earlier, red oaks and red maples are moving into the Boundary Waters Canoe Area Wilderness, the moose herd is declining, sales of ice-fishing shacks and snowshoes are down, and people who fish fret that a warmer climate could limit the growth and reproduction of the state's popular cold-water fish, such as walleye, northern pike, and trout. Coping with climate change will create colossal challenges not only for the people of Minnesota, but for the people of the United States and the entire world.[8]

Minneapolis Tribune

Sunday
January 11, 1976
Volume CIX
Number 222

1A Final

50c by carrier

325 cars in 4 snowy pileups

Staff Photo by Bruce Bisping

Drivers and emergency-vehicle crews walked through a knot of damaged cars north of 46th St., part of Saturday's pileup on Interstate Hwy. 35W.

35W wreck involves 250 autos; 17 people hurt

By George White

At least four chain-reaction traffic accidents piled up hundreds of vehicles on snow-slicked Interstate freeways in the Twin Cities area Saturday. More than 325 cars were involved.

About 250 vehicles were involved in clusters of chain-reaction pileups on northbound lanes of Interstate Hwy. 35W in south Minneapolis. Police on the scene said 17 injured people were taken to hospitals. No fatalities were reported.

Police said most of the collisions were fender-benders from which motorists drove away. However, 50 vehicles were seriously damaged and had to be towed.

(More reports, pictures, Page 1B.)

The string of accidents stretched from 35th St. in Minneapolis to 62nd St. in Richfield.

"I've never seen anything more chaotic," said one policeman, looking down on the freeway from the top of the 46th St. ramp.

The collisions began to spread at about 9 a.m. The state Highway Patrol and Minneapolis police blocked entrance ramps and closed the northbound lanes until vehicles were towed or driven away. The freeway was reopened at 12:30 p.m.

The highway department reported three other pile-ups Saturday morning. One involved 40 vehicles in the westbound lane of Interstate Hwy 694 between Main St. and the Mississippi River in Fridley.

According to the Highway Patrol, there were no injuries. It took three hours for authorities to clear that area.

About 20 vehicles collided on northbound Interstate Hwy. 35W between the Hiawatha and Washington Av. exits. There were no injuries and the area was cleared in a half-hour, the highway department reported.

The westbound lane of Interstate Hwy. 494 was the scene of a 25-car pile-up. It occurred between Penn and Xerxes Avs. S. on the Bloomington-Richfield boundary. There were no reports of injuries.

The accidents occurred while three inches of snow fell in the area. Visibility was poor and blowing, drifting snow made lanes indistinguishable, police said. Heavy traffic contributed to the problem.

"There was an unusual amount of traffic this morning," said Trooper Earl Erlander. "People also weren't ready for these conditions after a

Accidents continued on page 6A

African summit on Angola is divided

Associated Press

Addis Ababa, Ethiopia
An African summit called to bring peace to Angola split into pro-Soviet and neutral factions Saturday in a dramatic, name-calling opening session.

Holden Roberto, the National Front (FNLA) leader, and Jonas Savimbi, leader of the National Union (UNITA), listened impassively as Mozambique President Samora Machel called them "agents of South Africa" and "traitors to Africa." South African troops have reportedly sided with them in the Angolan civil war.

Machel said only the Soviet-supported Popular Movement (MPLA) could repel South African aggression and rule Angola. Without mentioning the United States by name, he condemned what he said were "last-minute attempts to send diplomats around Africa to rally anti-Popular Movement feeling.

President Leopold Senghor of Senegal, also speaking at the first emergency summit in the 12-year history of the Organization of African Unity (OAU), replied, "Those who condemn the alliance with South Africa would be honest only if they also condemned the alliance with Russia and Cuba." He said the OAU should work for peace talks among the Angolan rivals, followed by free elections.

Spectators were unable to recall any recent OAU summit in which divisions among African states had come out into the open so clearly. Usually OAU members thrash out their differences in private and make a public show of amity.

Sixteen presidents and prime ministers were present in the packed Africa Hall conference center. MPLA leader Agostinho Neto did not attend.

Most of the 21 OAU members that

Nebraska

Candidates prepare to dispel

People who live in the battleground of the elements must learn to adjust to the changing seasons anew each winter.

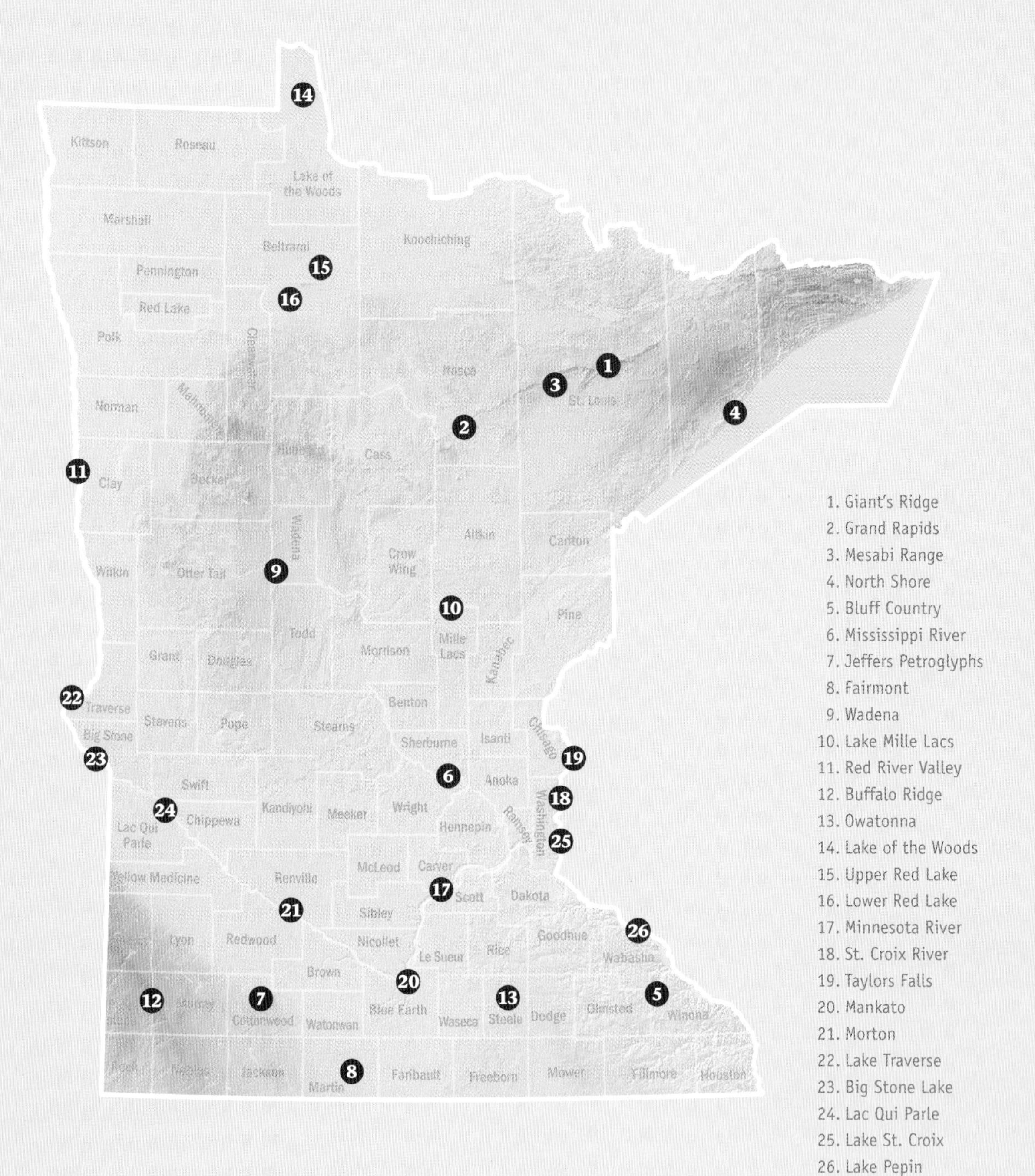

Kittson
Roseau
Lake of the Woods
Marshall
Beltrami
Koochiching
Pennington
Red Lake
Polk
Clearwater
Itasca
St. Louis
Norman
Mahnomen
Cass
Clay
Becker
Lake
Cook
Wadena
Aitkin
Carlton
Wilkin
Otter Tail
Crow Wing
Pine
Todd
Grant
Douglas
Morrison
Mille Lacs
Kanabec
Traverse
Benton
Stevens
Pope
Stearns
Big Stone
Sherburne
Isanti
Chisago
Swift
Anoka
Kandiyohi
Meeker
Wright
Chippewa
Lac Qui Parle
Hennepin
Ramsey
Washington
McLeod
Carver
Yellow Medicine
Renville
Scott
Dakota
Sibley
Lyon
Redwood
Nicollet
Goodhue
Le Sueur
Rice
Wabasha
Brown
Blue Earth
Waseca
Steele
Dodge
Olmsted
Winona
Cottonwood
Watonwan
Rock
Nobles
Jackson
Martin
Faribault
Freeborn
Mower
Fillmore
Houston
1. Giant's Ridge
2. Grand Rapids
3. Mesabi Range
4. North Shore
5. Bluff Country
6. Mississippi River
7. Jeffers Petroglyphs
8. Fairmont
9. Wadena
10. Lake Mille Lacs
11. Red River Valley
12. Buffalo Ridge
13. Owatonna
14. Lake of the Woods
15. Upper Red Lake
16. Lower Red Lake
17. Minnesota River
18. St. Croix River
19. Taylors Falls
20. Mankato
21. Morton
22. Lake Traverse
23. Big Stone Lake
24. Lac Qui Parle
25. Lake St. Croix
26. Lake Pepin

2

The Shape of the Land

Minnesota has gentle topography, with few hills and even fewer dramatic scenic vistas. Ancient glaciers modulated the land surface of the state by scraping away some of the underlying bedrock and then, as the glaciers melted, by depositing the debris they had carried from areas farther north. Glacial deposits of varying thickness blanket most of the state. Solid bedrock, which is exposed at the surface only in the northeastern and the southeastern corners of Minnesota and in a few areas in the southwest, is responsible for some of the state's most picturesque topography and scenery.

Northeastern Minnesota is part of the Canadian Shield, the anvil on which geology has forged the North American continent.[1] For the last 1.8 billion years, the Canadian Shield has been one of the most stable pieces of the earth's crust, but for the first half of geologic time northeastern Minnesota was a hotbed of geologic activity, where molten rock (magma) from deep within the earth was trying to force its way upward.

Some magma pushed to the land surface and erupted in spectacular volcanoes or oozed across the surface in massive lava flows. Some magma never reached the surface, but cooled and crystallized deep inside the earth and thrust up massive mountain systems. Earthquakes shivered the rocks from time to time. In calmer periods, rain-fed streams eroded the rocks and deposited sediment in low-lying areas. Eventually the weight compacted and cemented the sediment, creating sedimentary rocks. Heat and pressure subsequently transformed some sedimentary rocks into tougher metamorphic rocks.

The Land of 10,000 Lakes has gentle topography, with few spectacular scenic features.

The geologic activity concentrated some of the metallic minerals in the rocks into ore deposits large enough to be worth mining. The great ranges of iron ore are the best known, but miners have also dug ores of copper, nickel, and other metals from the rocks of northeastern Minnesota. Additional ore bodies may still lie waiting to be discovered, though no one can even guess whether they will be economical to mine. The deposits of iron ore have stimulated intensive research in northeastern Minnesota, and geologists know far more about the state's complex geology than they know about areas of similar geology in other parts of the world.

Light-colored veins in dark igneous rock may contain minerals worth mining.

Around 1.9 to 1.8 billion years ago, northeastern Minnesota was as mountainous as the Rocky Mountains are today. (Today's Rocky Mountains are a mere 65 million years old.) Hundreds of millions of years of erosion have removed all trace of these ancient Minnesota mountains except for their tough old roots, which are exposed in Giant's Ridge, a semicontinuous ridge of granite, 200 to 400 feet high, that extends nearly one hundred miles northeast from Grand Rapids along the north side of the Mesabi Range.

Huge cranes lift blocks of granite out of a quarry near Cold Spring.

Some of the magma that formed these mountains might have come from beneath the present basin of Lake Superior. The basin sagged when the magma flowed from beneath it; it is now a syncline, or geologic trough, between higher areas of tough old rocks on either flank. Erosion from the surrounding uplands deposited sediments in this trough, and a river probably flowed through it to the northeast until glaciers scoured out the present basin. Today the tough old rocks form majestic bluffs that tower hundreds of feet above the North Shore of the lake, and short, swift postglacial streams that notch the bluffs have picturesque waterfalls where they tumble over especially resistant rock formations.

During the last 2 million years glaciers have scoured the uplands of northeastern Minnesota. They scraped away the topsoil and gouged out the less resistant rock formations to create rockbound basins that now hold lakes of sky-blue water. Glacial ice rounded and smoothed the bare rocky knobs that rim the lakes. The plant life grows in the thin soil that has accumulated in the rocks' cracks and crevices. Northeastern Minnesota is unusual because the ocean has never covered it.

Throughout geologic history most of the United States has been subjected to repeated episodes of broad but gentle subsidence and uplift. Seas have covered large areas during the periods of subsidence. The layers of sediment deposited on the floors of these seas became consolidated into horizontal layers (strata) of sedimentary rock when the land was uplifted again. Such cyclical transgressions of the oceans between 550 and 350 million years ago formed the rocks in the bluff country of southeastern Minnesota, which is the only other part of the state where solid bedrock is at or close enough to the surface to influence the shape of the land.

In southeastern Minnesota, the lowest sedimentary rock strata, which were deposited first and thus are the oldest, are predominantly sandstone; the younger strata toward the top are predominantly carbonates (limestone and dolomite). The carbonate rocks resist physical erosion, and they caprock steep sandstone bluffs, but they dissolve fairly easily. Percolating groundwater also dissolves the carbonate rocks beneath the surface. The area is honeycombed with caverns and subterranean passageways; sinkholes dimple the land where the underlying rock has been dissolved and the ground has collapsed.

Glacial ice apparently did not cover the bluff country near the Mississippi River in southeastern Minnesota during the last several glaciations. Heavily weathered glacial deposits that are much older than the most recent glacial epoch extend nearly as far west as Interstate 35. When the last glacier melted, it left vast plains devoid of plant life in south central Minnesota. The wind sweeping across these plains

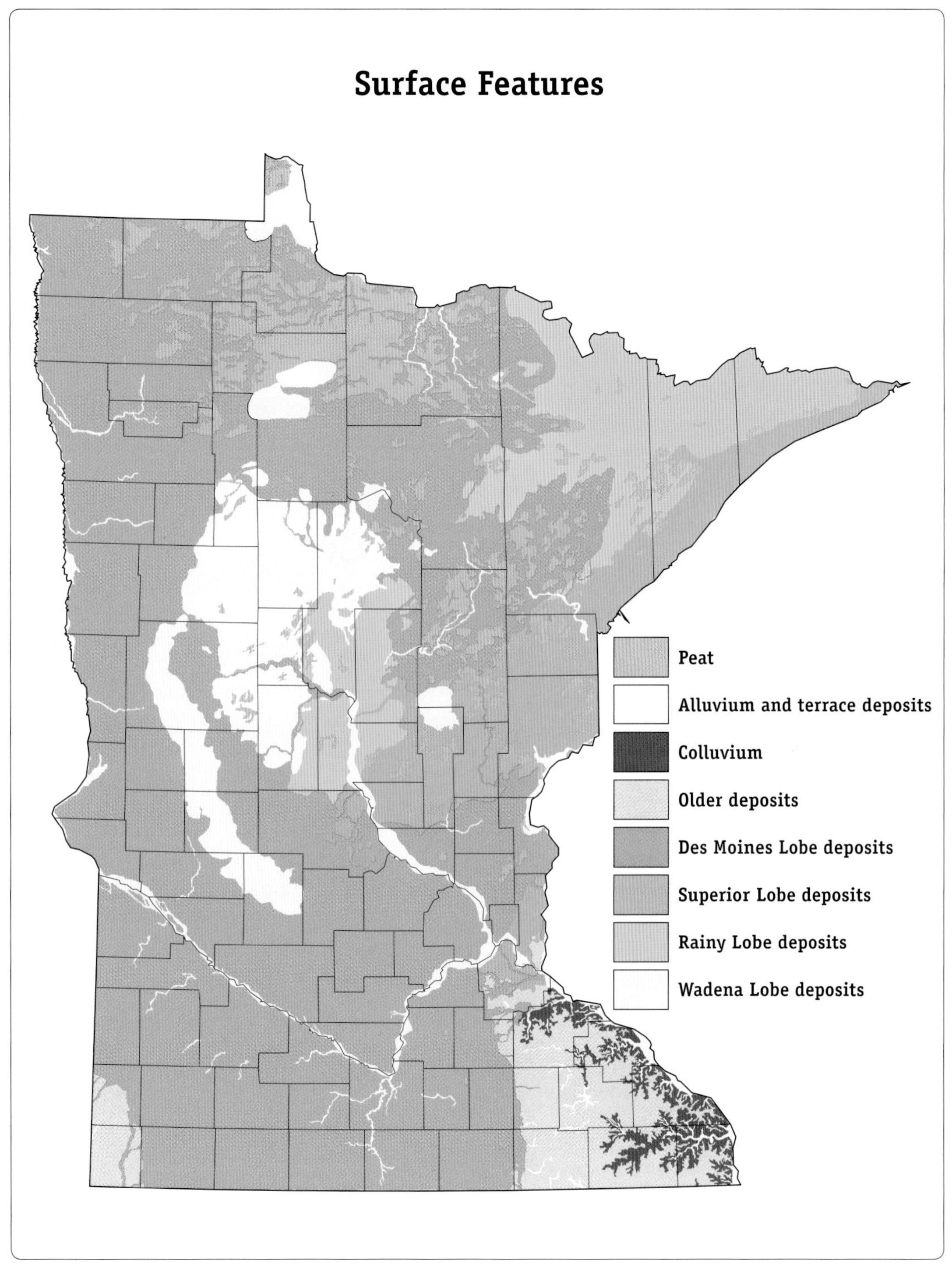

Glaciers have been responsible for most of the surface features of Minnesota.

picked up fine particles of dust and swirled them eastward in great dust storms that deposited blankets of windblown dust (called "loess"), which form fertile soils over much of the surface of these older glacial deposits. The tributaries of the Mississippi River have eroded deep valleys that separate the level upland areas. The hilly stream-eroded topography of southeastern Minnesota is like a museum that shows what most of Minnesota would be like today if glacial ice had not remodeled it.

Minnesota enjoyed a long quiet period after the rocks of the southeastern corner were formed. Then, about 100 million years ago, a period of mountain building began in the western part of the continent, and extensive warm, shallow seas crept eastward into the western part of the state. The rocks that were formed in these seas are buried beneath later glacial deposits. They make little mark on the land, but they were formed in the age of dinosaurs. These giant reptiles probably stalked the area that is now Minnesota. So far no one has discovered any dinosaur fossils in the state, but one can always hope.

The present land surface of nearly all of Minnesota is the product of deposition and erosion by glaciers. Glaciers form when summers are too cool to melt the snow

Snow-flecked fields on level uplands are dissected by deep dark wooded valleys in southeastern Minnesota's bluff country.

that fell during the previous winter. Eventually the snow compacts and recrystallizes into ice. Ice is plastic, and when a mass of ice becomes thick enough, the edges begin to bulge outward in great tonguelike lobes tens of miles wide and several hundred to a few thousand feet thick. A sheet of ice more than a mile thick covered much of Minnesota during the last glacial epoch.

The last 2 million years have had many colder periods, when enormous ice sheets have formed over central Canada and lobes of ice have bulged southward into the United States. A warmer interglacial period, when the ice sheet melted, ended each period of glaciation, but then a new ice sheet formed when the climate turned colder once again. At its maximum extent, glacial ice reached about as far as the present courses of the Missouri and Ohio rivers. Each glacial lobe slid over and obliterated or buried the features formed by earlier lobes. All the land-surface features of Minnesota, except in the very far southwestern and southeastern corners, were formed by the most recent glaciation, which peaked about 20,000 years ago.

Glaciers eroded by plucking and scouring. Meltwater from the ice trickled into cracks in the underlying rock and froze. These fingers of ice remained attached to the glacier as it moved on, and they plucked chunks out of the solid bedrock. The rocks frozen in the bottom of the glacier enabled it to act like coarse sandpaper, scouring and polishing the bedrock across which it passed. In places these rocks have cut long, parallel scratches (called striations) in the smooth ice-scoured bedrock. The hard Sioux quartzite rocks at the Jeffers Petroglyphs site in Cottonwood County have splendid examples of glacial striations, which reveal the direction in which the glacier was moving.

On a larger scale the ice scoured out the least resistant rock formations across which it moved and gouged out bare bedrock basins that have subsequently filled with water to become rockbound lakes. The Boundary Waters Canoe Area Wilderness in the northern part of northeastern Minnesota's Superior Upland is a superb example of an area of bare, ice-scoured bedrock speckled with many lakes.

"Drift" is the general name for all material deposited by glaciers. The bluff country of southeastern Minnesota, for example, is part of an area known as the Driftless Area; the area has no ice-deposited material because glacial ice never reached it, at least during the most recent glaciations. When a glacier finally melts, it dumps all the material it has eroded in irregular piles of unsorted debris that ranges in size from boulders to fine rock powder. This heterogeneous mixture of rock fragments of all sizes is called "till." The type of bedrock eroded by the ice determines the composition of the till it deposits; careful analysis of till may indicate the direction from which the glacier came.

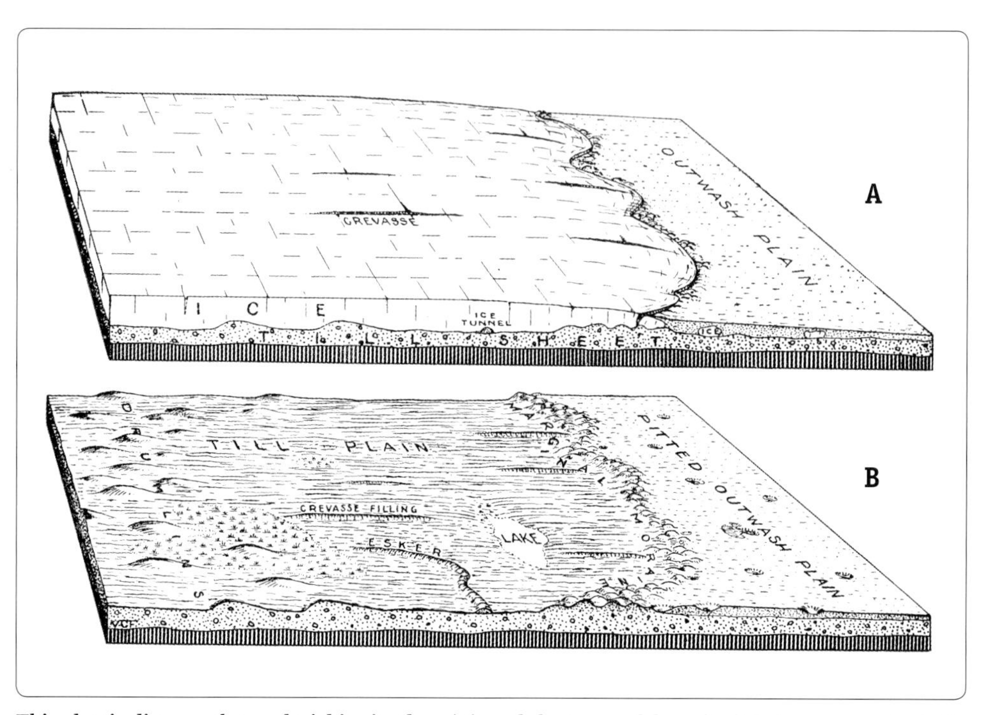

This classic diagram shows glacial ice in place (A), and the types of deposits it left after it melted (B).

Large streams of water flowed away from the glaciers that covered Minnesota when they started to melt. These streams picked up small particles of sand and silt. When the water lost velocity, it deposited these sediments in horizontal layers in which all the particles are roughly the same size. These water-sorted sediments are called "outwash deposits." Sometimes outwash deposits buried irregular blocks of ice that were separated from the glacier. When these ice blocks finally melted, they left depressions that have become filled with water to form lakes with attractive sandy shores.

Outwash deposits have sandy soils that drain quickly but lack essential nutrients for some plants. These loose, sandy soils are well suited to growing root crops, such as potatoes, if they are fertilized adequately, but they are susceptible to wind erosion and drought. The soils warm up quickly in the spring; near urban markets they are good soils for growing early vegetables.

Outwash deposits near cities are attractive areas for large-scale residential development. The land is level and poses no obstacles or barriers, and the well-drained, sandy soil helps to keep basements dry. Outwash deposits near urban areas often are mined for sand and gravel because running water sorted the particles by size when it deposited them.[2]

When glaciers melted, their deposits formed features known as "moraines," "till plains," "drumlins," and "eskers." End moraines are linear belts of choppy topography that were formed when the lead edge of the ice remained stationary for tens, perhaps hundreds, of years. The mass of the ice continued to move forward, but the leading edge remained in the same place because the ice was melting as fast as it was advancing. The glacier was like a giant conveyor belt that kept dumping piles of till where the edge of the ice was melting; over time it built up a belt of hummocky but not hilly topography, with slopes that are steep but short.

The slopes of moraine areas are so steep that they would be eroded if they were cultivated, and much of this land is used today for woodland or pasture. Such areas are better suited to dairy farming than to growing cash crops. Hundreds of lakes dimple some moraines, because the piles of till buried blocks of ice that eventually melted to form enclosed basins. Other lake basins were formed when the helter-skelter deposition of till blocked their outlets. The interesting topography and lake vistas of moraines near cities make them attractive as affluent residential areas.

When Minnesota's glaciers stagnated, then haphazardly melted and dumped a thin blanket of till over a large area, they formed "till plains." Till plains have undulating topography of low swells and swales. Till plains sometimes make excellent agricultural areas, but their surfaces are so irregular that they must be drained before they can be cultivated. Today most till plains are laced with networks of open drainage ditches and subsurface drainage systems.

Some till plains have swarms of "drumlins," or elongated oval hills fifty to one hundred feet high and half a mile long. Drumlins have steep snouts facing the direction from which the ice came, and long, streamlined tails. "Eskers" are low, winding ridges that look like abandoned railroad embankments. They were formed when streams of meltwater deposited water-sorted sediments as they flowed through tunnels in stagnant ice. Eskers are good sites for sand and gravel pits.

The uneven deposition of glacial till meant that till plains have large numbers of shallow depressions, or potholes, that collected water to form lakes. The postglacial growth of grasses and sedges has filled many of these lakes and turned them into marshes or peat bogs. Till did not completely fill some of the deeper preglacial valleys, which remain lower than the areas on either side. Today these buried valleys are marked by chains of lakes, such as Lake Calhoun and Lake Harriet in western Minneapolis, the line of lakes starting at Vadnais Lake north of St. Paul, and the north-south lines of lakes in Martin County near Fairmont. In these valleys huge blocks of ice were buried in the deposits of glacial till, and the lake basins were formed when these buried blocks of ice melted.

The dark wooded area at the top is a glacial moraine near Apple Valley, and the level outwash plain at the bottom is a desirable place for large-scale residential subdivisions.

Moraines have hummocky topography.

Morainic topography has slopes too steep for cultivation and is best left in woodland or pasture.

Myriad lakes dimple many moraines.

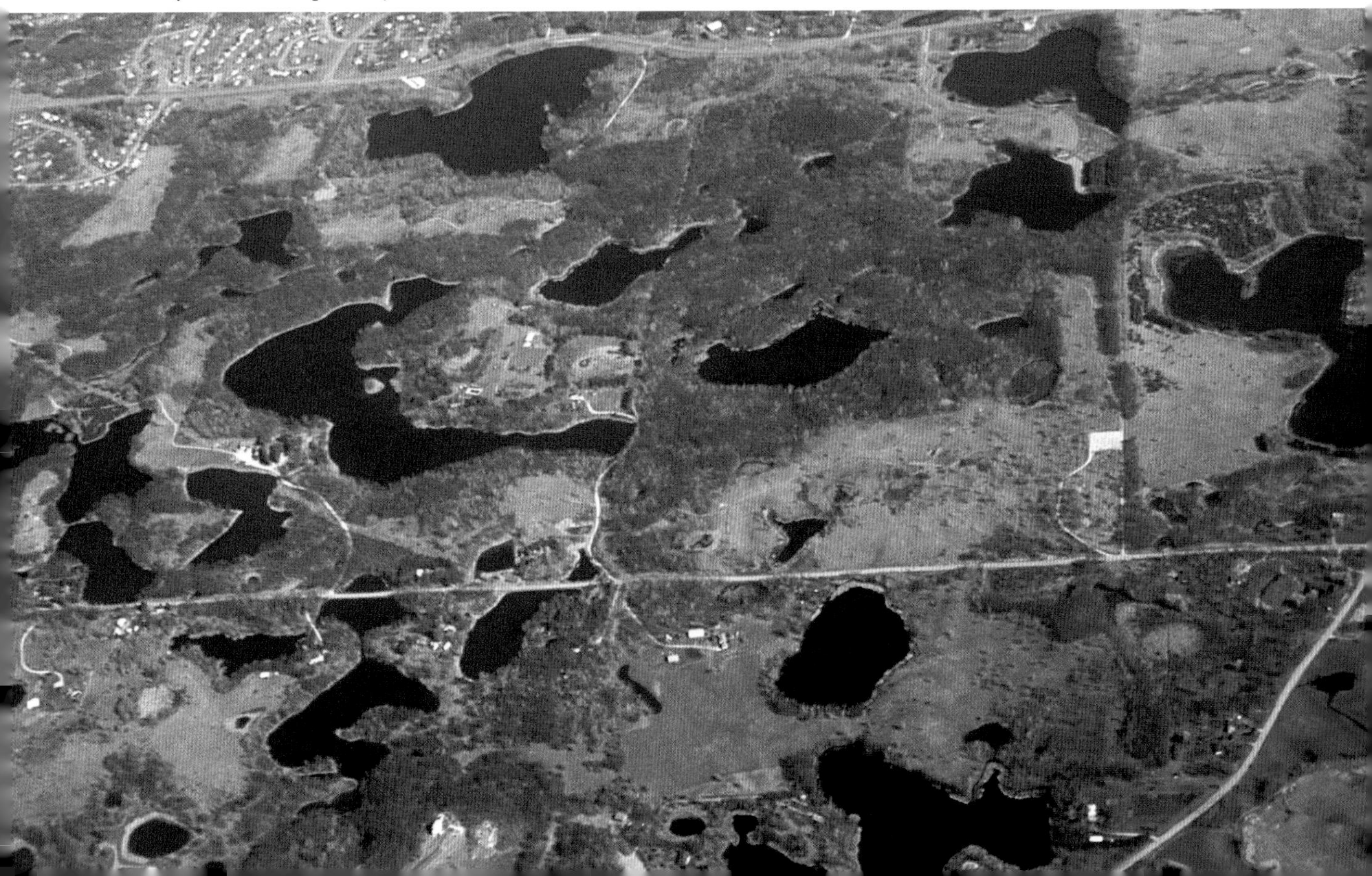

Till plains have undulating topography of light-colored swells and darker swales.

Till plains need drainage ditches to remove their excess water.

Light-colored vegetation is filling a small lake and turning it into a bog. The rings of trees are on the former shoreline of the lake.

THE DISTRIBUTION OF GLACIAL LANDFORMS in Minnesota is extremely complex because huge lobes of ice bulged southward across the state many times, and from different directions, over a period of more than 60,000 years. The different glacial landforms often shade imperceptibly into one another, sometimes over a distance of several miles.[3] They rarely have sharp boundaries. In broad outline, though, four major lobes—named the Wadena, Rainy, Superior, and Des Moines—were responsible for most of the present land-surface features of the state.

About 75,000 years ago, the Wadena Lobe moved southeastward into northern Minnesota, then recurved southward through the center of the state. It deposited fine, silty, gray till consisting mainly of particles of limestone from the Winnipeg Lowland. It formed the impressive Alexandria Moraine, which has an exceptionally large number of small lakes, and a fine swarm of drumlins east of it near Wadena. The western edge of the Alexandria Moraine coincides fairly closely with the boundary between prairie and woodland areas in Minnesota.

Then, about 30,000 to 20,000 years ago, the thin Rainy Lobe and the much thicker Superior Lobe advanced side by side into northeastern Minnesota. Their brown and reddish till consisted of coarse, sandy fragments of the tough old rocks of the Superior Upland. The soils derived from this till, which are of only average quality, are used mostly for dairy farming. The Rainy Lobe scoured the Superior Upland, and the Superior Lobe formed the thick, hummocky, boulder-strewn St. Croix Moraine. A later advance of the Superior Lobe formed the moraine that cups Mille Lacs.

About 14,000 years ago the massive Des Moines Lobe pushed across Minnesota from the Red River Valley into central Iowa. It deposited fine-textured, silty, gray to brown till consisting mainly of shale particles from Canada and North Dakota. It formed an undulating till plain that now has the finest farmland in south central and southwestern Minnesota, but it had large numbers of shallow swales that had to be drained before it could be cultivated.

The western edge of the Des Moines Lobe was the Coteau des Prairies, a smooth upland that rises 400 to 600 feet higher than the level till plain to the east. In Minnesota the Coteau is also known as Buffalo Ridge. Its eastern edge is so straight and sharp that it looks like a plateau whose topography is controlled by the underlying bedrock. Well borings have found only several hundred feet of glacial deposits beneath it, but many hillsides have tantalizing outcrops of the tough old Sioux quartzite.

Buffalo Ridge is a desirable site for developing windpower, because strong winds whip across the treeless prairie upland. As a result, Minnesota has become the nation's fourth largest generator of wind energy, behind only California, Texas, and Iowa. The countryside northwest and southeast of Lake Benton looks like a giant pincushion, with a new forest of wind generating-turbines. In 2005 the Buffalo

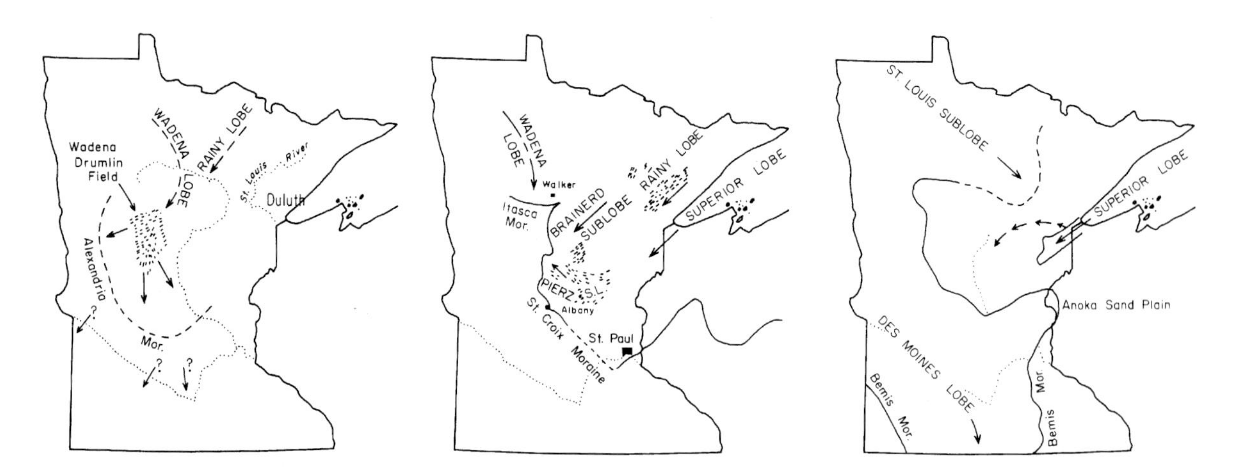

Four glacial lobes are responsible for most of the present land-surface features of Minnesota.

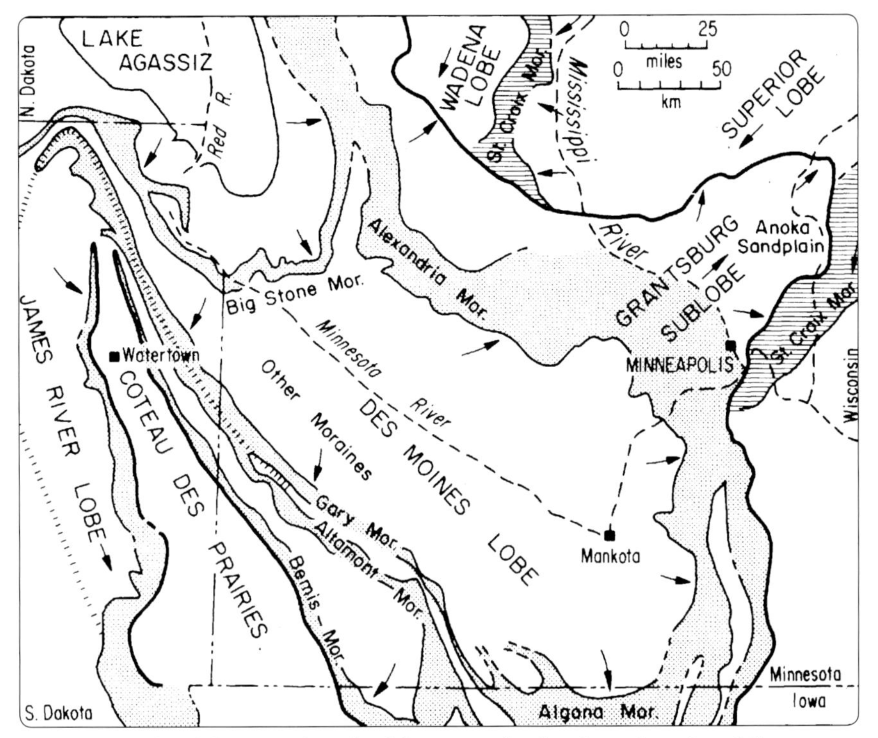

The Des Moines lobe created much of the present land surface of southern Minnesota.

Ridge had nearly 700 turbines that were up to 260 feet high and with blades up to 250 feet long.[4] Each tower has an interior service ladder.

The turbines cost around half a million dollars apiece, and the owners on whose property they stand collect $750 to $4,000 per turbine per year in rent for land that is fairly marginal agriculturally.[5] Lincoln County is the poorest county in the southern part of the state, so the wind turbines are a welcome source of income. The blades are idle for about two-thirds of the time, but when the wind speed reaches nine miles an hour, they automatically rotate into it and start to drive the generator that produces electricity. Wind power is an environmentally cleaner source of power than coal, oil, or nuclear fuel, and the power company has offered environmentally conscious customers the privilege of paying more for it, even though no one can identify the source of the electricity that enters an individual house.

The western edge of the Des Moines Lobe formed several parallel moraines along the eastern edge of the Coteau, and the area to the west has much older glacial deposits that are old enough to have developed a good stream drainage system. The eastern edge of the Des Moines Lobe overrode the Alexandria and St. Croix moraines; the Grantsburg Sublobe bulged northeastward past the Twin Cities into northwestern Wisconsin, where it created many small lakes and high enough hills to sustain several ski areas. South of the Twin Cities the Des Moines Lobe formed a morainic belt that runs through Owatonna into Iowa. East of this belt are older glacial deposits that have developed a stream drainage system, and the bluff country in the far southeastern corner of the state has rocky, stream-dissected topography that probably remained ice-free during the past 2 million years.

Glaciers like the Des Moines Lobe produced prodigious quantities of meltwater when they finally started to melt. Some of it collected as ponds and lakes in the low places that resulted from uneven glacial deposition, but far larger water bodies were dammed for a thousand years or so by the great ice sheet that was ever so slowly melting northward. Northern Minnesota naturally drains northward to Hudson Bay or eastward into Lake Superior, but ice sheets blocked these outlets. The water level behind these ice dams gradually rose higher and higher until it found outlets through which it could overflow to the south.

The largest ice-dammed lake was glacial Lake Agassiz, which was actually a vast inland sea greater in size than all five modern Great Lakes combined.[6] At one time or another Lake Agassiz covered most of southern Manitoba and western Ontario, as well as the northwestern corner of Minnesota. The eastern arm of Lake Agassiz in northern Minnesota was fairly shallow, and the floor of the former lake is now one of the world's largest uninterrupted wetlands, with many raised bogs and large areas

Wind turbine towers near Hendricks.

of peatland. Lake of the Woods, Upper Red Lake, and Lower Red Lake are in deeper basins on the former lake floor.

In the Red River Valley, Lake Agassiz was more than 300 feet deep. Waves and currents reworked the sediments that were deposited in the lake, filled in the irregularities on the lake floor, and created a featureless surface that is one of the flattest areas on the face of the earth today. Wind-driven waves built low beach ridges of sand and gravel that can be traced for miles along the eastern shore of the former lake. Originally these shoreline features were level, but today they are a few feet higher toward the north because the northern part of the continent has been rising slowly since the weight of the ice sheet has been removed.

Eventually the waters of Lake Agassiz rose so high that they overflowed at Browns Valley in western Minnesota and drained southward into what is now the Minnesota River. Such overflow outlets of ice-dammed lakes are called "glacial spillways." The St. Croix River, which forms part of the eastern boundary of Minnesota, was a spillway for glacial Lake Duluth, formed by meltwater from the Superior Lobe that collected in the Lake Superior basin. The river cut a deep gorge through an ancient lava flow to form the picturesque St. Croix Dalles near Taylors Falls.

The Lake Agassiz Plain is one of the levelest areas in North America.

For several thousand years, until the ice dam melted back and the lake found lower outlets farther north, the waters of Lake Agassiz flowed through the Minnesota River spillway across Minnesota. The original valley of the river was formed by water from the melting Des Moines Lobe. The river flowed straight southeastward across the western half of the state, but at Mankato it turned sharply northeastward on easily eroded sandstone rocks along the western edge of an upland underlain by tougher limestone strata.

The overflowing waters of Lake Agassiz transformed the Minnesota River valley into an impressive flat-bottomed, steep-sided trench one to five miles wide and 75 to 200 feet deep in the till plain deposited by the Des Moines Lobe. In a few places the river has cut down to the ancient basement bedrock, which is some of the oldest rock known on earth. The gneiss quarried near Morton, for example, was formed more than 3.5 billion years ago.

At one time the Minnesota River valley was even deeper than it is today, because episodes of erosion and deposition have alternated as the volume of water in the river and its load of sediment have increased or decreased. The river has cut its valley deeper, then filled it with sediment, then cut deeper again, not once but several times. The broad, level terraces perched well above the present floodplain are remnants of former floodplains.

Tributary streams once plunged over the edge of the Minnesota River spillway in waterfalls, but these streams have cut V-shaped notches back into the till plains on the adjacent uplands as the waterfalls have eroded their way upstream. Some of these streams have carried so much sediment into the spillway that they have dammed the river and created lakes. In western Minnesota they have formed Lake Traverse, Big Stone Lake, and Lac Qui Parle. In the eastern part of the state the Mississippi River has built a sediment dam across the St. Croix River to form Lake St. Croix; downstream, the Chippewa River has brought enough sediments from Wisconsin to dam the Mississippi and form Lake Pepin.

The Falls of St. Anthony in downtown Minneapolis show how waterfalls erode upstream.[7] Earlier these falls were at the present site of Fort Snelling, where the Mississippi River tumbled over the brink of the Minnesota River spillway. Resistant Platteville limestone caps the falls, but the Glenwood shale and the St. Peter sandstone beneath the limestone are easily eroded. Water churning over the falls eats away the lower formations, and chunks of the undermined limestone eventually fall into the river and are carried downstream.

The Falls of St. Anthony have migrated upstream slowly but steadily since Father Hennepin first saw them in 1680. In 1888 Newton H. Winchell, the Minnesota state geologist, used the retreat of the falls to make the first scientific estimate of postglacial time. He plotted the successive locations of the falls, as shown on early maps, and figured that they were moving upstream at an average rate of about five feet per year. Now the falls have been fixed in concrete by the U.S. Army Corps of Engineers, and their upstream migration has ended.

Glacial activity has been responsible for most of the surface features of modern Minnesota, except for the bluff country of the southeast. Glacial ice scoured the tough old rocks of the northeast. The Red River Valley of the northwest was formed on the floor of an enormous inland sea that was dammed by ice. The level ground of the southwest is a till plain deposited by the Des Moines Lobe, and the complex topography of the center and east is the product of many ice advances and retreats.

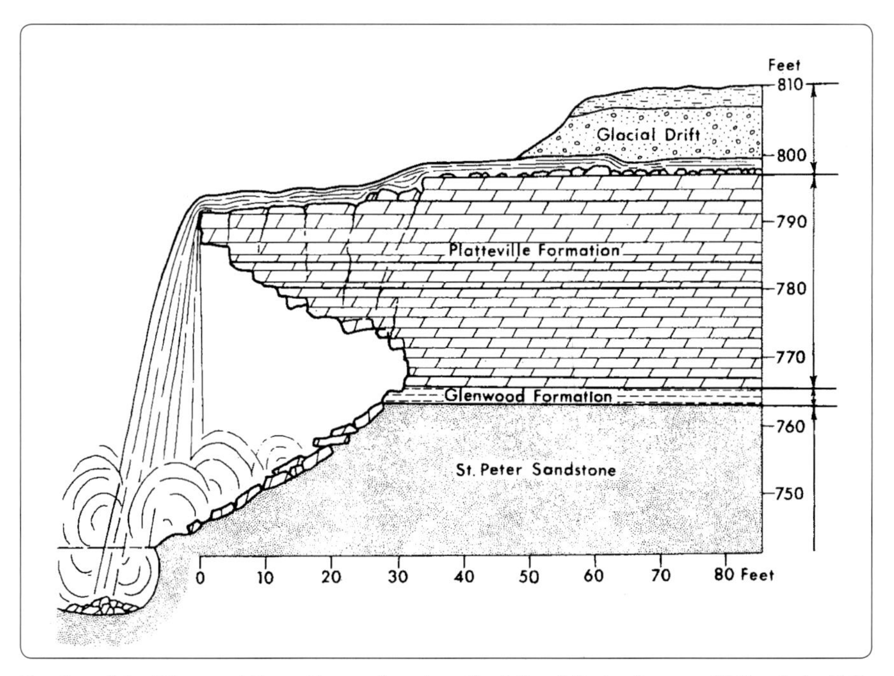

Erosion of the Glenwood Formation undermines the Falls of St. Anthony and Minnehaha Falls (shown above), forcing them to migrate slowly upstream.

Kittson
Roseau
Lake of the Woods
Marshall
Beltrami
Koochiching
Pennington
Red Lake
Polk
Clearwater
Itasca
Norman
Mahnomen
St. Louis
Hubbard
Cass
Clay
Becker
Wadena
Aitkin
Carlton
Crow Wing
Wilkin
Otter Tail
Pine
Todd
Grant
Douglas
Morrison
Mille Lacs
Kanabec
Traverse
Stevens
Pope
Benton
Stearns
Sherburne
Isanti
Chisago
Big Stone
Swift
Anoka
Chippewa
Kandiyohi
Meeker
Wright
Washington
Lac Qui Parle
Hennepin
Ramsey
Yellow Medicine
Renville
McLeod
Carver
Scott
Dakota
Sibley
Lyon
Redwood
Nicollet
Rice
Goodhue
Le Sueur
Wabasha
Brown
Murray
Cottonwood
Watonwan
Blue Earth
Waseca
Steele
Dodge
Olmsted
Winona
Rock
Nobles
Jackson
Martin
Faribault
Freeborn
Mower
Fillmore
Houston
1. Prairie Coteau Scientific and Natural Area
2. Bluestem Prairie Scientific and Natural Area
3. St. Cloud
4. Mankato
5. Faribault
6. Mississippi River
7. Cannon River
8. Minnesota River
9. Nerstrand Big Woods State Park
10. North Shore
11. Lake Superior
12. Boundary Waters Canoe Area Wilderness
13. Itasca State Park
14. Kellogg-Weaver Dunes Scientific and Natural Area

3

Evolving Ecosystems

Minnesota has three major biomes within its narrow elevation range of 1,700 feet. These biomes, or large areas with ecological communities of similar plants, animals, climate, and soils, are the tallgrass prairie, the eastern deciduous forest, and the northern coniferous forest. Types of vegetation within the biomes vary with temperature, precipitation, topography or landform patterns, soils, and land uses. The complex ecosystems are not static but have evolved since glaciers wiped the slate clean of earlier vegetation.

The locations of Minnesota's biomes have shifted with variations in temperature, amount of precipitation, and other environmental variables. The plants and animals of the biomes have also changed. Species responded to environmental alterations individually rather than as groups or assemblages of organisms.

Most of Minnesota was covered many times by glaciers. When the ice sheets retreated from the northern part of the state for the last time by 11,000 years ago, the newly exposed land was devoid of vegetation and soils. The bluff country of southeastern Minnesota, where the ground surface was not covered with ice during the recent glacial period, might have been an ice-age refuge for plants and animals if the climate near the ice margin was not too harsh and the soils were not frozen.

Fossil pollen, preserved in layered lake and bog sediments, records changes in vegetation over the past several thousand years. Paleo-environmental reconstructions suggest that vegetation and climate were relatively stable from 21,000 to 17,000 years ago. They changed rapidly from 16,000 to 8,000 years ago but were fairly stable from 7,000 to 500 years ago. The biggest changes during this period

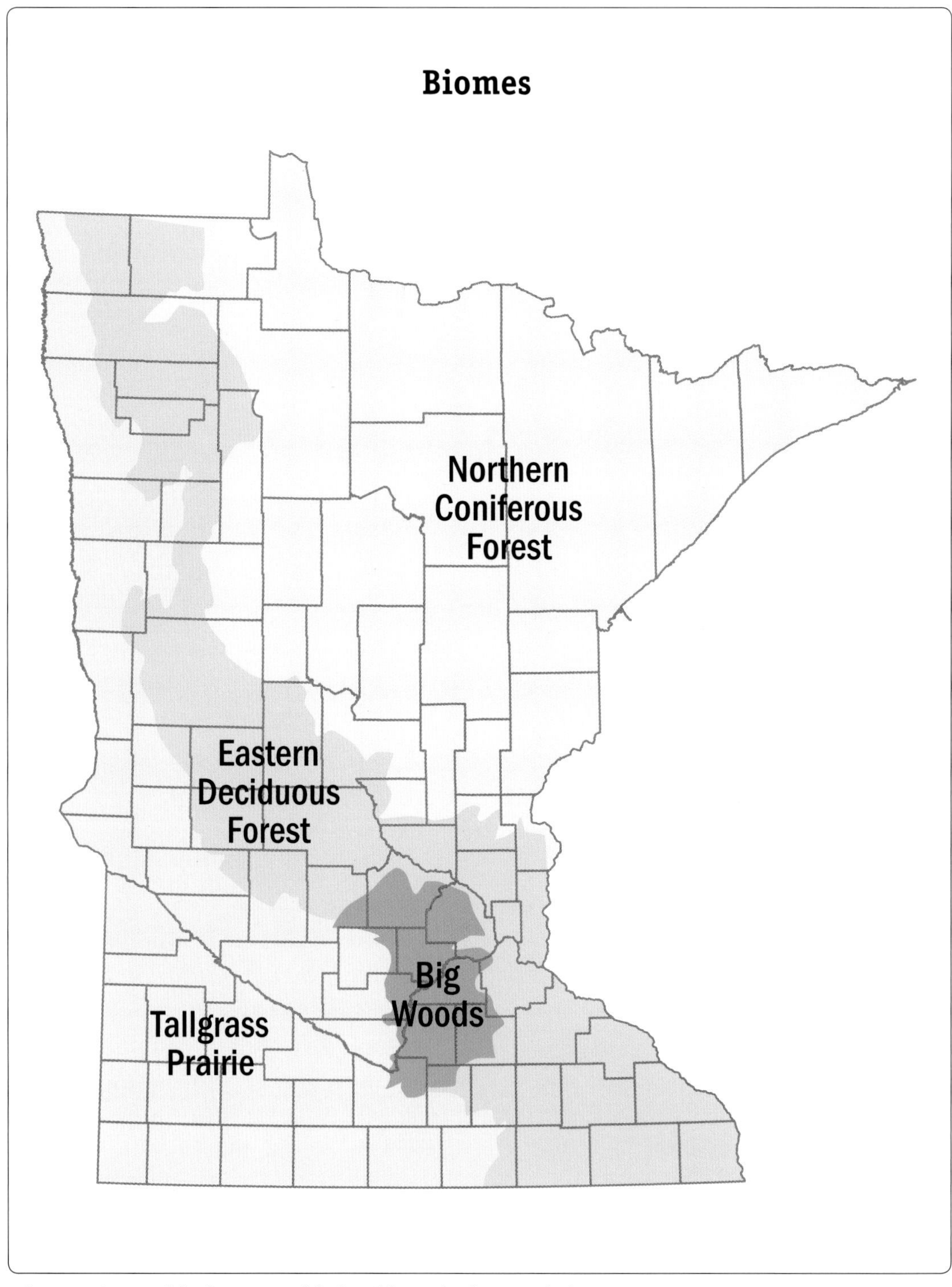

Minnesota is one of the few states with three biomes in close proximity.

were the eastward expansion of the prairie when the climate was drier and the westward movement of the deciduous forest during periods of greater precipitation and possibly less fire. European settlers began changing the vegetation dramatically in the mid-1800s.[1]

We do not know which plant species were the first to arrive in the wake of the melting ice sheet, but the initial colonizers probably were wind-dispersed tundra plants that could tolerate poorly developed soils and a harsh, variable climate. Present-day tundra vegetation in Canada includes primitive life forms such as algae and low-growing plants of the tundra. Soils began to develop in the ice-deposited glacial till or mix of particles, windblown rock flour (loess), and clay that settled to the bottom of glacial lakes. Mosses, algae, and lichens may have been early life forms that helped build up the substrate in which subsequent plants rooted.[2] Many tree species took several thousand years to return to Minnesota from the Appalachian Mountains and southern North America where the trees probably survived during periods of glaciation.[3]

Some seeds, such as those of aspen and other poplars, are feathery and lightweight and can easily be blown long distances. Others, such as acorns, are not easily dispersed because they are heavy and fall close to the parent tree. Seed palatability and likelihood of dispersal by animals, seed shape, germination requirements, environmental tolerance, and genetics may also influence the rate of spread of a plant species.

Some woody plants had traits that enabled them to survive better than others. Species such as white spruce, black spruce, and tamarack have a shape and physiology that can withstand short growing seasons and heavy snowfall. Aspen and paper birch establish soon after disturbance, when the ground has abundant sunlight. The nitrogen-fixing bacteria on the roots of plants such as alder absorb atmospheric nitrogen and convert it to a form that can be taken up and used by the plants growing on infertile soils. Special fungi help tree roots absorb nutrients. With a warmer climate came jack pine, which can tolerate nutrient-poor, well-drained soils. By 8,500 years ago, Minnesota was warm and dry enough that oak woodland, or scattered oaks among grasses and wildflowers, covered much of the southern part of the state. By 4,000 years ago, white, red, and jack pine were well established in northern Minnesota.

Tamarack is a needleleaf deciduous tree with golden foliage in autumn before the needles are shed for the winter.

Aspen and birch are two tree types that often grow up after wind, fire, or logging has removed the former vegetation.

THE FIRST DETAILED INFORMATION about Minnesota's vegetation was collected by the U.S. Public Land Office between 1847 and 1907. Surveyors measured each square mile of land and marked the section corners by identifying "witness" trees by species and size. They sometimes kept notes of the topography and vegetation along section lines, as well as the farming potential, location of wetlands and water bodies, and evidence of fire or wind blowdown. From these U.S. General Land Office survey notes, Francis J. Marschner in 1929–30 compiled a remarkable map titled *The Original Vegetation of Minnesota.*[4] The map documents the state's biomes at the time of European settlement, but today's land cover differs from the vegetation of Minnesota's early days of statehood.

Marschner's map shows Minnesota's three bands of vegetation that extend from northwest to southeast: prairie, deciduous forest, and coniferous (needle-leaf) forest. The width of each band's transition zones or "ecotones" varies with climate, soils, topography, and fire. The transition from prairie to deciduous forest to coniferous forest is abrupt in northwestern Minnesota, but the prairie-forest border in southeastern Minnesota, northeastern Iowa, and southern Wisconsin is not as sharp.

As recorded by surveyors, the prairie covered one-third of the state, extending diagonally from northwestern to southeastern Minnesota. The vegetation is

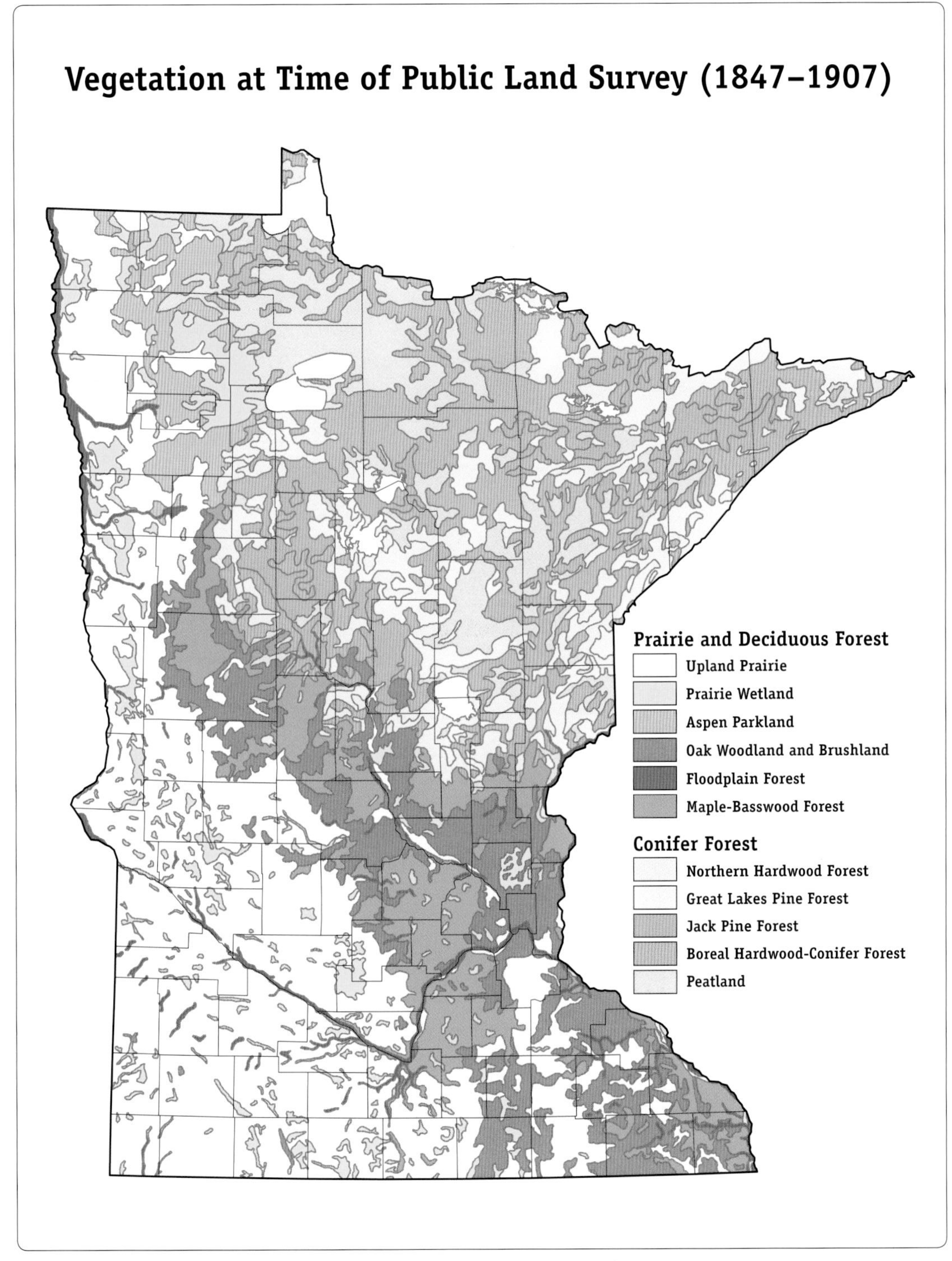

Minnesota's vegetation at the time of the Public Land Survey (1847–1907).

Nonwoody plants such as grasses and wildflowers flourish in Minnesota's prairies, where fire historically prevented trees from growing.

influenced by a decrease in precipitation from east to west and a decrease in temperature from south to north. The vegetation zones correlate well with climatic regions. The grassland has low winter precipitation, occasional major summer droughts, and a continental source of air.

Frequent fires maintained the nonwoody vegetation. The leaves of the plants burned, but the grasses and flowers resprouted from their roots after fire. Climate alone cannot explain the great expanse of prairie in the United States. A combination of natural conditions and human activities probably affected the vegetation on a local to regional scale.

Topography, fire disturbance, and soil moisture influence the prairie-forest ecotone. The amount of precipitation that falls in the grassland roughly counterbalances the amount of water that can be lost to the atmosphere through evapotranspiration, or evaporation from the ground and transpiration from plants. Trees and shrubs grow in wetter settings such as depressions and stream valleys, where the soils have sufficient moisture and where fire is inhibited. For this reason, fingers of forest poke into the prairie lowlands.

The prairie vegetation on the fertile, well-developed soils of the uplands included grasses, such as big bluestem and Indian grass, and wildflowers. Little bluestem and sideoats grama were grasses that tolerated the dry, nutrient-poor conditions on sandy uplands. The low-lying, wetter soils supported plants such as prairie cordgrass, sedges, and rushes. Many prairie soils (known as Mollisols) were rich in nutrients because of the chemical composition of the windblown glacial deposits or loess in which they formed, the high concentration of partially decomposed organic matter from leaves that are incorporated into the soils, and the minimal loss of nutrients downward through the soil because the annual rainfall of approximately twenty-five inches is relatively low.

While settlers seeking to farm prairie land did not have to clear trees, sod busting, or cutting through the thick mat of roots, was no easy task. Most of the original prairie has been plowed under, with a few prairies remaining in cemeteries, along railroad rights-of-way, and in protected areas such as the Prairie Coteau Scientific and Natural Area (SNA) in Pipestone County in southwestern Minnesota and the Bluestem Prairie SNA in Clay County in northwestern Minnesota.

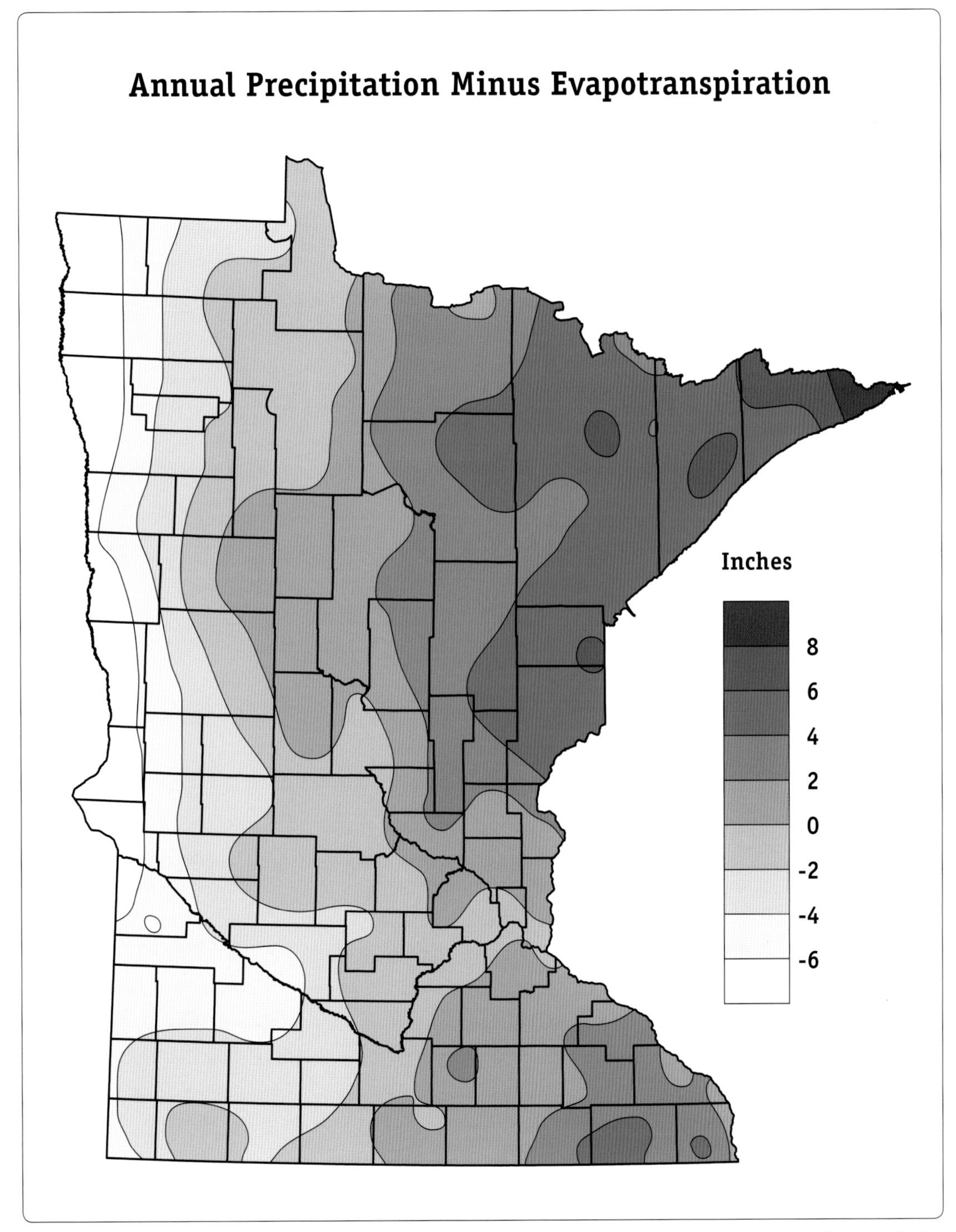

In the prairie in western Minnesota, the demand for moisture (potential evapotranspiration) typically exceeds the amount of rainfall (precipitation), which limits the growth of trees.

Oak savannas are a transitional vegetation type, with species from open prairies and closed forests. Sparsely distributed trees surrounded by grasses and wildflowers are adapted to the frequent fires that maintain the openness of the oak savanna.

Frequent fires and insufficient soil moisture kept trees out of the prairies, except where species such as willows and cottonwoods grew along streams and rivers. The oak woodland or oak savanna developed in the transition zone between prairie and deciduous forest. The soils were wet enough for trees, and the oak species resisted fire. Bur oak, for example, survived fires because of its thick, corky bark and its ability to resprout from its roots when fire killed young stems. Oak woodlands had continuous grass cover under a tree and shrub canopy of 30 to 70 percent. These open parklands covered 10 percent of Minnesota at the time of European settlement, but less than 1 percent remains. Many of the remaining tracts of oak woodland are sandy and unsuitable for agriculture.

MINNESOTA'S BIG WOODS WAS PART OF THE deciduous forest biome. The Big Woods was named Bois Grand by French settlers in the seventeenth century, and the name has stuck, even though little of the forest remains. The region extended from south of St. Cloud to Mankato to Faribault and Northfield. The Mississippi River lies to the northeast, the Cannon River is along the southeastern border, and the Minnesota River cuts through the area that was forest. Prairie and oak savanna lay to the east and west of the Big Woods.

The maple-basswood forest of the Big Woods contained elm, basswood, sugar maple, red oak, and white oak. Settlers cleared the trees for agricultural fields and maintained woodlots for firewood, fence posts, and lumber. In the spring the farmers collected sap from sugar maples and boiled it down to make maple syrup, an important locally produced sweetener.

The biomes have developed over the past several thousand years, and the Big Woods is young in comparison. Fossil pollen and charcoal fragments from lakes suggest that the Big Woods was established about 650 years ago when trees grew up in the former prairie. The moist, woody vegetation may have reduced fires, especially in areas with landforms, lakes, or rivers that served as firebreaks, allowing the forest to grow denser.[5]

Plant geography in the Big Woods has changed significantly since the European settlers arrived, largely because of human activities, including farming. Now less than 10 percent of the Big Woods region is forest, whereas 60 percent is agricultural land, and 14 percent supports urban development. Nerstrand Big Woods State Park, about forty-five miles south of the Twin Cities, protects one of the last large remnants of the central hardwood community. Nerstrand Woods is only 1,280 acres. Much of the former forest is now horse pastures, house lots, and fields of corn and soybeans.[6]

Nerstrand Woods receives approximately thirty inches of rain per year. The rolling hills were created by glaciers, which dumped unsorted rock fragments at this end moraine. Meltwater carved valleys as the glaciers receded, and more recently streams have eroded them. The glacial till overlies sedimentary rocks deposited about 500 million years ago. Oak savannas were on the margins and in pockets in fire-prone areas. The Big Woods constituted 80 percent of the region; the remaining 20 percent was prairie, lakes, and woodlands. The canopy composition has shifted from oak to sugar maple, basswood, elm, green ash, and ironwood.

Conservationists are concerned that development pressure may decrease biological diversity in the Big Woods region. Forest fragmentation favors invasive bird species such as the brown-headed cowbird over native species such as the scarlet tanager. Clearing land for housing construction, timber extraction, and farming is a threat for Minnesota's sole federally endangered species, the dwarf trout lily. This species, which probably evolved from the white trout lily no more than 9,000 years ago, grows only in about 600 acres of woodland in the Cannon River watershed in

Spring wildflowers in the Big Woods complete their life cycle during a short window of opportunity when abundant sunlight reaches the forest floor before the leaves of the canopy trees are fully open.

Rice, Goodhue, and Steele Counties. It does not grow naturally anywhere else in the world. It spreads by underground bulb rather than seed, so a flood or other major disruption is needed to deposit an uprooted plant or bulb on a slope downstream to establish a new colony.[7]

Big Woods remnants are especially beautiful early in the growing season when the wildflowers are blooming. These spring ephemerals take advantage of the available light and bloom before the canopy trees have leafed out. Late April or early May is the time to admire the dwarf trout lily. Its mottled leaves look similar to those of the white trout lily and yellow trout lily, but the rare plants are distinguished by the shape and pale pink color of their blossoms.

In northern Minnesota the deciduous forest grades into coniferous forest. The trees are not exclusively cone-bearing needle-leaf species but are adapted to the snowy winters and short growing season. The coniferous forest is really a mosaic of northern hardwood forest, pine forest, boreal forest, and peatlands that formed on extensive plains left behind by glacial lakes. The vegetation types vary with land position, soils, underlying glacial deposits, fire frequency, and effects of wind, insect outbreaks, and other natural disturbances. Wetlands rather than forests are common where the drainage network has not yet developed on ice-sculpted land and former lake bottoms.

Itasca State Park has one of Minnesota's largest remaining old-growth northern coniferous forests.

At the time of the land survey in the last half of the nineteenth century, a northern hardwood forest of sugar maple, basswood, yellow birch, and red oak dominated highlands protected from fire along the North Shore of Lake Superior, but many of the other forests of northern Minnesota were coniferous. Red pine, white pine, jack pine, and white spruce were important species of the coniferous forest, but the species composition has changed since the logging era.

The abundance of jack pine at the time of the government survey indicates that fires were common. Jack pine can tolerate droughty, nutrient-poor soils, but it needs heat from fires or sunlight after the forest has been cleared to unstick the resin that holds the seeds in the cones. The seeds are then broadcast so the trees can reproduce.

Park managers have reintroduced fire to the pine forest at Itasca State Park. They hope that prescribed burns will remove dense underbrush and invasive species, allowing seeds of red and white pine to germinate.

Red pine, also known as Norway pine, is the state tree. Red pine and white pine were once abundant in many parts of the northern forest. The logging industry left only about 26,000 acres of tall-pine forest untouched in the Boundary Waters Canoe Area Wilderness, one of the few areas in the upper Great Lakes with extensive old-growth forests.[8] The other large remnant of old-growth red and white pine forest is at Itasca State Park in Becker, Clearwater, and Hubbard counties.[9] The park was established in 1891 and now protects about 5,500 acres of unlogged pine forest and another 2,000 acres of old-growth hardwood and wetland forest. Some 500,000 people visit Itasca State Park each year, but only a few travel beyond the headwaters of the Mississippi River to appreciate the vegetation diversity. Park visitors who walk a short distance into the woods from the southern edge of Wilderness Drive can stand in awe at the base of Minnesota's largest red pine and the park's largest white pine and try to imagine what the forest looked like before the lumberjacks went to work. The understory includes Minnesota's state flower, a tall orchid called the showy lady slipper that blooms in July.

The forest that has regenerated after the logging era consists of tree species native to the area, but in different proportions. Today the park is heavily managed in an attempt to grow the future pine forest. Park staff and volunteers oversee prescribed burns; remove unwanted vegetation, especially in the understory; plant desired species; and attempt to minimize the effects of browsing by deer.

The showy lady slipper is Minnesota's state flower.

The bluff country of southeastern Minnesota is a mosaic of vegetation types related to the soils and varied topography. Running water rather than ice has carved the layers of sedimentary rock, exposing steep bluffs capped by resistant rock. In the valleys, sandy outwash from melting glaciers and windblown loess provide a variable surface in which soils form. The uplands were wooded, the fertile valleys were cleared for farmland, and the sandy outwash plain was dry prairie and oak savanna that could be irrigated or planted with desert crops such as melons. Kellogg-Weaver Dunes Scientific and Natural Area in Wabasha County near Kellogg is an example of sand prairie and dry oak savanna. Average rainfall of thirty-four inches per year is fairly high, but the loose, sandy soils are so well drained that little moisture is available for plants, and the vegetation resembles a short-grass prairie more than a tallgrass prairie.

Vegetation patterns in bluff country vary not only with soils but also with the shape of the land surface. The cliff faces along the bluffs have unique microenvironments. Cool air seeps out of underground caves to cliff faces, supporting boreal plants and animals that are usually found in cooler climates to the north and west, or live here and nowhere else in the world. Some snail species and perennial plants are listed as federally threatened or endangered due to the rarity of their habitat and the possibility that people would disturb them.

Plants are continually adjusting to the ever-changing climate, to altered disturbance patterns, and to human modifications of the environment. Myriad conservation groups throughout Minnesota are attempting to restore the state's vegetation to its appearance at the time European settlers arrived. Volunteers and employees plant grasses, cut down trees, burn prairies, fence out deer, remove "exotic" species, and alter drainage patterns. Some critics wonder why the goal is to re-create the landscape of 150 years ago rather than of some other phase of the evolving vegetation, such as 650 years ago when the Big Woods formed. Is it possible to remove the human imprint? If we want to return to "natural" conditions, why not reconstruct the vegetation at the time of the continent's early human settlers some 12,000 years ago?

The world has changed since the surveyors walked the section lines in the 1800s. Climate, atmospheric chemistry, soils, agents of disturbance (such as plowing, draining, and logging), herbivores and pathogens, species composition, and the land surface itself are different. We do not know the former conditions exactly, and even in the absence of human activity, the vegetation of 150 years ago would have changed in ways we cannot know. Vegetation restoration is expensive, challenging, and time-consuming, but people hope that it will lead to healthier ecosystems and more aesthetic scenery.[10]

Kittson
Roseau
Lake of the Woods
Marshall
Beltrami
Koochiching
Pennington
Red Lake
Polk
Clearwater
Itasca
Lake
St. Louis
Norman
Mahnomen
Cass
Clay
Becker
Wadena
Aitkin
Carlton
Crow Wing
Wilkin
Otter Tail
Pine
Todd
Mille Lacs
Morrison
Grant
Douglas
Kanabec
Traverse
Stevens
Pope
Benton
Big Stone
Stearns
Sherburne
Isanti
Chisago
Anoka
Swift
Washington
Kandiyohi
Meeker
Wright
Chippewa
Lac Qui Parle
Hennepin
Ramsey
Renville
McLeod
Carver
Yellow Medicine
Scott
Dakota
Sibley
Lyon
Redwood
Nicollet
Goodhue
Rice
Le Sueur
Wabasha
Brown
Murray
Cottonwood
Blue Earth
Steele
Olmsted
Watonwan
Waseca
Dodge
Winona
Rock
Nobles
Jackson
Martin
Faribault
Freeborn
Mower
Fillmore
Houston
1. Lake Pepin
2. Grand Portage
3. Pigeon River
4. Rainy River
5. Duluth
6. St. Louis River
7. St. Croix River
8. North West Fur Company post
9. Snake River
10. Red River Valley
11. Pembina
12. Lake Mille Lacs
13. Traverse des Sioux
14. Mendota
15. Lower Sioux Agency
16. Redwood Falls
17. Birch Coulee
18. Fort Ridgely
19. New Ulm
20. Mankato
21. Red Wing

4

Indians, Voyageurs, and Croupiers

In 1659, when Pierre Esprit de Radisson and Médard Chouart des Groseilliers became the first Europeans to explore the area now known as Minnesota, the land was occupied by people who called themselves Dakota (Sioux). The Dakota fashioned arrowheads, spears, knives, axes, hoes, and other tools of stone or of animal bones. They made pots of clay for cooking and storing food, and they wore clothing and moccasins made of animal hides.

The Dakota were hunter-gatherers who moved with the seasons but returned to the same favored spots year after year.[1] They wintered in hide-covered conical tipis at sheltered places in the deep woods. The tipis were easy to set up and take down, and the poles could be rolled up in the hide to form a bundle that dogs could drag. After the spring thaw the group moved to the sugar bush, where women tapped maple trees to make sugar and men could fish, trap muskrat and beaver, and hunt ducks, geese, and other small game with spears or bows and arrows.

In late spring, groups moved to a fertile place near a river or lake, where women could plant corn, beans, squash, and other crops; the men hunted and fished. Dakota lived in rectangular lodges of poles covered by birch bark that they had to repair each year. The women and children collected wild berries, roots, herbs, and medicinal plants while their corn was growing, and when it was ripe, they dried it for winter use. Summer was also a time for pleasures, dances, and ceremonies.

In early fall Dakota moved to the lakes where wild rice grew. They paddled their canoes through the marshes, bent the stalks of the plants over the canoe's sides, and beat the grass heads to make the grain fall into the bottoms of the canoes. Wild rice was a staple of their diet. Late fall was the time of the big hunt, when men killed enough deer, buffalo, and other large animals to feed their families through the winter. Women dried the meat to preserve it and cured the hides. People often went hungry if the big hunt was not successful.

In the early 1700s another Indian group, the Ojibwe or Anishinaabe (Chippewa), began to move into northern Minnesota from the east. In skirmishes over the next century, they drove the Dakota out of the pine woods into the hardwoods and out onto the prairies. Some Dakota had already started to inhabit the prairies before the Ojibwe intrusion. The Ojibwe lived in dome-shaped wigwams formed of bent poles covered with birch bark or hides, and they used birch bark canoes, but their lifestyle was similar to that of the woodland Dakota. The Dakota and Anishinaabe named many of the lakes and rivers in Minnesota.

Indians began trading furs to people from France and England almost as soon as Europeans arrived in North America.[2] The fur trade flourished because Europeans wanted things the Indians could produce, and the Indians wanted European goods. Europeans wanted furs, especially beaver, because beaver hats symbolized wealth and power. They were willing to pay high prices for them. In return, the Indians wanted guns, axes, knives, kettles, textiles, and other manufactured goods that made their lives easier.

The early fur trade was organized by Indians, who were trappers as well as traders. By the 1640s, the Huron had developed a fur-trading network on the Great Lakes, and by 1670 some were trapping and trading as far west as Lake Pepin in southeastern Minnesota. The English expected the Indians to bring their furs to trading posts at the edge of the settled area where the English or other Euro-Americans lived, but the French were more aggressive. They took over the transportation of trade goods and furs from the Huron, but they still depended on Indians to trap and hunt for them, and to feed them.

Travel overland was so slow and difficult that the fur trade relied almost completely on travel by canoe on rivers and lakes, including the Great Lakes. French fur traders used the network of Indian routes on the Great Lakes, but they lost it to the English North West Company after the Treaty of 1763, in which France ceded Canada to Britain. Montreal, at the head of ocean navigation on the St. Lawrence River and gateway to the Great Lakes, was the headquarters of the fur trade. Agents in Montreal imported trade goods from Europe, packed them in watertight bales that weighed around ninety pounds, and hired French-Canadian voyageurs to paddle the canoes that carried them westward to the great transshipment posts.

Voyageurs were very strong but short in stature; few stood more than five foot six, because taller men could not fit comfortably in canoes.[3] Most did not need to know how to read or write. They could paddle tirelessly for eighteen-hour days and carry two ninety-pound bundles over long portages at a dogtrot pace that left other travelers puffing and panting. They liked to laugh and joke, and they accepted hardship and danger with equanimity.

Fur companies typically gave each voyageur two blankets, two shirts, two pairs of pants, and a plug of tobacco when he was hired. The voyageurs wore bright-colored sashes and knit woolen caps, and let their hair grow long as protection against the hordes of mosquitoes and flies that still are the bane of the north country. They subsisted on two meals a day of boiled corn seasoned with strips of pork; at night they rolled up in their blankets and slept on the ground beneath their overturned canoes.

The smallest birchbark canoes used by Indians and traders on small waterways weighed less than 300 pounds. They were made from, and could be repaired with, forest materials that grew everywhere. Strips of bark from birch trees were sewn together with the supple thonglike roots of red spruce trees, then were calked with melted gum from pine trees. The canoes had to be calked frequently, and passengers hardly dared move in them for fear of breaking the gum.

Medium-size North canoes were shorter, twenty-five feet long, and needed only six or eight voyageurs. They could carry about 3,700 pounds but were light enough for two men to haul over a portage. These canoes were used throughout the interior west.

Large Montreal canoes, employed on the Great Lakes and largest rivers, were thirty-five to forty feet long and required perhaps a dozen paddlers. These canoes were six feet wide at the middle and could carry up to ninety bales, each of which weighed ninety pounds. The voyageurs sat on the cargo bales for hours on end as they paddled. They sang lively songs in time with the rhythm of their paddles, speeding along at a clip of four to six miles an hour. They stopped regularly to smoke their pipes and measured distances in the number of pipes smoked or the ground they covered between smoking breaks.

In May, brigades of two to thirty canoes loaded with trade goods left Montreal for the two-month trip to the west. In July they rendezvoused at major transshipment posts, such as Mackinac and Grand Portage, with hundreds of traders and voyageurs from small interior posts. They unloaded their bundles of trade goods, which were repacked for carriage to the interior posts, and they packed furs from the interior posts to carry back to Montreal.

The canoemen of the Great Lakes were called pork-eaters *(mangeurs de lard)* because they lived on corn boiled with pork fat. They rendezvoused with the winterers *(hivernants)* who spent the winter in the woods at interior posts and lived on what they could buy from Indians or take from the land. The voyageurs who paddled and portaged were the manual laborers of the fur trade. They were supervised by the educated gentlemen, traders, and clerks, who had to be able to read, write, and keep careful records.

Five principal fur trade routes led to Minnesota: from Hudson Bay south to the Red River Valley; from Grand Portage up the Pigeon River and down the Rainy River to the Lake of the Woods; from Fond du Lac (southwest of modern Duluth) up the St. Louis River and portage to the headwaters of the Mississippi; from northwestern Wisconsin up the Brule River and portage to the St. Croix and down to the Mississippi; and from Green Bay across Wisconsin by the Fox and Wisconsin rivers to Prairie du Chien, and then up the Mississippi.

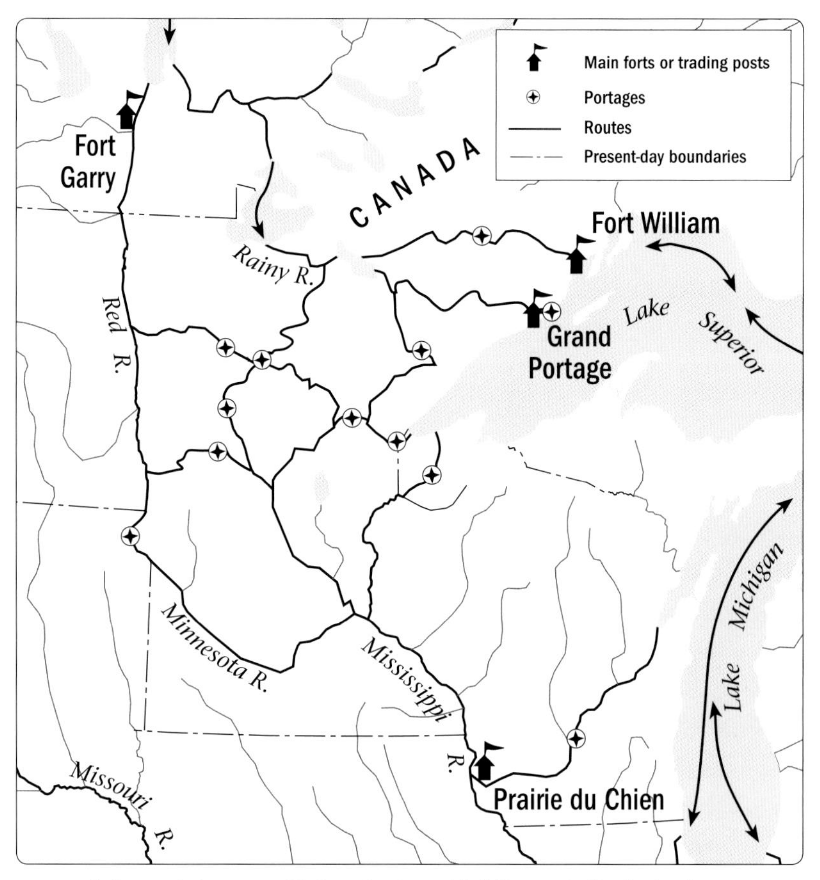

Lakes and rivers provided routes to and in Minnesota for fur traders.

Grand Portage, at the very northeastern tip of the state, may seem remote to modern Minnesotans, but it was the first important commercial center in the state. This portage was the easiest connection between the Great Lakes and the vast waterway system leading north to Hudson Bay and west to the Rocky Mountains. By 1733 Grand Portage had become a major rendezvous point, where goods from Montreal were exchanged for furs from interior posts.

In 1760 the British built a small fort at Grand Portage. Competing traders built trading posts, which had to be stockaded for protection against drunken brawls and fighting, because some traders used rum to lubricate trade. Eventually the traders merged to form the North West Company, which built its substantial headquarters at Grand Portage in 1780.

A gatehouse was the principal entry to the reconstructed Grand Portage National Monument.

The Grand Portage complex consisted of some sixteen wooden buildings, including a business office, storehouses, shops, living quarters for traders and clerks, and a large mess and meeting hall that could seat 100 people.[4] The complex was enclosed by a ten-foot-high palisade of wooden stakes for secure storage of goods rather than for defense against attack. Outside the palisade were a canoe yard, where canoes were built, and camping areas for voyageurs and Indians.

An artist's conception of how Grand Portage might have looked in its heyday.

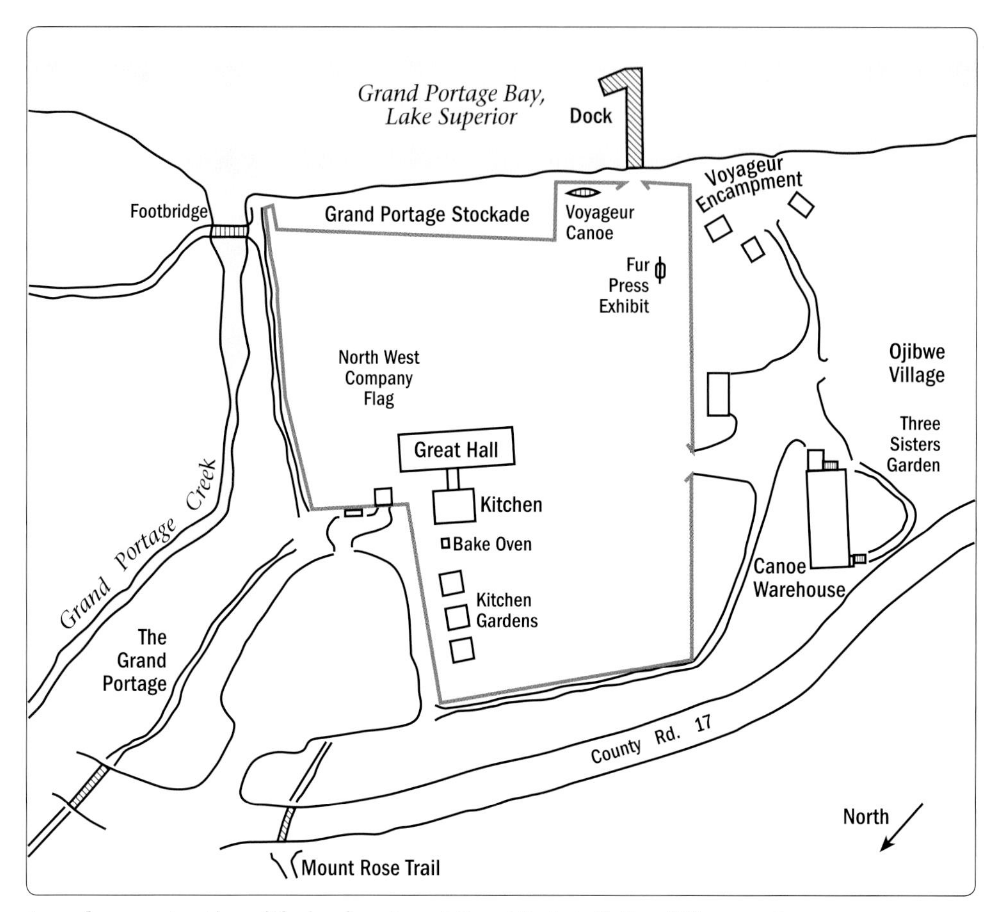

An early reconstruction of the headquarters at Grand Portage National Monument.

The annual rendezvous in July was a busy time. Hundreds of traders, voyageurs, Indians, and tough men from the lonely interior posts renewed their friendships. The voyageurs were paid their annual wages at the rendezvous. Many drank them up at "company-store" prices and wound up in debt to the company. Clerks repacked trade goods from Montreal in bundles for the interior posts and rebaled furs from the interior for shipment east. The rendezvous ended with a great ball in the mess hall, when the men danced with young Indian women to the music of fiddles, flutes, and bagpipes.

The North West Company used Grand Portage as its main western base until 1803, when it moved to Fort William (now Thunder Bay) in Canada to remain under the British flag. The buildings at Grand Portage gradually crumbled away and disappeared. In 1854 the Lake Superior Ojibwe signed a treaty with the United States turning over much of their land, including the site of the post; it remained unused for almost 70 years. In 1922, the Minnesota Historical Society assumed management of the area and began a decades-long process of excavation, planning, and restoration efforts. In 1958 a reconstructed site was turned over to the National Park Service and designated the Grand Portage National Monument. The post has been further reconstructed for accuracy. Visitors to the site learn about the fur trade and view the reconstructed Great Hall, kitchen, and a warehouse. Park rangers in period costumes organize a variety of special events, interpretive programs, and craft demonstrations, including building birchbark canoes. The nine-mile portage is now a year-round trail for hikers, backpackers, and cross-country skiers.

During the years of the fur trade, some Montreal men went up the portage to the Rainy River to meet traders from the distant Athabasca country in Canada. They officially became "Nor'westers" when they crossed the height of land east of Rainy Lake, from which water flows north to Hudson Bay. Nor'westers were entitled to wear plumes in their caps. They probably liked the excitement of shooting the rapids on the Rainy River. In 1960 the Underwater Research Project of the Minnesota Historical Society sent divers to search the riverbeds downstream from dangerous rapids where canoes loaded with trade goods might have capsized. The divers recovered kettles, axes, spears, knives, buttons, and a variety of other metal goods, many of which are now on display at the national monument.

After the rendezvous the traders, clerks, and voyageurs returned to their lonely winter posts on interior lakes and rivers. Many of the small interior camps were temporary, used only for a season or two. The voyageurs were remarkably skilled axmen. They could build a post in an impressively short time, and posts were abandoned as easily as they were constructed. Historian Grace Lee Nute has mapped the sites of more than 150 trading posts in Minnesota, and more still wait to be discovered.[5]

Fur traders built their posts near the winter camps of Indians and on rivers or lakes for ease of travel and for fishing, because fish were a staple of the winter diet. The first job of the trader after he had selected a site was to build storehouses and enclose them with a palisade of vertical logs, because the traders were leery of Indians and allowed only a few at a time inside the stockade. The trader, who remained at the post, sent his helpers out to hunt deer and beaver for fresh meat, and to trade with the Indians. He bartered with Indians for wild rice and maple sugar, and some posts had gardens with potatoes, beans, and other vegetables.

The reconstructed North West Company fur post in Pine City is a fine example of a small interior wintering post. It is based on the post built by trader John Sayer and his men in 1804. Sayer, two clerks, and eight voyageurs paddled up the Brule River from Lake Superior, portaged to the St. Croix, and went down the St. Croix

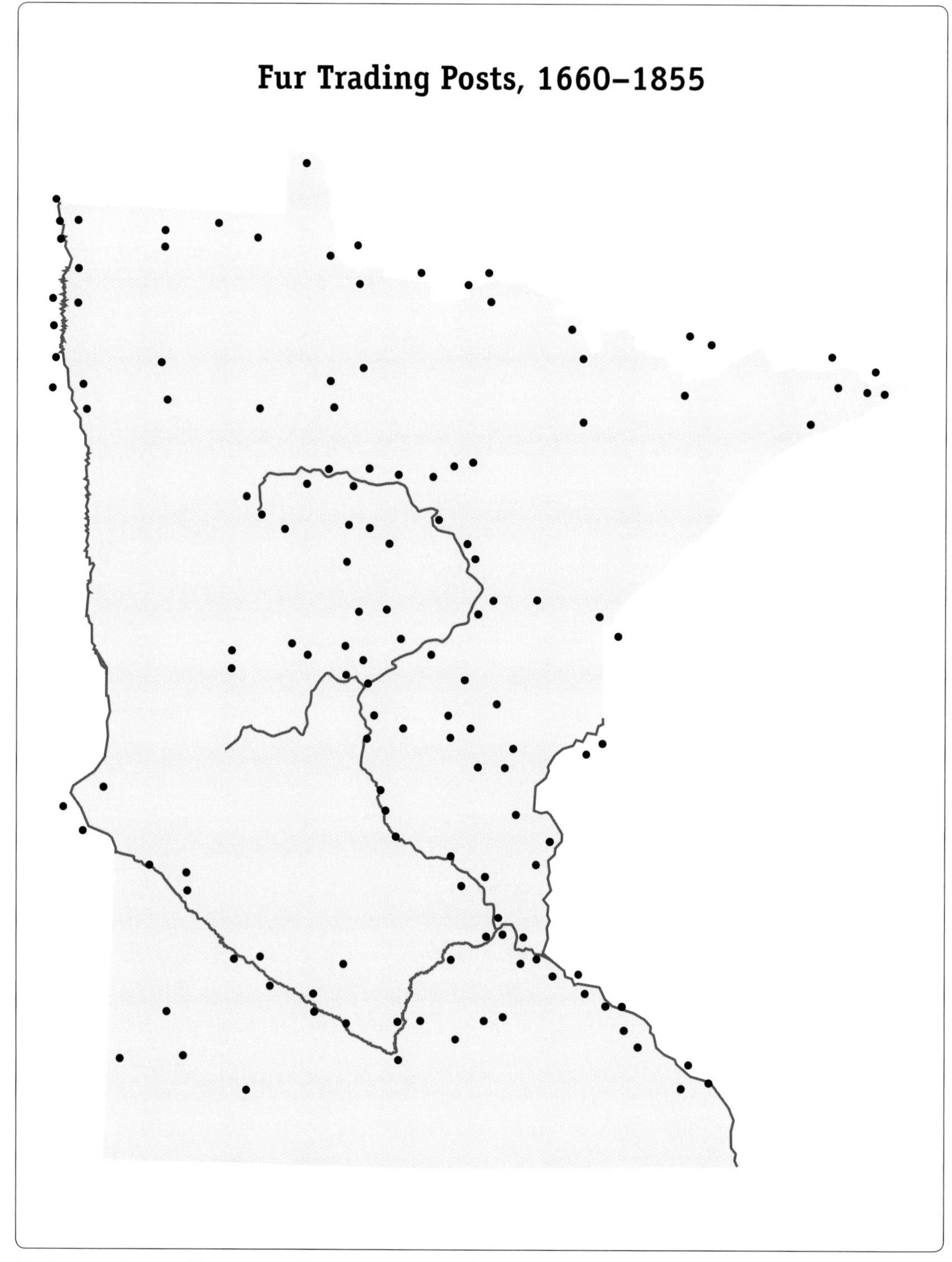

Most over-winter trading posts in Minnesota were near rivers and streams.

A sketch of the reconstructed North West Company fur-trading post near Pine City.

and up the Snake. Sayer picked a site near the winter camp of the local Ojibwe after talking to the chief, and his men built the post in only six weeks.

The post is palisaded with vertical poles and has bastions at opposite corners. The long log building is divided into six rooms, with short bunks against their back walls. Two were bunk rooms for the voyageurs, one was the quarters of the clerks, and one was the home of Sayer, his Ojibwe wife, and their children. The fifth room was a secure storeroom for trade goods and furs, and the end room was the "store" where Sayer did business with Indians. Costumed interpreters tell modern visitors about the life and activities of the post.

John Sayer, like many traders, for comfort and convenience married a woman from the tribe with which he traded. She could be an interpreter and teach him the language, and her family ties with the tribe were important. She could also perform domestic duties, such as making canoes, snowshoes, and moccasins. The children of Euro-American men and Indian women, sometimes called Métis, frequently entered the fur trade.

Métis who settled in the Red River Valley at the edge of the prairie grassland became renowned buffalo hunters. They hauled their furs and hides to St. Paul in two-wheeled Red River carts made entirely of wood. The ungreased axles of these carts screeched so loudly that they could be heard for miles. Each year from the

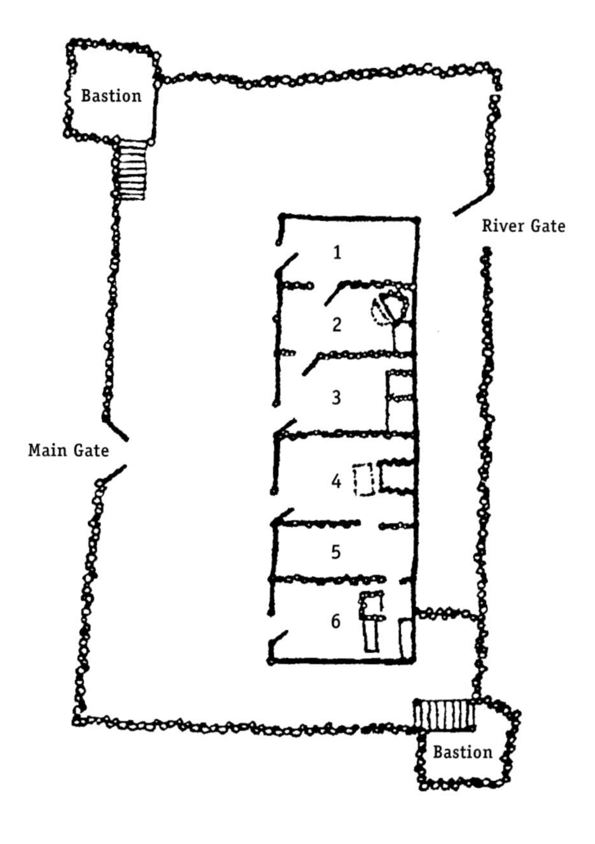

A palisade with corner bastions encloses the reconstructed North West Company fur-trading post near Pine City.

1830s to the 1860s, trains of 100 or more carts pulled by oxen made the two-month trip from Pembina to St. Paul. Their trails became important routes to the northwest.[6]

Americans moving up the Mississippi River from St. Louis and Prairie du Chien were slow in getting into the fur trade, but eventually they ousted the British. British troops had captured many American posts during the War of 1812 but were forced to return them by the Treaty of Utrecht in 1815. In 1819 the U.S. Army began to build Fort Snelling at the confluence of the Mississippi and Minnesota rivers. The fort created an island of safety for fur traders, and Mendota became a major base for the fur trade, but the fur trade was already on its last legs. Beaver had fallen out of fashion, replaced by a variety of lesser furs. The fur trade continued on a smaller scale, but times were tough for fur traders and Indians alike.

The North West Company fur-trading post, surrounded by its stockade.

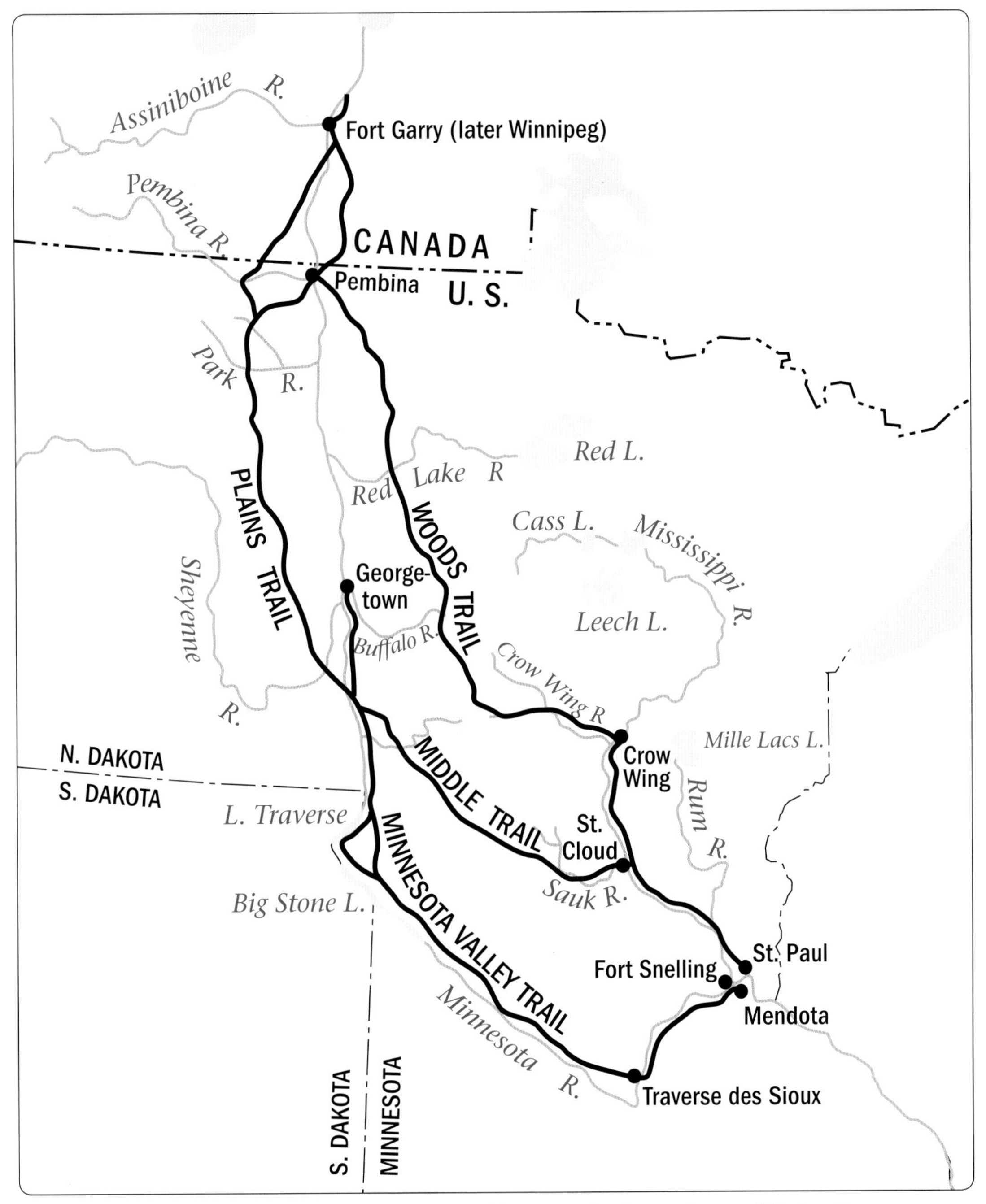

The Red River trails connected the Mississippi River with streams flowing north toward Hudson Bay.

Lumbermen and settlers were already starting to cast greedy eyes on Minnesota. They wanted the government to open land for white settlement by taking it from the Indians. In 1837 the United States negotiated a treaty with the Ojibwe and Dakota for the triangle of land between the Mississippi and St. Croix rivers as far north as Mille Lacs; in 1851 it acquired most of southern Minnesota from the Dakota in treaties signed at Traverse des Sioux and Mendota. The Dakota retained a ten-mile-wide strip of broken, hilly land along the Minnesota River that Euro-Americans did not covet, but in 1860 they were forced to cede the portion of this strip north of the river. By 1866 the Ojibwe had also signed treaties ceding most of their land in northern Minnesota.

Until 1871, the policy of the United States was to treat Indian tribes as though they were independent sovereign states, like England or France, and to make formal treaties with them. Government officials staged elaborate treaty-signing ceremonies, made grandiloquent speeches, and presented uniforms, medals, and flags to Indian leaders. The treaties they signed promised annuities of cash and food to the Indians, agreed to pay Indian debts to traders, established reservations, and guaranteed the preemption rights of Euro-Americans who were already living on the land. After 1871, the policy of treaty-making between the United States and tribal nations ended, but treaties already enacted continued in force, and tribal sovereignty, although altered, continues.

Some treaty provisions that were ignored or forgotten for years have created major confrontations when they have been revived. For example, some treaties reserved to designated tribal groups the right to hunt game and to fish on the lands they had ceded. In recent years some Indian groups have asserted that poverty and high unemployment on their reservations have forced them to exercise these treaty rights in order to survive, and the courts have upheld their cases, because treaties are the "supreme law of the land," according to the U.S. Constitution, and are binding legal documents. Some non-Indians have severely harassed Indians who choose to exercise their treaty rights.

After Indians signed treaties, incompetent government agents and dishonest traders frequently treated them badly, and even well-intentioned agents tried to force Indians to forsake their traditional hunting and become farmers. Relations between Indians and traders were based on credit: traders gave the Indians goods in the fall, and the Indians paid off their debts with furs in the spring. Indians had no way of checking the books the traders kept. They suspected traders of cheating them, and times were so tough that they frequently ran up hefty debts. The traders even grabbed off some of the annuity payments the Indians had been promised when they signed the treaties. The payments were usually late.

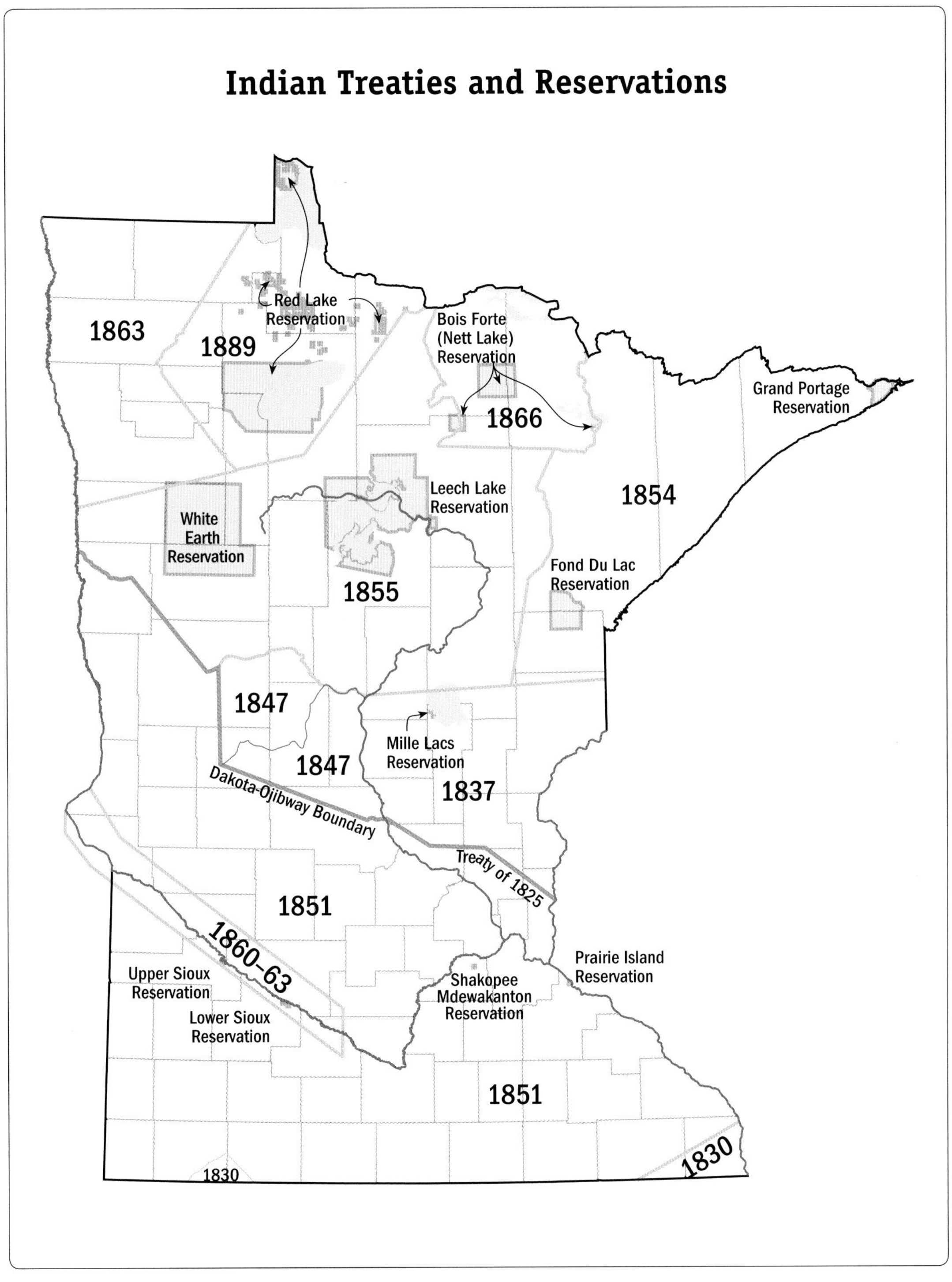

Major Indian land cessions and contemporary reservations.

In 1862 Indian relations with Euro-Americans in Minnesota reached a crisis. The previous year had been particularly harsh for reservation Indians, and the Dakota at the Lower Sioux Agency near Redwood Falls found themselves near starvation. The agent refused to give them food from the warehouse, even though it was well stocked, until the government's annuity payments had arrived, an event which was late, as usual. When the Indians protested that they were hungry, an arrogant young trader named Andrew Myrick reportedly said, "If they are hungry, let them eat grass."

On August 17, four young Dakota men killed five white settlers. Knowing that the settlers would punish them, some Dakota proposed taking the offensive: since many white men were away fighting the Civil War, this could be their last chance to take back their land. After debate, Dakota who chose to fight made a surprise attack on the Lower Sioux Agency, killing the agent, traders, and Indians and looting and burning buildings. One of the first people killed was Andrew Myrick. When his body was found, his mouth had been stuffed full of grass.

While some Dakota protected white settlers, others attacked isolated farms, killed settlers, plundered their belongings, burned their houses and barns, and made off with their horses and cattle. One woman who had been captured by the Dakota said she knew her husband had been killed when she saw an Indian riding his horse. Terrified settlers fled to Fort Ridgely, which the Dakota unsuccessfully attacked twice. Dakota also failed in their attacks on New Ulm and other settlements.

A hastily mobilized army defeated the Dakota at the battle of Wood Lake on September 23, 1862, and the six-week conflict ended when many Dakota fled west or to Canada. Although 303 captured Dakota were sentenced to be hung, President Abraham Lincoln approved that verdict for fewer. Thirty-eight were hanged in a public square in Mankato on the day after Christmas, an event that was the largest mass execution in American history.

Most of the rest of the Dakota people from Minnesota, who had had nothing to do with the war, were nonetheless transported by force to bleak reservations in South Dakota and Nebraska. A few were allowed to stay, or came back to stay, on small parcels of land near Granite Falls, near Redwood Falls, at Shakopee, and at Prairie Island north of Red Wing, but Dakota population and lands in the state were greatly diminished. Although the state's Ojibwe had stayed out of the war, in 1867 government officials tried to force all of them to move to the White Earth Reservation. White settlers wanted to force all Indians onto reservations in remote areas on land that no one else wanted and then to forget about them.

Before long, however, white men seeking profits turned their attention to the

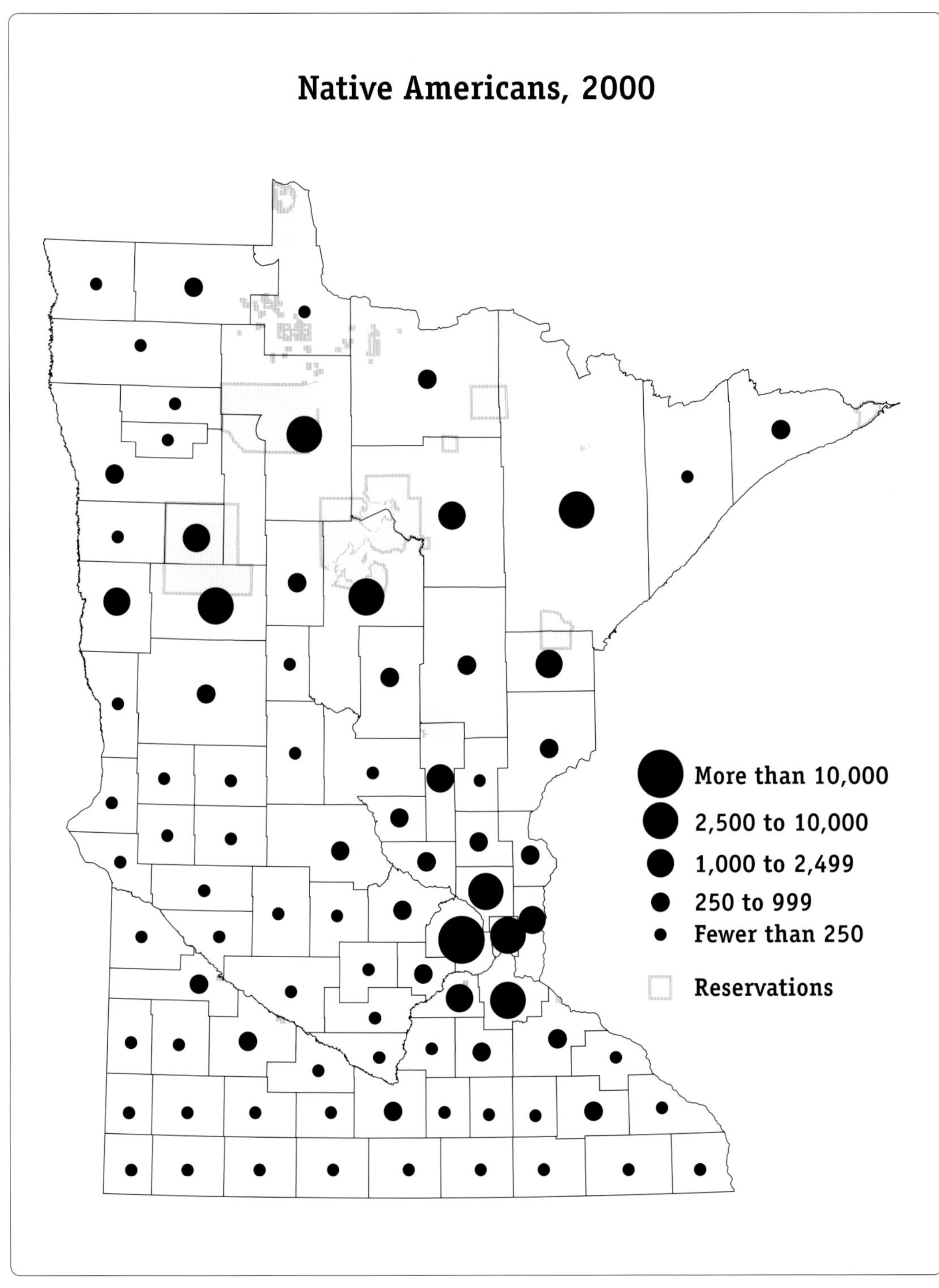

Most Native Americans lived on reservations or in the Twin Cities in 2000.

reservations. Lumbermen wanted access to the forests on the reservations, and farmers complained that the Indians were wasting good farmland by failing to use it properly. Their "seasonal round" required extensive grounds for hunting and gathering, and the entire tribe held the land of the reservation in common.

Some people believed that the Indians would become "proper" farmers if they had their own parcels of land. In 1887 the Dawes Severalty Act allotted 160 acres to every head of household on the reservation. The government did not deed these allotments to individuals but held title to them "in trust" for twenty-five years to keep the Indians from selling them.[7] Most of the lands nevertheless wound up in the hands of whites. All the land of the reservation that was "surplus" to the acreage needed to make these allotments was made available for public sale, and white people bought most of it.

By 1934 it was obvious that the policy of allotment had failed. Indians owned, for example, only 10 percent of the large White Earth Reservation. The Indian Reorganization Act of 1934 returned to tribal ownership all remaining "surplus" land that had not been sold and extended the trust period indefinitely. Because the federal government holds the land in trust for its American landowners, it is exempt from state and local laws and taxes. Some lands are collectively owned by tribal nations, and others are heirship lands owned by individual tribal members.

For half a century, Indians could not afford to buy off-reservation land. Reservations were frequently physically isolated areas of extreme poverty in an ocean of national wealth.[8] Most reservations had an extremely weak economic base, and many band members relied on hunting and gathering for their food. Frequently they had to leave the reservation to find jobs, and many of those who remained were trapped in a vicious cycle of social problems.

After World War II Indians moved in larger numbers to cities, and by 2000 (the most recent year for U.S. census data), only one-quarter of Minnesota's Indians lived on reservations. Another one-third lived in the Twin Cities, and many traveled back and forth between the cities and the reservation. In the cities many Indians live in substandard low-income housing and cope with discrimination and high unemployment. Unemployment rates were and are higher than official statistics suggest because people are not counted as unemployed when they no longer look for work.

A few Indian entrepreneurs have made valiant efforts to start small businesses on reservations, but success is difficult because they have been hobbled by insufficient capital, lack of business experience and advice, inadequate technical assistance and infrastructure, and poor Internet service.[9] Few Indians own enough land to

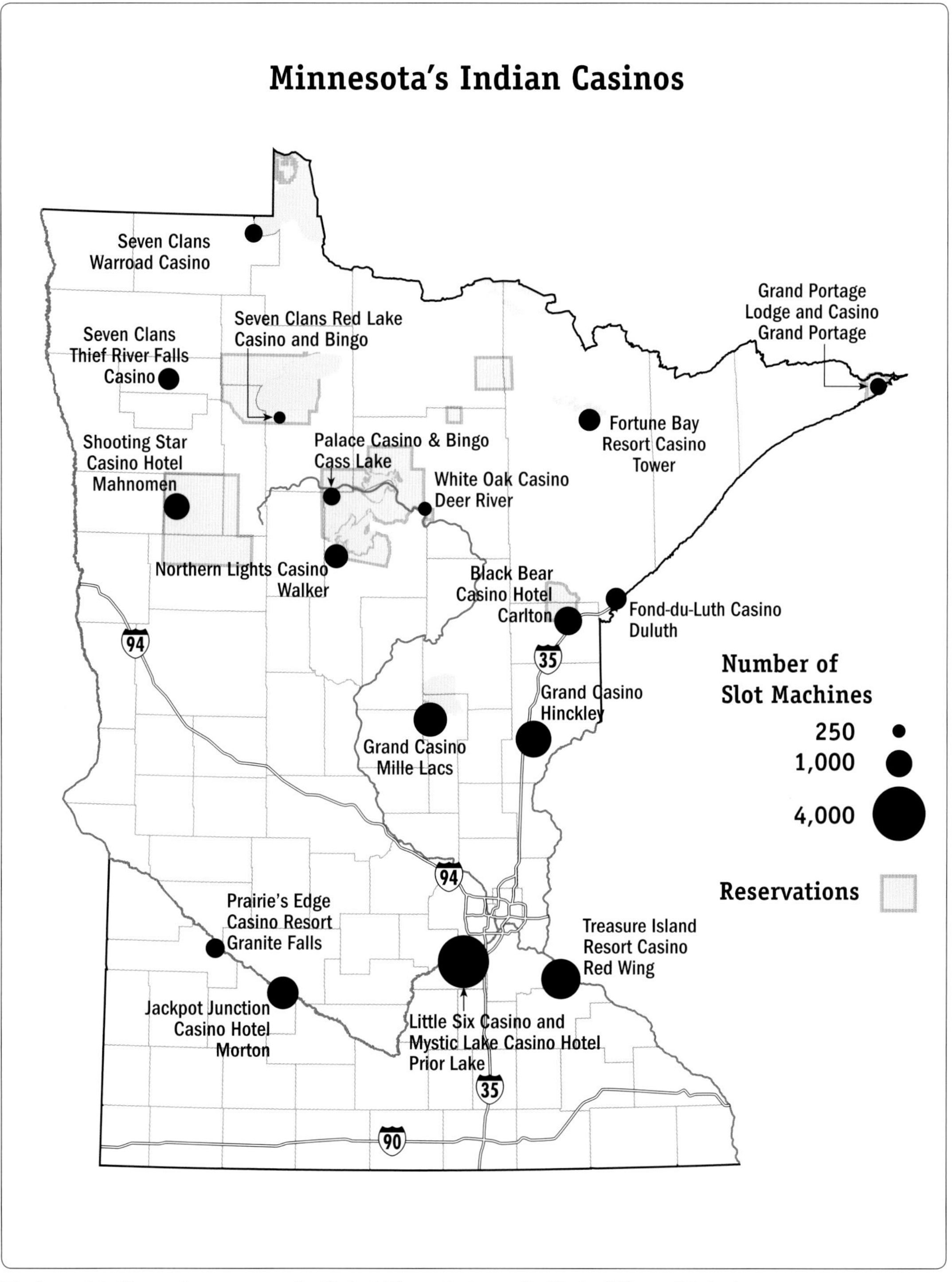

The largest Indian casinos are near the Twin Cities or between the Twin Cities and Duluth.

serve as collateral for business loans, and tribal law takes precedence on Indian land, which makes bankers reluctant to lend money to businesses on reservations.

A big break came with the 1988 Indian Gaming Regulatory Act, which authorized tribes in any state in which gambling is permitted to negotiate compacts with their states that would allow them to establish gambling casinos. Minnesota's Indian gaming compacts permit slot machines and blackjack games, but do not allow roulette, poker, or craps. Minnesota's compacts (unlike Wisconsin's compacts) have no termination date, so Minnesota has not been able to change the compacts and demand tax payments from the casinos when they prospered.

Tribes have developed eighteen casinos in Minnesota. Although the bands initially hired outside management firms, Indians have gradually taken over the operations. Four casinos that are easily accessible from the Twin Cities have been extremely profitable, but those in more remote areas are less successful. The tribes are not required to reveal their income because of confidentiality agreements with the state, but in 2004 Governor Tim Pawlenty estimated that they make a profit of around $1.4 billion a year.[10] Most casinos are on trust land, so they pay no state or local taxes.

Rows of slot machines along the sides flank the lines of blackjack tables down the center at Jackpot Junction Casino.

For many patrons, few of whom are Indian, the popular casinos are an inexpensive entertainment. A day at the casino can cost a careful player less than a day at the ballpark or going to a concert or play. Casinos are surrounded by vast parking lots that are crowded even on weekdays. Special areas are set aside for limousines, buses, motorcycles, and snowmobiles, and campground areas are available for recreational vehicles.

Casinos are laid out like shopping malls, easy to enter but hard to exit. The impatient tinkling casino sound greets visitors when they arrive, and lights flash everywhere. Long banks of slot machines flank lines of blackjack tables. The scene is eerily reminiscent of a computer lab, with rows of people staring at small screens. Few of the hundreds of seated visitors laugh, smile, or talk, but many smoke, because casinos are among the few remaining public places in Minnesota where one can still smoke a cigarette indoors. Many slot-machine users are mature women, with pails of coins at their sides. Few young people are to be seen. Attendants patrol the floor with trays of free soft drinks. Few croupiers are Indians, even though the casinos give employment preference to tribal members.

Casinos compete intensely with each other. They stage elaborate advertising and marketing campaigns, run bus service from distant cities, and have frequent-player programs. They offer inexpensive all-you-can-eat buffets, restaurants, and high-rise hotels with competitive room rates. Some aspire to be entertainment destination resorts, featuring lounges with big-name entertainers, convention centers, golf courses, and marinas. Some have playgrounds and child care centers, although the casinos' patrons are more likely to need wheelchairs.

Some tribes have dispersed their casino profits, while some have invested them in reservation infrastructure. The Mystic Lake Casino near Shakopee reportedly has made payouts of more than $900,000 a year to each tribal member.[11] Tribes concerned about the health, education, and welfare of their members have built schools, hospitals, museums, and water and sewage treatment plants. Some have made voluntary payments to local governments for road construction and maintenance. Some tribes have tried to buy back non-Indian land within their reservations.

Some of the poorer tribes would like to buy land near cities, place it in trust, and build casinos on it. Their proposals must be approved by the governor. Most local governments have opposed such plans vigorously, because they would lose valuable tax base. Duluth, however, has welcomed the Fond du Luth Casino, which is in a former department store in the heart of town.

Casinos are highly controversial, and they have spent large sums of money to curry public favor. Critics point to the social costs of gambling, while advocates stress their benefits for economically distressed areas and people and their value as entertainment centers. Politicians facing budget deficits wish to change the casinos' tax-exempt status. Economists disagree about how to measure the costs and benefits of casinos accurately. They agree, though, that they are neither cure-alls nor catastrophes.

Kittson
Roseau
Lake of the Woods
Marshall
Beltrami
Koochiching
Pennington
Red Lake
Polk
Clearwater
Itasca
Norman
Mahnomen
St. Louis
Lake
Hubbard
Cass
Clay
Becker
Wadena
Aitkin
Carlton
Crow Wing
Wilkin
Otter Tail
Pine
Todd
Grant
Douglas
Morrison
Mille Lacs
Kanabec
Traverse
Stevens
Pope
Stearns
Benton
Big Stone
Sherburne
Isanti
Chisago
Swift
Anoka
Kandiyohi
Meeker
Wright
Chippewa
Lac Qui Parle
Hennepin
Ramsey
Washington
McLeod
Carver
Yellow Medicine
Renville
Scott
Dakota
Sibley
Lyon
Redwood
Nicollet
Le Sueur
Rice
Goodhue
Wabasha
Brown
Pipestone
Murray
Cottonwood
Watonwan
Blue Earth
Waseca
Steele
Dodge
Olmsted
Winona
Rock
Nobles
Jackson
Martin
Faribault
Freeborn
Mower
Fillmore
Houston
1. Red River
2. Lake of the Woods
3. Mississippi River
4. Lake Superior
5. Duluth
6. Northwest Angle
7. Mankato
8. St. Croix River
9. St. Cloud
10. St. Peter
11. Pembina
12. Minnesota River
13. Big Stone Lake
14. St. Croix Falls
15. Nerstrand Big Woods State Park
16. Hibbing

5

Dividing the Land

The cession of Indian lands to the United States was only the first step in opening them up to white settlement. The land had to be surveyed, land offices had to be opened, and public sales had to be announced before title to land in the public domain could be "alienated," or transferred to private citizens.[1]

The land that was to become Minnesota was acquired by the United States in three stages.[2] The area east of the Mississippi River was part of the Old Northwest, which Britain ceded to the United States at the Treaty of Paris in 1783, ending the Revolutionary War. The area west and south of the Mississippi was acquired as part of the Louisiana Purchase in 1803. The area drained by the Red River and its tributaries, which drain northward to Hudson Bay, lay outside the Louisiana Purchase but was ceded by Britain after the War of 1812 in the Convention of 1818, which fixed the northern boundary of the United States at the forty-ninth parallel from the Lake of the Woods to the Rocky Mountains.

The Treaty of Paris was based on seriously flawed maps. It stipulated that the northern boundary of the United States should go from Lake Superior to Long Lake, and thence by the water route to the Lake of the Woods, but there is no Long Lake in the area. The British insisted that it was the estuary of the St. Louis River at Duluth, but the Americans claimed that it was Thunder Bay at the mouth of the Kaministikwia River in Canada. Not until 1842 did they finally compromise on the present boundary along the Pigeon and Rainy rivers.

The Treaty of Paris also said that the boundary should run through the Lake of the Woods to its northwest corner and then proceed due west to the Mississippi

River, an impossibility, because the Mississippi is almost directly south of the lake. The Convention of 1818 solved this problem by running a line south from the northwest corner of the lake to the forty-ninth parallel, which endowed Minnesota with the curious little topknot known as the Northwest Angle.

The other boundaries of the state, which seem immutable today, evolved only slowly and contentiously. In 1846, when Iowa was admitted to the Union, Iowans wanted its northern boundary drawn through Mankato, at 44° 10′ North, but Congress did not want such an unreasonably large state and drew the present boundary at 43° 30′ North. In 1848 the new state of Wisconsin wanted the entire Mississippi River as its western boundary, but once again Congress demurred and drew the line along the St. Croix River, with a line north to Lake Superior.

In 1857, when Minnesota was approaching statehood, there was vigorous debate about whether the new state should extend north-south or east-west. The farmers in the south wanted a wide state reaching west to the Missouri River, with its northern boundary at 45° 30′ North, just south of St. Cloud. The fur traders in St. Paul wanted a tall state that went all the way north to the international border.

Boundaries of Minnesota

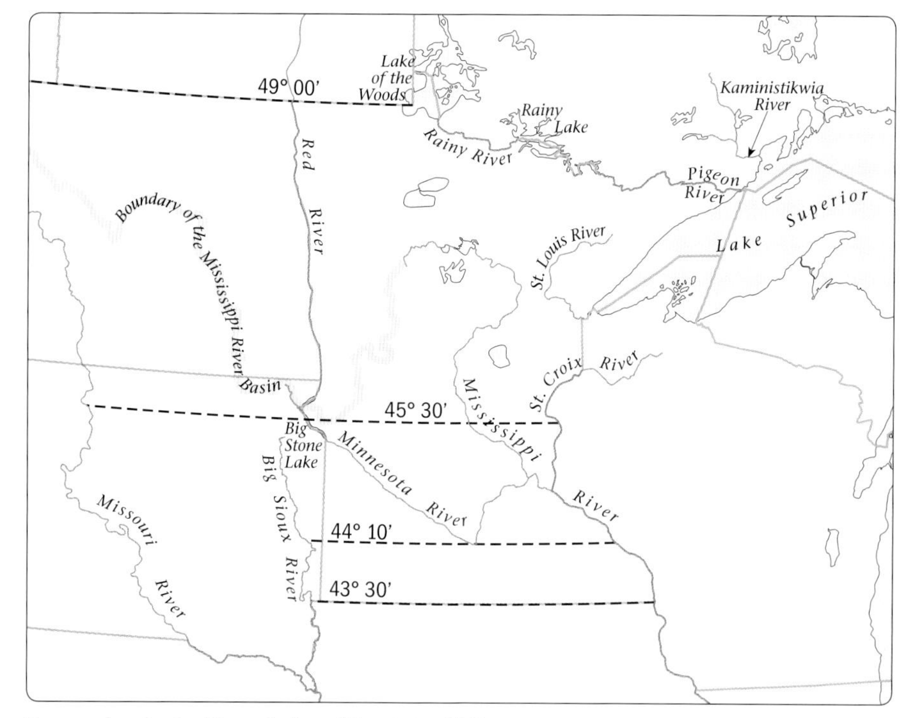

Proposed and actual boundaries of the state of Minnesota.

The advocates of an east-west state decided to buttress their case by getting the territorial legislature to pass a bill moving the state capital from St. Paul south to St. Peter, which would be more central to the state they advocated. A legislator named Joe Rolette, a colorful fur trader from Pembina, was supposed to take the bill to the governor for his signature, but instead he disappeared with it until it was too late for the governor to sign it into law. Nobody believed Joe's story that he had been kidnapped, and it turned out that the bill was unconstitutional anyhow.

Congress passed the enabling act that allowed Minnesota to become a state in 1858. Congress, the final authority for determining the state's boundaries, favored the north-south idea. In the northwest the Red River was an obvious boundary. Some people thought the western boundary should continue south along the Big Sioux River in South Dakota, but instead Congress simply drew a line straight south from the source of the Minnesota River at Big Stone Lake to the Iowa line.

LONG BEFORE STATEHOOD, PEOPLE WERE chomping at the bit to buy land in Minnesota, but first it had to be surveyed. Almost immediately after the Indian cession of 1837 the commissioner of the General Land Office had sent surveyors into the area between the St. Croix and Mississippi rivers. By 1847 enough of this land had been surveyed so that it could be sold, and the first land in Minnesota was sold at the public land office in St. Croix Falls, Wisconsin.

Land in Minnesota, as in the rest of the Midwest, was surveyed in accordance with the Ordinance of 1787, which provided that the public domain should be divided into a uniform grid of squares with boundary lines following the cardinal compass directions.[3] The survey system starts at a baseline that runs straight east and west, and a principal meridian that runs straight north and south. The surveyors drew both east-west township lines parallel to the base line at six-mile intervals, and north-south range lines parallel to the principal meridian at six-mile intervals.

The grid of township lines and range lines created six-mile squares called "townships," which were numbered north and south of the baseline, and west and east of the principal meridian. The meridians converge northward because they eventually meet at the North Pole; thus, every fourth township line was a correction line, where the spacing of range lines was adjusted.

Each township was divided into thirty-six plots of one square mile (or 640 acres) that are called "sections."[4] Sections were numbered from 1 to 36, with 1 in the northeast corner and 6 in the northwest corner, then zigzagging back from 7 to 12 in the second tier of sections, and so on to section 36 in the southeast corner.

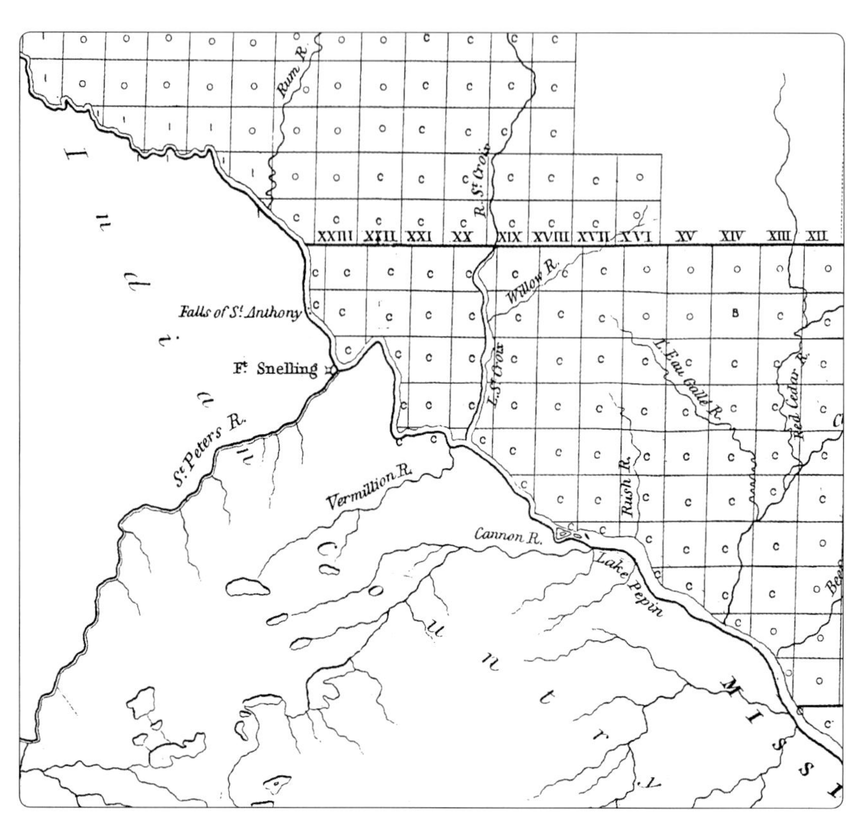

The 1850 annual report of the Commissioner of the U. S. General Land Office marked the area west of the Mississippi River as "Indian Country." The townships east of the river that had been completely surveyed were labelled "C."

Sections could be divided into four "quarters" of 160 acres, quarters could be divided into 40-acre "forties," and so on. This system uniquely identifies every parcel of land in the entire state. For example, the NE¼ of the NW¼ of the SW¼ of Sec. 31, T29N, R22W, is the site of the state capitol in St. Paul, and a couple more quarterings could put you right on the desk of the governor.

The surveyors were required to mark the corner of each section by chopping prominent blaze marks on the two or three closest trees. The surveyors' notes describing these "bearing trees" are a valuable record of the contact vegetation of the state. Francis J. Marschner produced the landmark vegetation map of Minnesota in 1929–30 by laboriously working his way through the endless pages of surveyors' notes and mapping what they said about the bearing trees or markers at each square mile in the entire state.

The original survey system is still evident in the modern map of Minnesota. Away from the major rivers, most county lines follow survey lines, and most survey

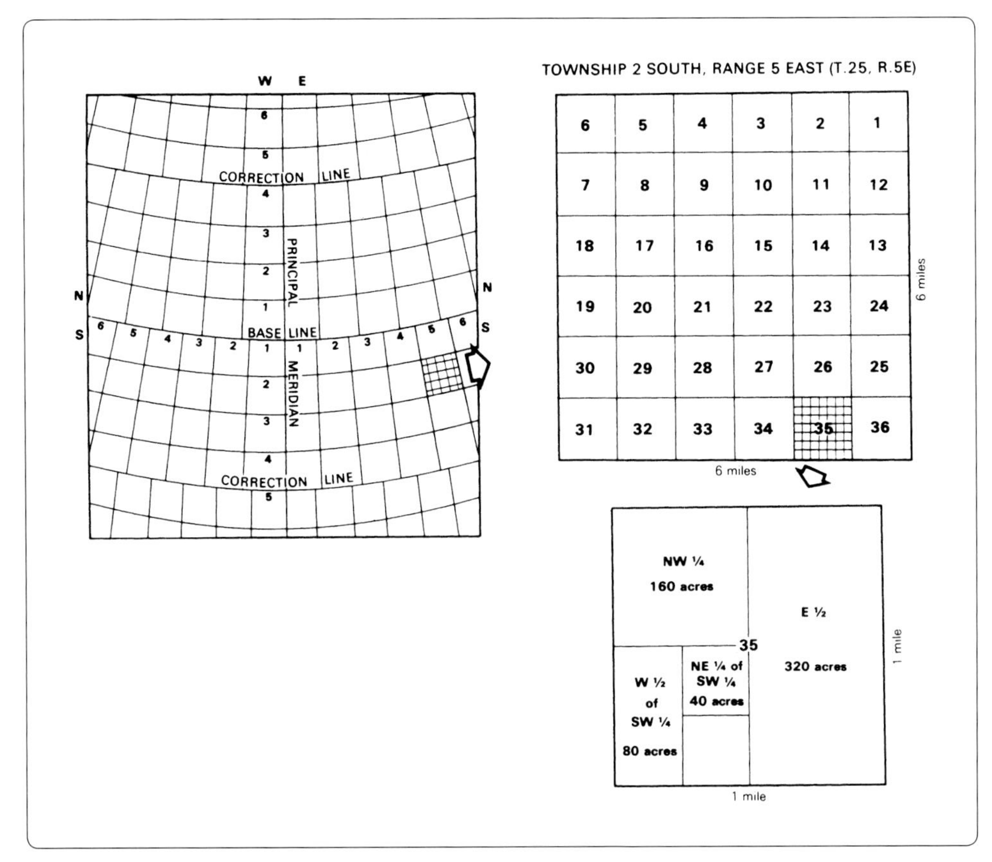

The Public Land Survey divided the land into townships, sections, and fractions of sections.

townships are now civil townships. Country roads run along most section lines to form a checkerboard road network, and roads that run north and south make awkward right-angle jogs when they cross correction lines.

After the land had been surveyed, it was sold by public auction at land offices that were moved westward with the spread of occupance when most of the land near them had been sold. Forty acres was the smallest unit that a land office could sell, but some owners of steep, wooded land subdivided it into five- or ten-acre woodlots, which they sold to farmers on the adjacent treeless prairie uplands. The Nerstrand Big Woods State Park northeast of Faribault was created by amalgamating many such small woodlots.

The land was sold for a minimum price of $1.25 an acre, but direct cash purchase was only one of many ways in which people acquired public land. Some buyers paid with scrip, which was a document entitling its owner to a specified acreage of land. In the 1850s lumbermen acquired much of the large acreage they needed in the area

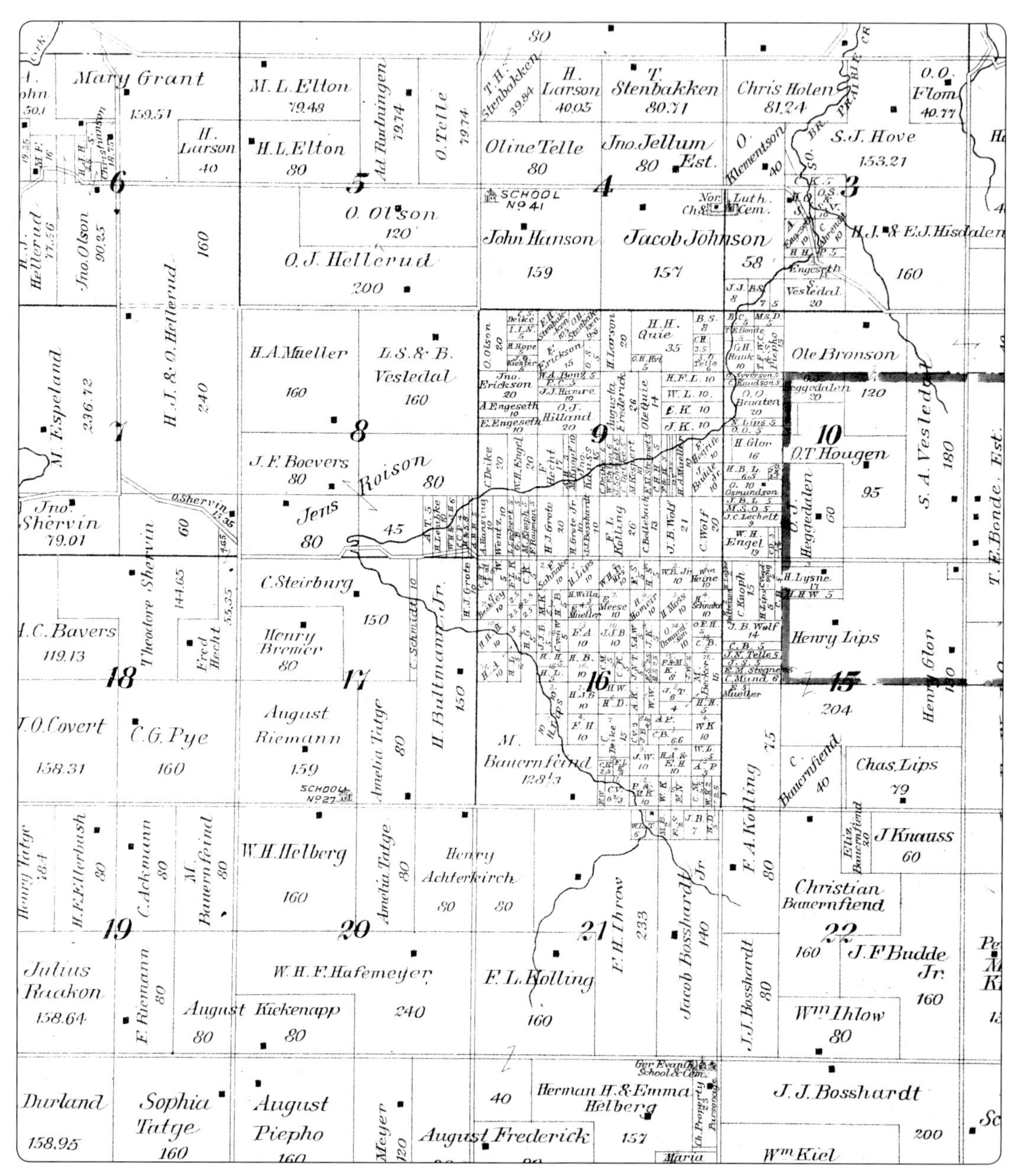

Farmers with 160-acre farms on the treeless prairie uplands bought five- or ten-acre lots in steep wooded areas. These former woodlots, shown in the fine detail of this plat map, have been amalgamated into Nerstrand Big Woods State Park.

between the Mississippi and St. Croix rivers by using land warrants the United States had given to veterans of the Mexican War. These warrants could be used to pay for land anywhere. Veterans in the East, who had little use for them, were only too happy to sell them for as little as ten cents an acre to land brokers, who in turn sold them to lumbermen.

In 1862 the Morrill Act founded land-grant colleges and universities by giving each state 30,000 acres of land for each of its senators and representatives in Congress. This land was to be sold to endow agricultural colleges. The states in the East that had no public domain land were given scrip that could be used in other states. The states sold their scrip for as little as fifty cents an acre, and eager buyers in Minnesota snapped it up. There were also other forms of scrip, but they were less significant.

The federal government gave the state 17 million acres, one-third of its entire area, which it could use or sell for worthy purposes, such as paying for the construction of schools, public buildings, railroads, and other internal improvements. For example, sections 16 and 36 in each township were granted to its inhabitants for schools. The state as well as the federal government made huge land grants to railroads. The railroad received all odd-numbered sections in a strip six miles wide along its intended route, and it could select "in lieu" land within fifteen miles to compensate for any of this acreage that had already been sold.

The federal government also gave the state 4.7 million acres that surveyors had identified as swampland. It assumed that the state would construct levees, dig drains, and pay for them by selling the improved land, but Minnesota is not Louisiana. Because the swamplands of glaciated areas are so widely distributed and so irregular, massive reclamation projects were not feasible, and the state sold its swampland at greatly reduced prices to people who had to drain it themselves. The proceeds from these sales went into the state's general fund. A few surveyors were suspected of having classified some of the best farmland as swampland and then coming back to buy it.

The two principal land laws that affected the settlement of Minnesota were the Preemption Act of 1841 and the Homestead Act of 1862. The preemption law allowed a settler to obtain patent to 160 acres of land by living on it for a specified length of time, cultivating it, and paying the minimum price of $1.25 an acre. It protected bona fide settlers against land speculators because it gave them the right to buy land at the minimum price before it was offered for public sale, where wealthy speculators could bid up the price.

Preemption encouraged squatting because it gave first land-claiming rights to people who were already living on it. An amendment to the act in 1854 allowed

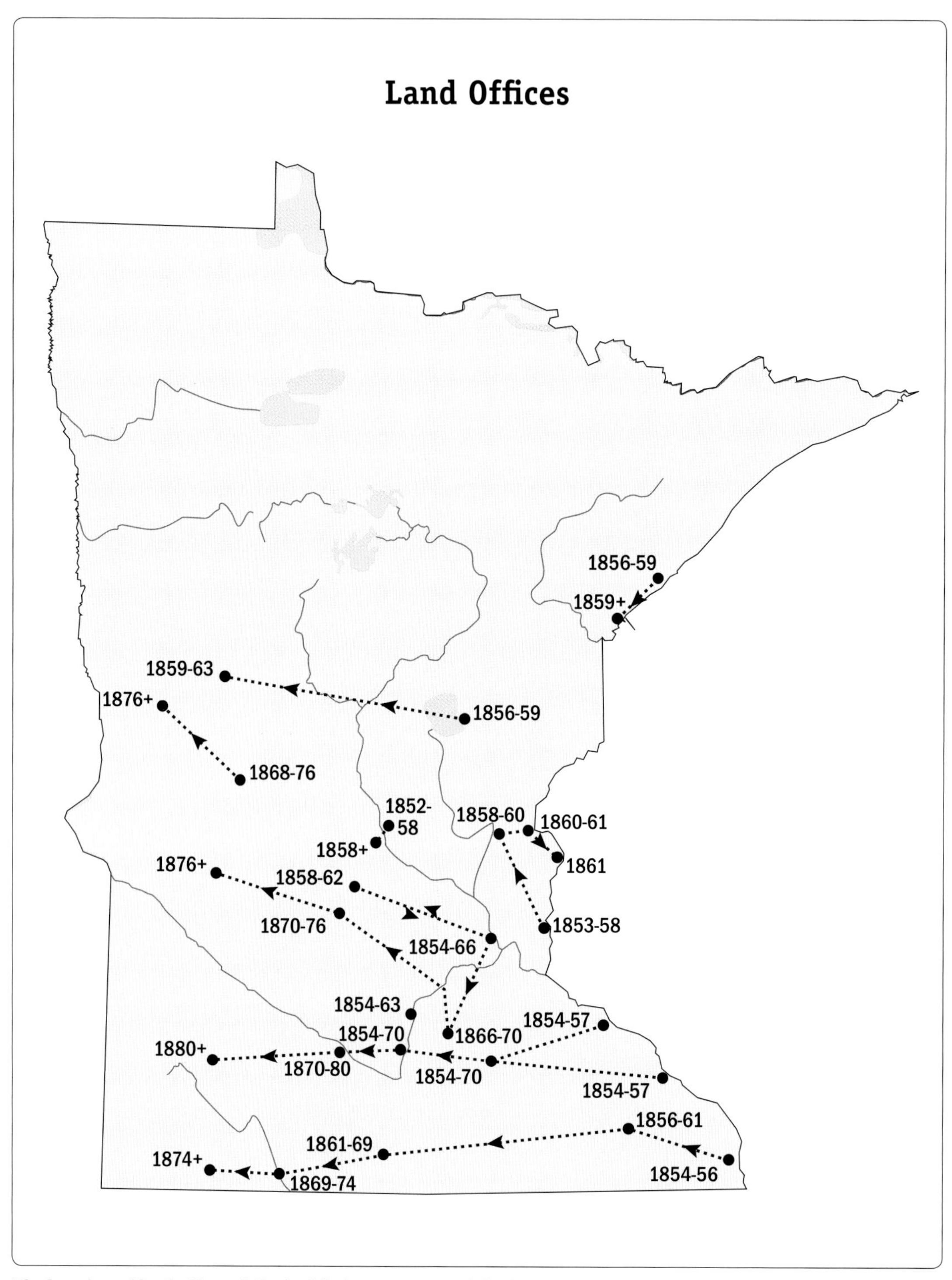

The location of land offices shifted with the movement of the frontier.

settlers to preempt land even before it had been surveyed, although their claims often had to be adjusted after the survey had been completed. This amendment triggered a huge land rush, and the eight land offices in Minnesota sold 5.3 million acres, or 10 percent of the entire state, between 1854 and 1857. In 1855 alone some 20,000 acres, or more than half of the present city, were preempted where Minneapolis now stands.

The Homestead Act of 1862 gave 160 acres of free land to any head of household who would live on it for five years. This act culminated seven decades of rancorous debate about whether the objective of public land policy should be to increase revenue or to encourage settlement. Frontier settlers always wanted free land, but the young United States needed revenue badly, industrialists in the East feared that free land would entice away their workers, and planters in the South suspected that small farmers would oppose slavery.

Congress could not pass the Homestead Act until the southern states seceded in 1861. The act had limited impact in southeastern Minnesota, where little public land remained, but it ensured the rapid settlement of the western part of the state. Even though the Civil War was underway, 9,500 homesteaders claimed more than 1.25 million acres of land in Minnesota between 1863 and 1865, and claims rarely dropped below 200,000 acres a year until 1889, when the public lands were closed.

The land laws of the United States were oriented toward agricultural settlement, and they posed problems for mine owners and lumbermen, who needed substantially larger acreages. For example, the famous Hull Rust Mahoning Mine north of Hibbing alone is larger than nineteen 160-acre homesteads. The mine owners and lumbermen felt they were forced to circumvent the laws in order to earn their livelihood. Their ploys to gain larger acreages have infuriated bureaucrats and many historians, but bureaucrats who tried to prosecute them could find no witnesses who were willing to testify against them. One can be incensed by these scofflaws, or admire their ingenuity.

From the very earliest days people who lived near unsold government land treated it as common property. They grazed livestock, cut hay, and took wood from it. Some lumbermen saw nothing wrong in continuing this practice, albeit on a vastly larger scale, and they kept cutting timber illegally on public land until they were arrested and fined. Some were not stopped even then.

Northeastern Minnesota was a cauldron of fraudulent land claims, both by mine owners and by lumbermen. Land office clerks and registrars were notoriously underpaid, receiving most of their income from the fees they collected. Some apparently were willing to accept "fees" from land agents to countenance their trickery. It was common for agents to hire dummy entrymen to file false claims and then to buy the land from them as soon as their entry had been completed. One enterprising land agent even paid a mailman to write on a series of blank claim forms the names of people on his route.

Few dummy entrymen ever went near their claims, but one who did took an ordinary fourteen-by-sixteen-inch wooden box, painted doors, windows, and a shingled roof on it, and swore at the land office that he had a good fourteen-by-sixteen board house with doors, windows, and a shingled roof.[5] A frustrated inspector who tried to track down entrymen with false claims reported that "one was driving a dray in Duluth, others were keeping saloons, more were working on the docks, and the rest were following their various occupations, if indeed they had any," but he added sadly, "The community is not in sympathy with us" in the attempt to extirpate fraud.[6]

It was easy simply to walk away from land that had been acquired fraudulently if it proved barren of minerals or after its timber had been cut. In this way much of northeastern Minnesota was abandoned, became tax-delinquent, and was returned to the public domain. The modern ownership map is a crazy quilt of land owned by lumber companies, mining companies, railroad companies, land companies, the state and federal governments, and private individuals.

By 1891 so many people had become alarmed by the problem of tax-delinquent and cutover timber areas that Congress passed a law to protect the remaining forest areas by authorizing the establishment of forest reserves from the public domain. These forest reserves evolved into national forests, which are mainly public land but have substantial acreages of private land within their boundaries. They may acquire additional land by purchase or by exchange to tidy up the ownership map. The state also used some of its public land to create state forests, state parks, and wildlife preserves.

Public forests are managed for sustained yield and multiple use. They offer timber for sale at competitive bidding, but they also emphasize recreation and wildlife protection. They are part of the public domain managed for the benefit of all citizens, not just a few individuals.

The way in which the lands of Minnesota were divided and acquired has left indelible marks on the landscape. Some of these marks are so sharp and bold that they are inescapable. Others are so nuanced and subtle that they require careful scrutiny and are most easily seen on maps, on aerial photographs, or from airplanes.

The Public Land Survey has given Minnesota its "checkerboard" of square and rectangular fields.

1. Itasca State Park
2. Mississippi River
3. St. Croix River
4. Stillwater
5. Snake River
6. Winona
7. Minneapolis
8. Rum River
9. Duluth
10. Moorhead
11. Aitkin
12. Grand Rapids
13. Cloquet
14. St. Louis River
15. Virginia
16. International Falls
17. Rainy River
18. Hinckley
19. Chisholm
20. Baudette
21. Bayport
22. Warroad
Kittson
Roseau
Lake of the Woods
Marshall
Beltrami
Koochiching
Pennington
Red Lake
Polk
Clearwater
Itasca
St. Louis
Lake
Norman
Mahnomen
Hubbard
Cass
Clay
Becker
Wadena
Aitkin
Carlton
Crow Wing
Wilkin
Otter Tail
Pine
Todd
Grant
Douglas
Morrison
Mille Lacs
Kanabec
Traverse
Stevens
Pope
Stearns
Benton
Big Stone
Sherburne
Isanti
Chisago
Swift
Anoka
Washington
Chippewa
Kandiyohi
Meeker
Wright
Hennepin
Ramsey
Lac Qui Parle
Yellow Medicine
Renville
McLeod
Carver
Scott
Dakota
Sibley
Lyon
Redwood
Nicollet
Goodhue
Le Sueur
Rice
Wabasha
Brown
Murray
Cottonwood
Watonwan
Blue Earth
Waseca
Steele
Dodge
Olmsted
Winona
Rock
Nobles
Jackson
Martin
Faribault
Freeborn
Mower
Fillmore
Houston

6

The First Great Industry

Lumbering was Minnesota's first great industry. It played a major role in the early growth of the state by providing the capital needed to develop the frontier.[1] Entrepreneurs made fortunes in the lumber business and invested their profits locally. They founded banks and financed railroads. They provided the capital to build flour mills, whose product found a ready market in their sawmill towns and logging camps. They endowed colleges, museums, and other cultural institutions. Without the capital created by the lumbering industry, Minnesota's economy would have grown far more slowly.

American lumbering began on the eastern seaboard. As eastern forests were felled, the industry moved slowly westward through the great coniferous forest that reached for more than a thousand miles from the coast of Maine to the prairies of northwestern Minnesota.[2] The lumbermen who arrived in Minnesota were looking for white pine, and they scorned as weeds all other species, such as spruce, fir, birch, and aspen. They were not interested in the oaks, maples, basswoods, and other hardwood trees of the Big Woods, but farmers were all too happy to have these trees for firewood, for fence posts, and for building houses and barns.

The virgin stand of red and white pine that is protected at Itasca State Park suggests how awesome the original forest must have been. Lumbermen felled trees that were up to 200 feet tall, 5 feet in diameter, and more than 250 years old. The wood is strong, easy to cut, and odorless, but so light that it floats on water. It seasons well, is slow to decay, and resists the ravages of time and weather.

Lumbermen traveled up the Mississippi River from St. Louis to the "pineries" of the St. Croix triangle, the name they gave to the land between the Mississippi and St. Croix rivers, almost as soon as the ink was dry on the 1837 treaty ceding the Indian land to the United States. The first sawmills were built at Marine on St. Croix in 1839 and at Stillwater in 1844; within ten years Stillwater had five more. These mills were powered by steam raised by burning wood scrap. The lumbermen borrowed money in New England to build them, and most of the early workers were from Maine. Stillwater was so remote that the lumbermen had to open banks and stores where their workers could buy food, clothing, and other goods.

The lumbermen acquired timberland, sent crews to log it, sawed the logs into lumber, and sold logs and lumber. They preferred to log in winter, because snow and ice facilitated dragging logs to streams. The first known logging crew, twelve men and six oxen, camped at the junction of the St. Croix and Snake rivers in the winter of 1837.

Logging camps were even more transitory than fur trading posts. Used no more than a winter or two, they were abandoned when lumberjacks had felled the trees within a reasonable walking distance, no more than five miles. At one time or another several thousand sites in Minnesota have been used for logging camps, but few traces of them remain.

In the fall the foreman brought a small crew into the woods, selected a site, and built a camp. The lumberjacks arrived in November or early December and worked for three or four months. The early logging camps consisted of single, windowless, one-room log shanties that measured perhaps twenty-five by forty feet. In the center of the room an open fire for cooking and for heat was never allowed to go out. On two sides of the room the men slept in a row, with their heads to the wall and their feet to the fire. Their beds were balsam boughs a foot thick laid on the ground.

In the opposite corner of the room the cook prepared the staple diet of baked beans, boiled salt pork, blackstrap molasses, and biscuits. The Maine men drank tea, but Swedes and Norwegians preferred coffee. Working in the bitterly cold woods gave the men ravenous appetites, and a good cook was essential for a successful camp. The working day started at four in the morning with a generous breakfast. The men were in the woods before dawn and worked until after dusk, because the winter days are short in Minnesota.

The foreman selected each tree that was to be felled and chopped a deep notch on the side to which it should fall. Choppers cut it off slightly above the notch on the opposite side. Swampers cut off the branches, and a barker chopped away the bark from the side on which it would slide when a team of oxen dragged it through the brush to the river. At the river a sawyer cut it into sixteen-foot logs. The stamp marker pounded the owner's mark, or brand, into the end of each log, and a swamper cut bark marks where they could be seen when the log rolled in water. Minnesota has more than 20,000 officially registered log marks.

In the spring the logs were floated downstream to the sawmills. The lumberjacks built dams on small streams to hold water for the spring drive and stacked logs on the frozen lakes behind the dams. In the early days they built earthen dams that they dynamited to start the drive, but later they built permanent dams with sluiceways they could open to let the torrent of logs and water come pouring through. They modified many streams by blasting boulders from their beds and widening narrow bottlenecks with dynamite.[3] The rushing logs widened streams by undercutting their banks, which deposited sandbars and mud banks in their channels.

The logs had to be separated by giant booms upstream from the sawmills when several companies drove on the same stream. The Stillwater boom, which was built in 1851, extended nine miles upstream. The boom was built by driving posts into the streambed and connecting them with floating logs chained together. At the boom all logs had to pass through a narrow channel where the catchmarker could identify the mark on each log and direct it to the boom pocket of its owner.

In spring the lumberjacks became "river rats" and drove logs downstream to the sawmills.

From the boom some logs went directly to local sawmills; others were lashed together with pegs and ropes into great rafts that were towed to sawmills as far down the Mississippi River as St. Louis, which was the first major market for Minnesota logs and lumber. The rafts of logs, and later lumber, were larger than four or five football fields. The largest, in 1896, was 270 feet wide and 1,550 feet long. A stern-wheel steamboat pushed the raft, and a small boat tied sideways at the front moved back and forth to guide it. At the peak of the lumber era, more than one hundred steamboats were in the rafting business on the Mississippi.

Winona, downstream on the Mississippi River in the southeastern corner of the state, was far from the white pineries, but like other river towns it became a major sawmilling center because it was close to a major market, the prosperous farms on the prairies to the west. Prairie farmers made the steep descent down through the bluffs to the Mississippi at Winona with wagonloads of wheat, then climbed back up to their homes on the treeless prairie uplands with their wagons full of lumber.

By 1852 Winona had become a major wholesale and retail center for lumber rafted down the river from sawmills upstream, and in 1855 the first of ten large sawmills was built in the city. These mills sawed logs that were rafted down the river from Minnesota; they also sawed logs floated down the Chippewa River from the pineries of northwestern Wisconsin.

Minneapolis was by far the largest sawmilling center in early Minnesota. The Mississippi and Rum rivers brought logs down to the Falls of St. Anthony, which was the largest waterpower site west of Niagara Falls. The first sawmill was built at the Falls of St. Anthony in 1848. The money to build it came from Boston; its machinery, its management, and most of its workmen were from Maine. Other men from Maine soon followed, and by 1856 the falls boasted no fewer than eight sawmills.

One of the sawmills at the Falls of St. Anthony in Minneapolis.

By 1870 Minneapolis had already become the leading center of the lumber business in Minnesota. It had thirteen major sawmills, eight water-powered and five steam-powered. It had a large number of subsidiary factories that used the by-products of the mills to make doors, sashes, blinds, and other building materials; barrels, boxes, crates, and other containers; furniture; wagons, carriages, sleighs, and other vehicles; and a host of other wooden products.

Minneapolis was home to the head offices and principal banks of the leading companies. It was a major distribution center at the hub of the railroad network that was spreading westward across the prairies. Lumber marketing companies in Minneapolis developed retail lumberyards in nearly every small town along the railroad lines. The lumberyard, which also sold coal, performed the same function on the treeless prairies as the farm woodlot, which had provided fuel and building material for farmers in wooded areas farther east.

THE COMPLETION OF RAILROADS from St. Paul to Duluth in 1870 and from Duluth to Moorhead in 1871 ushered in a whole new era in the lumber business because they opened up vast areas of pineland that were not accessible from rivers. Railroads began to replace river drives as the principal means of hauling logs to mills, and the lumbermen established scores of new sawmills with company towns at strategic sites along the railroad lines.

Some of these new towns became diversified and continued to grow, but others simply disappeared after the mill shut down. Aitkin flourished as a major supply base for logging camps in northern Minnesota. Trains from Duluth hauled supplies to Aitkin, and steamboats carried them to the camps along the Upper Mississippi River.

The lumber companies built railroads into areas that were not accessible from streams.

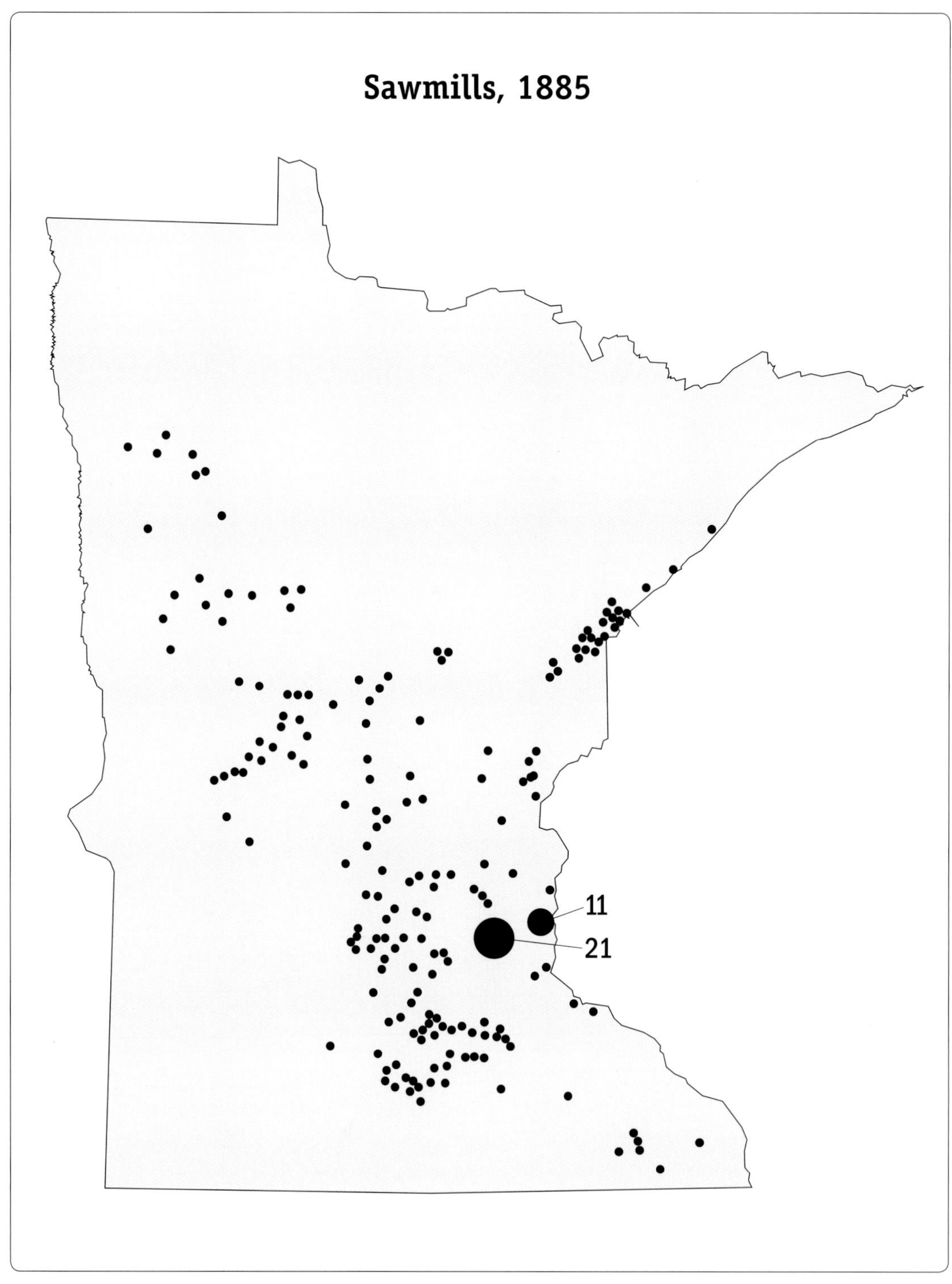

In 1885 most sawmills were at waterpower sites in the wooded areas.

The technology of logging and the logging camps themselves kept getting better and better. Horses replaced oxen for pulling sleds stacked with logs over icy roads to the river or railroad line; they, in turn, were replaced by crawler tractors. Axes gave way to crosscut saws, and they in turn to power saws. Machines reduced the necessity of hand tools, heavy lifting, and sheer brute strength, but winter work in the woods still was harsh and dangerous. Lumberjacks could easily be maimed or killed by falling trees or logs.

The Forest History Center near Grand Rapids has a fascinating replicated logging camp from the turn of the century, which was the peak of the white pine era in Minnesota.[4] It has a bunk room for about eighty lumberjacks, the building where the cook fed them, a large barn for the horses, and smaller specialized buildings. They are all built of log, which was typical before 1910; thereafter most camps were built of rough lumber with tarpaper sides and roofs.

Popular lore has glamorized the tough lumberjack life, with its colorful language and tall tales. (Most lumberjacks never heard of Paul Bunyan, whose legend was invented by an advertising man around 1910 and was greatly embellished in the 1920s.) At the end of the hard day or week working in the snow, most lumberjacks were much too tired to walk into town, even if one was near, during the working season. Most jacks went to work in the sawmills, or back to their farms, when they were paid off at season's end, but some earned a reputation for carousing in bars and brothels. Small towns were said to roll up their sidewalks when the lumberjacks were coming. Many sidewalks were made of boards held together with wire, and the jacks' cleated boots would have chewed them to splinters. (In 1887 alone the city of Minneapolis laid no fewer than sixty-seven miles of board sidewalks.)

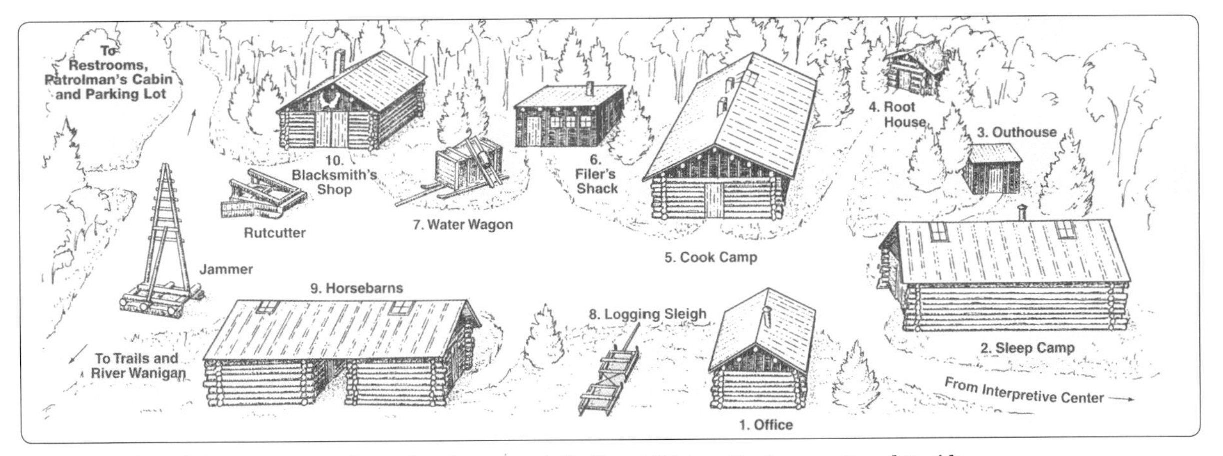

An artist's view of the reconstructed 1910 logging camp at the Forest History Center near Grand Rapids.

Lumberjacks slept in bunkhouses like this one in the reconstructed logging camp at the Forest History Center near Grand Rapids.

Because Duluth was a railroad hub, it became northeastern Minnesota's first major sawmilling center. Trains carried lumber westward to the farms on the treeless plains of the Red River Valley and beyond. They returned with wheat that was stored in the grain elevators at Duluth while it was waiting to be shipped eastward. Steam schooners also carried lumber eastward and returned with cargoes of coal, salt, and limestone. The construction of the railroads themselves required large amounts of lumber, as did the construction of docks and grain elevators in Duluth. Duluth was close to the forests whence came the logs for its sawmills, and it was at the head of the cheap all-water route to Tonawanda, New York, a suburb of Buffalo, which was the greatest wholesaling center for white pine in North America.

Cloquet, fifteen miles west of Duluth, also became a major sawmilling center. The St. Louis River drops 600 feet between Cloquet and Lake Superior. So many logs were smashed and broken in its falls and rapids that a sawmill was built above the falls in 1878. By 1896 the tumbling waters of the St. Louis River powered six large sawmills at Cloquet. The principal initial market for Cloquet lumber lay westward on the Great Plains, but its mills also began to ship lumber eastward when production in Duluth started to lag after 1903.

In its early days Duluth had to compete with established producers in Michigan who were closer to Tonawanda, but by 1890 Michigan had nearly been stripped of its pinelands. Saginaw mills sawed 36 percent of all the white pine in the Great Lake states in 1873, but only 16 percent in 1890. Michigan lumbermen began moving to Minnesota, and so did lumber merchants from Chicago, who were concerned about the declining supply of Michigan's forests.

The Michigan lumbermen had good business contacts with eastern markets, and they brought big ideas, big money, and big reputations to Minnesota. They built impressive sawmills in Minneapolis and in Duluth, and they acquired huge blocks of Minnesota pine lands. They stimulated the growth of Duluth, sparked a second wave of growth in Minneapolis, and launched the golden age of Minnesota lumbering, which lasted from 1890 to 1905.

This golden age was also nearly its terminal stage, because larger operations could cut down forests so fast. As a result the lumber business declined rapidly after its peak. By 1907 lumber had become so expensive that the construction industry began to shift to cement and structural steel. The St. Croix boom closed in 1914, and the last log was sawed at the Falls of St. Anthony in 1919.

The lumber business hung on a bit longer in the north. What was touted as "the largest, most modern and complete lumber plant in the world" began sawing at Virginia in 1910. It covered a square mile, was served by a network of logging railroads, and kept 3,000 lumberjacks busy cutting trees. A large sawmill complex was completed in 1910 at International Falls on an impressive waterpower site on the Rainy River. The Virginia sawmill closed in 1929, followed by the mill at International Falls in 1934; thereafter, the lumber business in Minnesota consisted of small operations sawing what lumbermen hitherto had despised as "weed" species.

As lumbering operations grew, lumbermen needed ever larger acreages of land. The ways in which some of them acquired and used land in Minnesota provide one of the more colorful episodes in the state's history. While some lumbermen were proper and circumspect, many were not, blatantly cutting timber illegally on public lands because they knew that the state did not have the resources to police the land adequately. Some unscrupulous lumbermen took every possible advantage of laws that were intended to encourage settlement. They acquired large acreages with military land warrants and agricultural college scrip that they bought for ridiculously low prices. Some hired transients to file fraudulent homestead claims that they abandoned after the timber had been cut.

The State of Minnesota lost millions of dollars in revenue in the sale of stumpage, or the right to harvest timber from the land without actually buying it. The land still belonged to the state after the timber had been cut from it. The state auditor was supposed to estimate the value of the standing timber before stumpage rights were sold, but he was not qualified to do the job properly and often accepted the absurdly low estimates of the lumbermen who wanted to buy it.

LUMBERMEN LEFT MINNESOTA with vast acreages of cutover land from which they had stripped the timber. Until the 1920s many people assumed that farmers would move in and cultivate the land after the timber had been removed and that lumbermen were rendering a service by preparing the land for agriculture. The lumbermen formed land companies to sell cutover land to unsuspecting immigrants, who were enticed by the state immigration office with extravagant promises.

At first some farmers could sell hay and produce to the logging camps, tend horses and oxen for a fee in summer when the camps were dormant, and supplement their meager farm income by taking winter jobs in the camps as lumberjacks. These opportunities disappeared when the camps closed down, and the farmers were forced to realize that the harsh climate and infertile soils of the coniferous forest simply are not suitable for commercial agriculture.

Many farms were abandoned, taxes rose on the few that remained, and in time they too had to be abandoned. The situation was exacerbated because taxes on standing timber were so high that lumbermen felt compelled to cut it as soon as possible instead of managing it wisely, and they felt no sense of responsibility for the land after they had logged it. They refused to pay taxes on their cutover land, and by 1934 nearly half of the land in northeastern Minnesota was tax-delinquent. Much of this land was incorporated into state and national forests.

The logging slash—limbs, branches, and leaves that the lumbermen had left on the tax-delinquent cutover lands—was a serious fire hazard. The state suffered disastrous fires that killed many people and burned vast areas at Hinckley in 1894 (350,000 acres), at Chisholm in 1908 (400,000 acres), at Baudette in 1910 (360,000 acres), and at Cloquet in 1918 (200,000 acres). For comparison, the area of Hennepin County is 356,000 acres.

The fires were so hot that they created firestorms. For example, the summer of 1894 was one of the driest on record, and fires were burning everywhere. On September 1 two major fires were roaring south of Hinckley. When they merged, their heat created a giant updraft. This updraft sucked in fierce blasts of strong winds that whipped the fires into a wall of flame that exploded over Hinckley and devastated the entire town. The heat was so intense that it buckled steel railroad tracks. The Cloquet fire of 1918 killed 453 people and destroyed 52,000 homes; the Hinckley fire killed almost as many people.

Most lumbermen fled Minnesota when the pinelands were gone. The lumber companies that remained had already been trying to figure out how to use the "inferior" species and second-growth trees that they previously had scorned. They built factories to make doors, sashes, boxes, and other wood products. Minnesota has two of the nation's three leading wooden window-making companies, Andersen Windows in Bayport near Stillwater, and Marvin Windows in Warroad, both of which were founded in 1904.

A new breed of lumberman began to turn trees into paper rather than lumber; by the 1920s, pulp and paper were beginning to replace lumber as the principal forest product of Minnesota. Paper mills had been constructed in Duluth in 1890, at Cloquet in 1898, and at Grand Rapids in 1901, but the development of new papermaking machines encouraged the construction of large new paper mills. Traditional lumbering and sawmills had wasted 70 to 80 percent of a tree, but the new paper mills used nearly all of it, and they could use all kinds of trees.

The grand rapids of the Mississippi River have been harnessed to provide power for the paper mill in Grand Rapids.

Wood consists of cellulose fibers held together by a gluelike substance called lignin.[5] Papermakers chip logs of pulpwood into small pieces and dissolve the lignin with strong chemicals to produce an amorphous fibrous mass of pulp. They wash, screen, bleach, and dry the pulp and run it through rollers that press it into continuous ribbons of paper. Paper machines, which are as long as a football field, roll out paper at speeds of up to thirty miles an hour.

Many of the modern pulp and paper mills of Minnesota are at major waterpower sites that once had sawmills.[6] They dominate towns like International Falls, Grand Rapids, and Cloquet. Next to the mills are huge woodyards with interminable stacks of 100-inch-long pulpwood logs and mountains of wood chips. The mills have had to install complex, expensive systems to prevent pollution of air and water by the strong chemicals they must use to dissolve the lignin. They produce high-quality coated papers for magazines and calendars, building board, cardboard, packaging, and a variety of other special paper products.

Aspen is a small, fast-growing tree that occupies more than one-third of Minnesota's timberland. Formerly it was underutilized, but it has become the primary species for making paper and also for making oriented strand board (OSB), which is now more popular than plywood for most structural panel and other construction

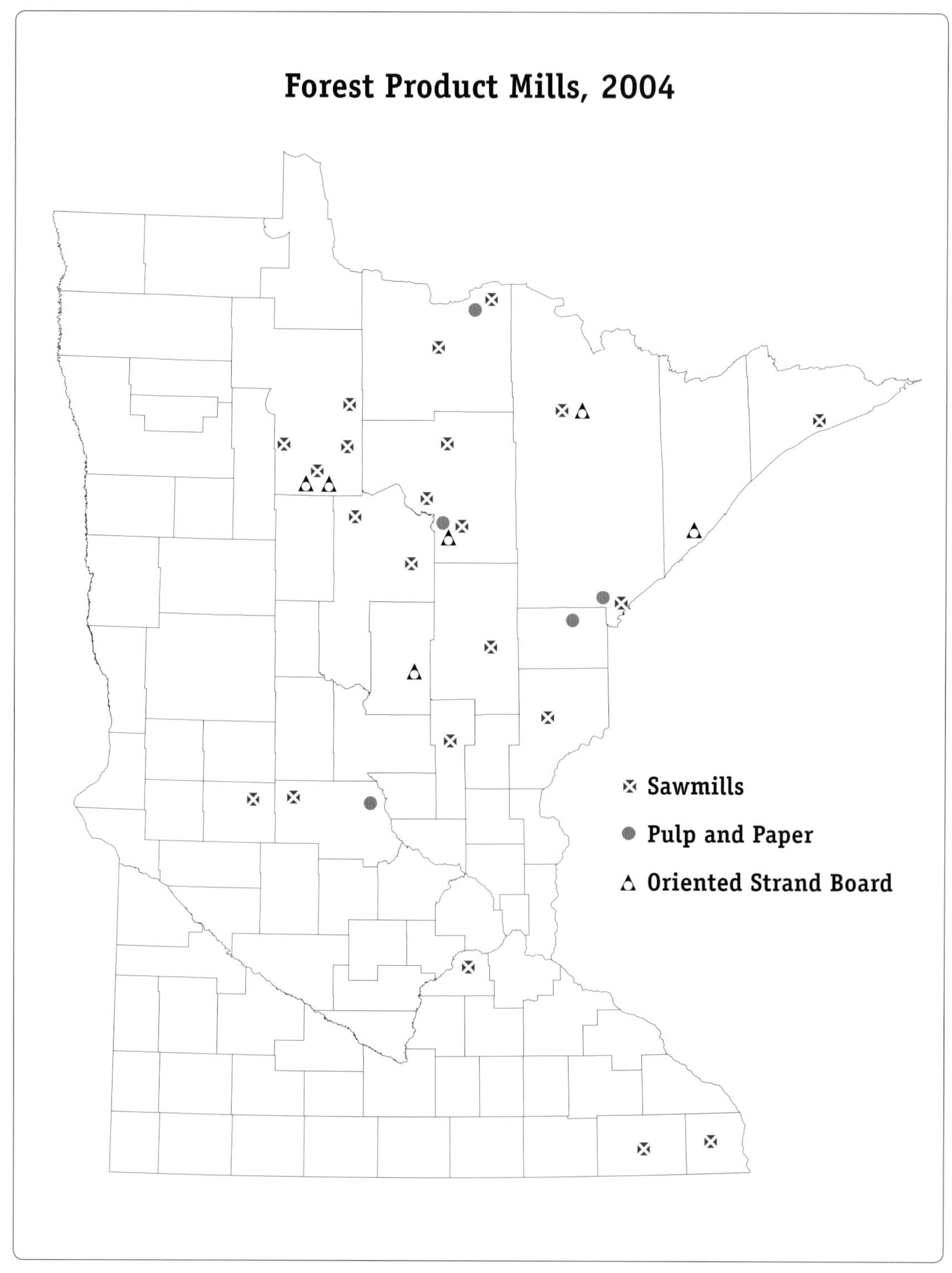

Modern forest products mills use all kinds of trees, and nearly all of the tree.

The paper mill at the international falls of the Rainy River dominates the city of International Falls.

Modern paper mills need prodigious quantities of pulpwood, which is stacked in yards near the mill.

purposes.[7] To make OSB, which first came on the market in 1978, strands of wood fiber are carefully aligned, layered, subjected to intense heat and pressure, and glued with resins containing water-resistant waxes. OSB production in North America increased from only 0.75 billion square feet in 1980 to 25 billion square feet in 2005.

Clear-cutting is the most efficient strategy for managing aspen forests because the trees quickly resprout from roots. Hunters like periodic clear-cutting because it creates the optimum habitat for deer and for ruffed grouse, the number one upland game bird in Minnesota, but some people object to it on aesthetic grounds, and forest managers must be sensitive to their complaints.

Paper companies own hundreds of thousands of acres of woodland, but they buy some of their pulpwood from independent loggers. They manage their forests efficiently for sustained yields, maintain hiking and snowmobile trails, and permit hunting and fishing. They replant land as soon as they have logged it, but their efforts to restore forests of white pine have been unsuccessful because nursery stock

A modern forest harvest operation. The feller-buncher cuts and bunches the trees, and the skidder drags them to the harvester, which in this sketch chips them instead of sawing them into pulpwood.

that was imported from Europe was infected with white pine blister rust, which causes deadly cankers on native white pine.

The large blocks of paper-company land have linked federal, state, and county land into vast wooded areas to which the public has had access for recreation, but stockholders are forcing paper companies to sell much of their land because they believe that too much of their capital is tied up in assets that are not producing income.[8] Paper companies are selling their land to investment companies that are subdividing it into smaller plots and selling or renting it to individuals who are closing it to the public. The increased fragmentation of the northern forest threatens key wildlife habitats, and the access roads built by the new owners have encouraged invasive species such as crows, raccoons, feral cats, and weed plants.

The feller-buncher has hydraulic shears that can snip off a tree as easily as scissors cut through paper.

The harvester saws trees into 100-inch pulpwood length and stacks the logs beside the trail.

Today an independent logger working with three men can cut as much pulpwood as thirty lumberjacks could cut in 1950. They can cut 100 to 150 trees an hour using a feller-buncher, which is a Bobcat with powerful hydraulic steel shears mounted in front of it at ground level. It leaves no stumps. The feller-buncher has steel claws that clamp the tree, while the shears can cut off a twelve-inch pine as easily as scissors cut through paper. The feller-buncher then twirls the tree like a baton and places it on the ground. A skidder picks up several trees and drags them to the harvester. The harvester has a long claw that picks up a tree and feeds it to a buzzsaw that cuts it into 100-inch logs. The harvester claw then piles the logs into a stack to be picked up by the truck that hauls them to the mill. The replacement cost of this equipment is several million dollars or more.

Minnesota's first great industry fueled early growth in the state. Lumbering began in the triangle between the Mississippi and St. Croix rivers and then moved to the northern half of the state where the construction of railroads enabled lumbermen to penetrate the state's vast pinelands. Peaking around 1910, sawmilling has been replaced by the modern pulp and paper business, whose enormous mills completely dominate the communities where they are sited.

1. Root River
2. Zumbro River
3. Wabasha
4. Cannon River
5. Red Wing
6. Big Woods
7. Mankato
8. Red River
9. Winona
10. Owatonna
11. Faribault
12. New Ulm
13. St. Cloud
14. Mesabi Iron Range
Kittson
Roseau
Lake of the Woods
Marshall
Beltrami
Koochiching
Pennington
Red Lake
Polk
Clearwater
Itasca
St. Louis
Norman
Mahnomen
Hubbard
Cass
Clay
Becker
Wadena
Aitkin
Carlton
Crow Wing
Wilkin
Otter Tail
Pine
Todd
Morrison
Mille Lacs
Kanabec
Grant
Douglas
Benton
Traverse
Stevens
Pope
Stearns
Sherburne
Isanti
Chisago
Big Stone
Swift
Anoka
Kandiyohi
Meeker
Wright
Washington
Ramsey
Hennepin
Lac Qui Parle
Chippewa
Yellow Medicine
Renville
McLeod
Carver
Scott
Dakota
Sibley
Rice
Goodhue
Lyon
Redwood
Nicollet
Le Sueur
Wabasha
Brown
Murray
Cottonwood
Watonwan
Blue Earth
Waseca
Steele
Dodge
Olmsted
Winona
Jackson
Martin
Faribault
Freeborn
Mower
Fillmore
Houston

7

Settlers Move In

In 1850 the only part of Minnesota Territory open for legal white settlement lay east of the Mississippi River, but in 1851 Governor Alexander Ramsey persuaded the Dakota to sell their land in southern Minnesota to the United States. By the Treaties of Traverse des Sioux and Mendota, the Dakota ceded their hereditary rights to more than 19 million acres of land, the entire southern half of the present state, and land-hungry white settlers began pouring in. It took time for the U.S. government to survey the land and establish land sale offices, but by 1857 the state had eight land offices, which had already sold more than 5 million acres.

Most of Minnesota's early white settlers probably came up the Mississippi River by paddlewheel steamboat, even though the river was frozen for some five months of the year, because travel by water was far easier than travel overland on crude or nonexistent roads. Paddlewheelers were essentially elaborate rafts that could float on only a few feet of water, and their captains used to joke that they could even run on a heavy dew. They were driven not by underwater screw propellers but by large revolving paddlewheels that churned on their sides or sterns.

Paddlewheelers stopped at landings all along the Mississippi. Visitors were impressed by the picturesque bluffs that loomed over both sides of the river valley, but the bluffs had little good farmland. Settlers pushed inland up the Root River valley across from La Crosse, Wisconsin, up the Zumbro River valley from Wabasha, and up the Cannon River valley from Red Wing. These rivers were not navigable, but their valleys gave access through the steep wooded blufflands to the Big Woods and the great prairies on the level uplands a few score miles to the west. A few

The Mississippi River was the first major route into Minnesota for white settlers.

Red Wing nestles at the foot of bluffs that overlook the Mississippi River.

Paddlewheel steamboats were like rafts that could float on only a few feet of water.

Paddlewheel steamboats lined the St. Paul waterfront around 1860.

paddlewheelers continued up the Minnesota River as far as Mankato, but for most of them the head of navigation was St. Paul, headquarters of the fur trade and jumping-off point for the Red River trails to the northwest.

The first visiting steamboat had reached Fort Snelling as early as 1823. In the 1850s and 1860s, eight to ten boats a day were arriving in St. Paul, but railroads quickly took the place of steamboats. In 1854 a railroad west from Chicago reached the Mississippi at Rock Island, Illinois, and passengers could travel from New York to St. Paul in the comfort of train and steamboat. The railroad celebrated the event by inviting 1,200 distinguished guests and a number of journalists to make the trip on seven chartered steamboats, and the resulting publicity helped to establish Minnesota firmly on the mental map of the nation.

The railroad from Chicago reached Dubuque, Iowa, in 1855, Prairie du Chien, Wisconsin, in 1857, La Crosse in 1858, and St. Paul in 1867. After the Civil War, railroad lines began to radiate from the Twin Cities hub, and merchants in the towns along the Mississippi River realized that they needed their own railroad lines in order to be competitive. They financed lines west from La Crosse and from Winona to bring wheat from the prairies east to the ports on the Mississippi, and to haul lumber and other necessities back to the farms on the prairies.

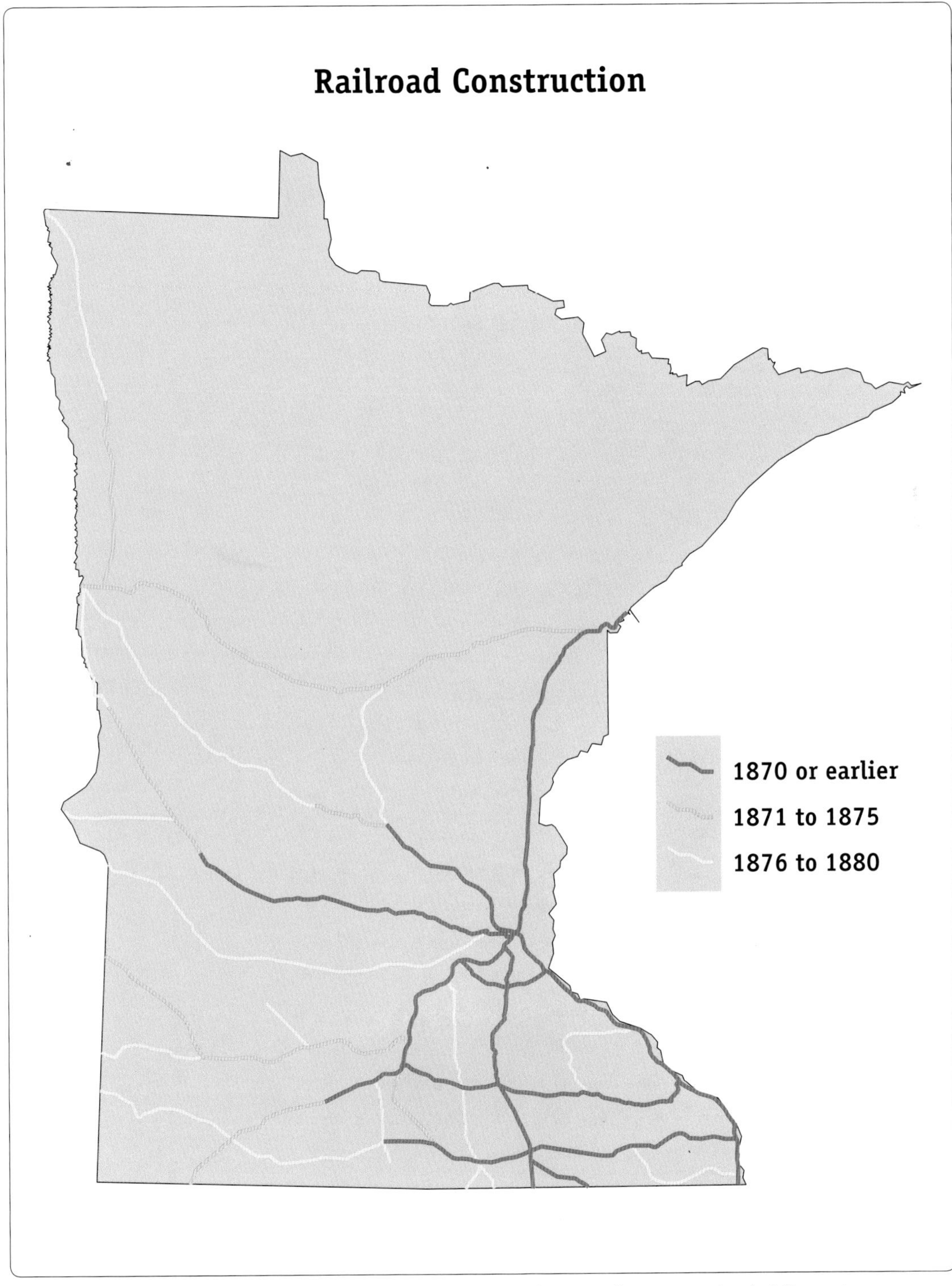

After 1865 railroads began to replace steamboats as the principal means of transportation in Minnesota.

The construction of railroad lines coincided with the spread of agricultural settlement.[1] The first lines were built through settled areas, but soon they extended beyond them and opened up new areas for settlement. The government granted land to railroads to help finance their construction, and the railroads encouraged settlement to generate more business. The railroads sold land to intending settlers and platted towns at regular intervals along their lines to serve as places where the farmers of the surrounding countryside could sell their products and buy the goods they needed.

The spread of agricultural settlement in Minnesota is mapped by plotting the first census year in which each township had a population of at least twelve persons per square mile of land area. This density could have been achieved by three families of four persons each on three 160-acre homesteads, which would have left one homestead in each section still waiting to be claimed and settled. Most agricultural areas of the state have remarkably similar histories of rapid early population growth followed by a long, slow, gradual decline.

Fillmore County in southeastern Minnesota, the first nonriver county in the state to be settled, illustrates this pattern, both at the township and at the county level. Three agricultural townships in the county peaked in population in 1870 and have been losing people ever since. The county overall reached its peak population in 1880 and has been losing population steadily since 1900. Across the state most agricultural areas were settled by 1900 and have been losing population since. The growth of the population of Minnesota in the twentieth century has been concentrated in its villages, towns, and cities.

In 1860, the first census year after Minnesota attained statehood in 1858, many townships along the Mississippi, Minnesota, and St. Croix rivers were already settled, as were the first upland townships west of the bluff country of southeastern Minnesota and townships south of the Twin Cities toward Owatonna. The entire southeastern corner of the state had effectively been settled by 1870. By 1880, settlement had pushed northwestward along the line of the old Red River trail at the edge of the Big Woods and the prairie, as well as along the Minnesota River valley. By 1890 and 1900, settlement had continued to expand to the northwest, but most of southwestern Minnesota was not effectively settled until 1900.

The census of 1860 reported that Minnesota had 172,023 residents, and the population of the state has continued to grow steadily ever since. It had so few people in 1850 that in-migration had to fuel its early growth, but the newcomers quickly started having children. The share of the state's people who were native-born Minnesotans grew from only one in five in 1860 to better than half by 1900, but until

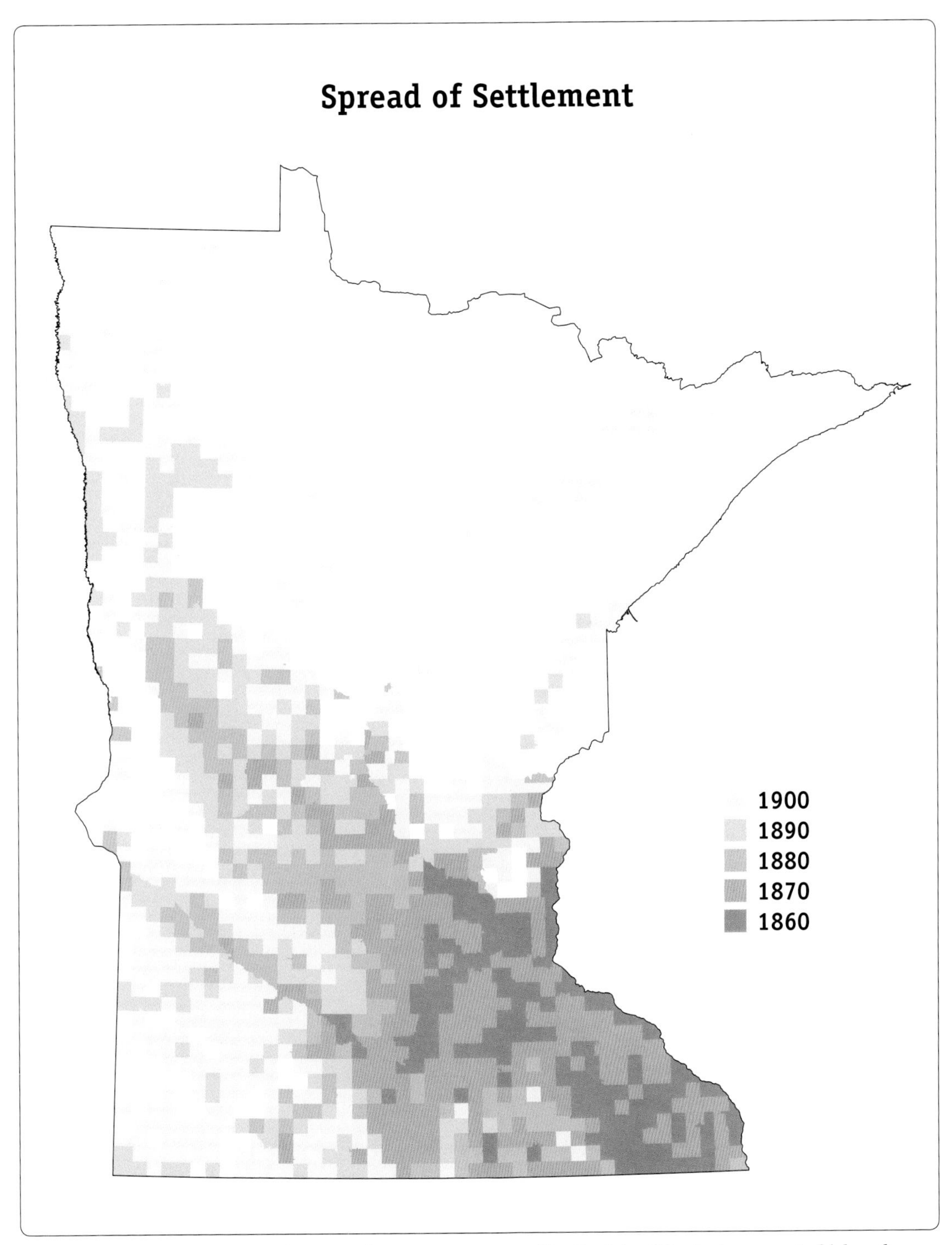

Settlement spread northwestward across Minnesota, as measured by the date of the first census at which each township had at least 12 persons per square mile.

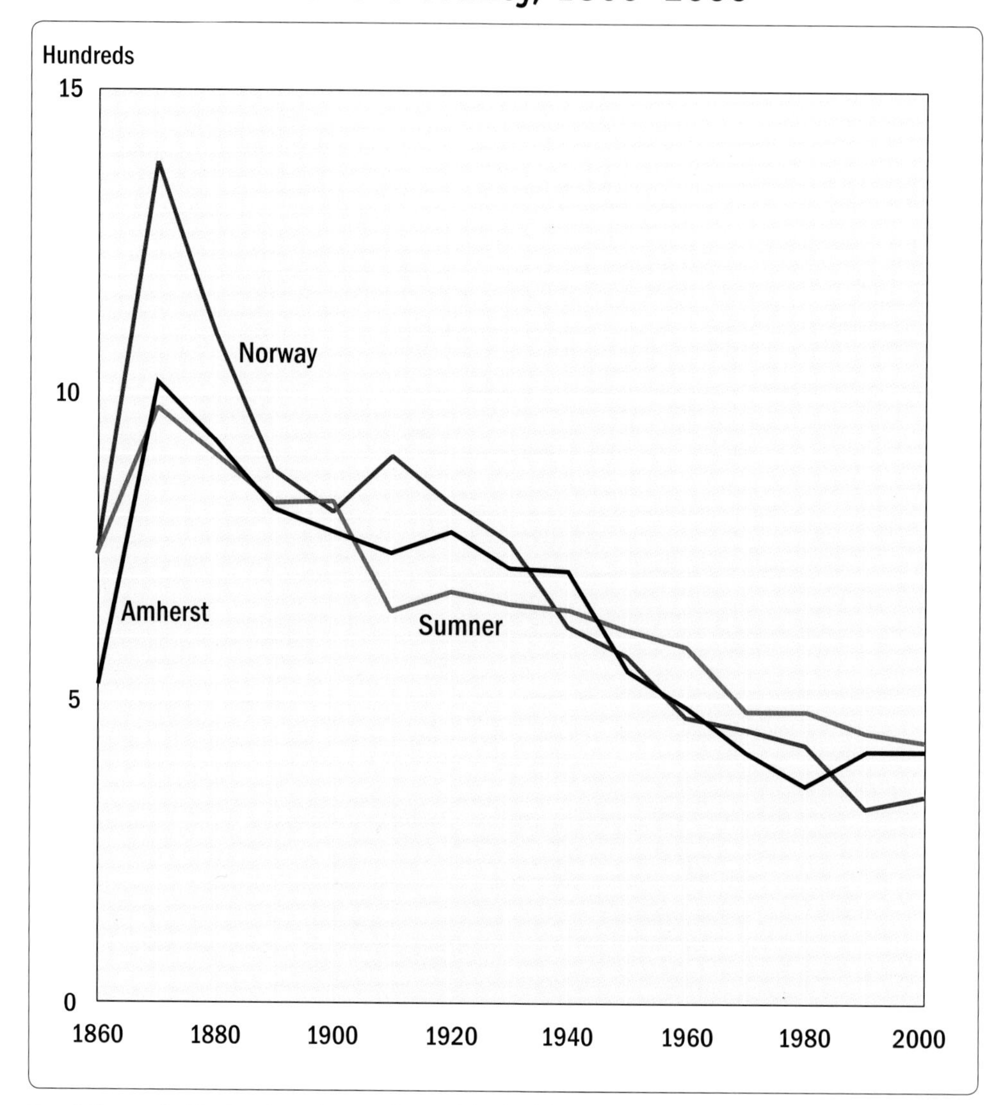

Population peaked in 1870 in typical rural townships in Fillmore County and has been declining ever since.

Population of Fillmore County, 1860–2000

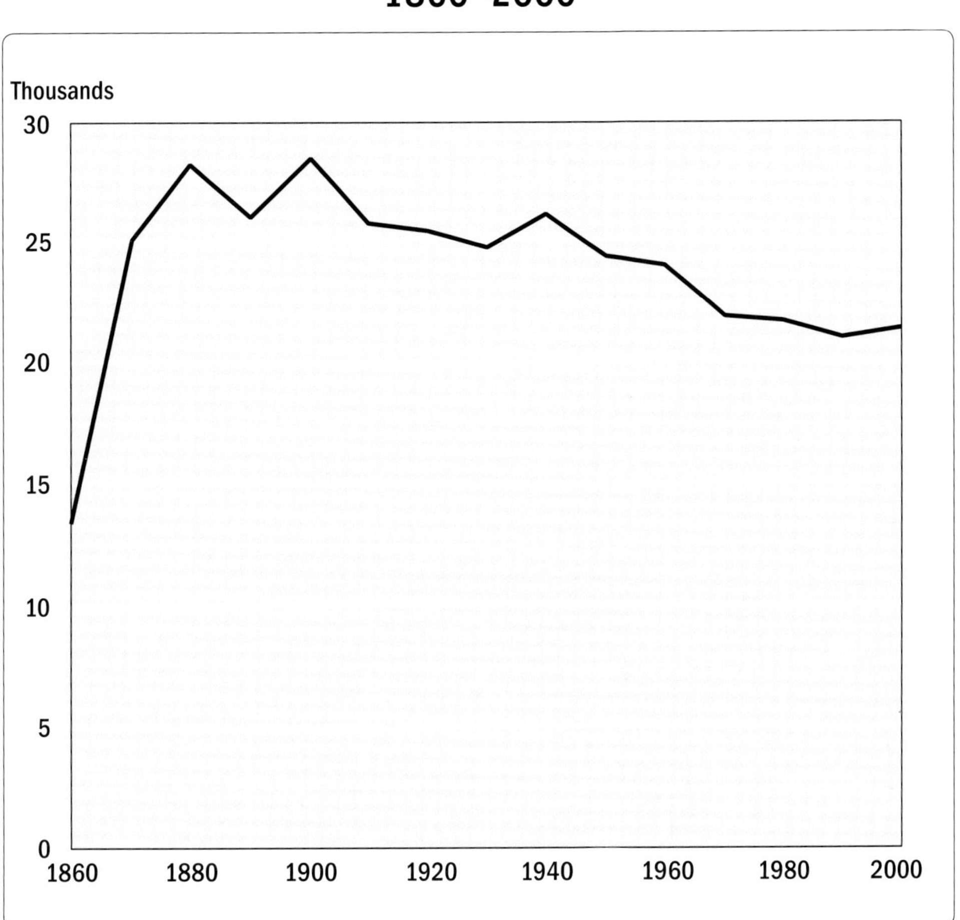

The population of Fillmore County peaked in the early days of settlement and has been declining subsequently.

1900 more than half of the people of Minnesota were in-migrants.[2] Between 1860 and 1910, one-third to one-quarter of the state's residents had come from countries overseas and a dwindling share had come from eastern North America.

Most immigrants to Minnesota were part of three great streams of migration from eastern North America: Yankees and New Yorkers (herein, "Yankees"), Middle Atlantic residents, and people from neighboring states. Other immigrants came from Ireland, Germany, Norway, Sweden, Austro-Hungary, and Finland.[3] More than one-third of the non-native residents of Minnesota in 1860 were products of the great migration stream that surged westward from New England, New York, and New France (Quebec) across Ontario, Michigan, and Wisconsin into Minnesota. The vanguard was men from Maine who had established lumbering centers on the rivers well before statehood. The size of this stream dwindled steadily after 1860, and it had become insignificant by the turn of the century, but it had a long-lasting impact.

Some Yankees were farmers who settled on the land, but many became townspeople who were eager to make money in business. They spoke the national language and had good commercial contacts with friends and acquaintances in eastern cities. They founded towns, opened stores and banks, and built factories. They felt superior to the non-English immigrants who farmed the countryside, and the dislike was mutual. They published newspapers, became civic leaders, and established the township as an important unit of local government.

The Yankees seemed incapable of minding their own business, and they eagerly imposed their values on others. They recognized the importance of education, founded schools and colleges, and recruited New England spinsters (unmarried women) to teach Yankee values to schoolchildren. They were loyal to the Congregational Church, formed temperance societies, and were fierce Republicans and abolitionists. They resented the power of the fur traders in St. Paul, who were Democrats. The Republicans wanted a farming state that extended west to the Missouri River, whereas the fur traders wanted a state that reached north to the Canadian border.

The Yankees who came to Minnesota were literate and conscious of their own importance. They left rich records, quite unlike the in-migrants from the older states of the Midwest and the South, who quietly settled on farms and were content to allow the Yankees to monopolize the spotlight. While written records of non-Yankees are sparse, census data show that neighboring states were a steadily increasing source of in-migrants as the size of the Yankee stream dwindled; between 1900 and 1910, nearly half of all new in-migrants to Minnesota were natives of Wisconsin and Iowa.

Immigrants to Minnesota from Ireland constituted nearly 10 percent of the non-native population in 1860, but thereafter few Irish immigrants found their way this far west. They were mostly unskilled laborers who did not know how to farm, although some were farm laborers. Most headed for the cities, where they worked in sawmills or in railroad construction. Leaders like Archbishop John Ireland tried to set them up on farms to distance them from the lure of urban saloons, but the colony he started in Rice and Le Sueur counties near Faribault did not last long.

Germans were the most numerous ethnic group to migrate to Minnesota, although Norwegians were always a close second. Some romantic souls claimed that the Mississippi River valley in southeastern Minnesota reminded German immigrants of the Rhine River valley in Germany, but the hardheaded immigrants were

Place of Birth

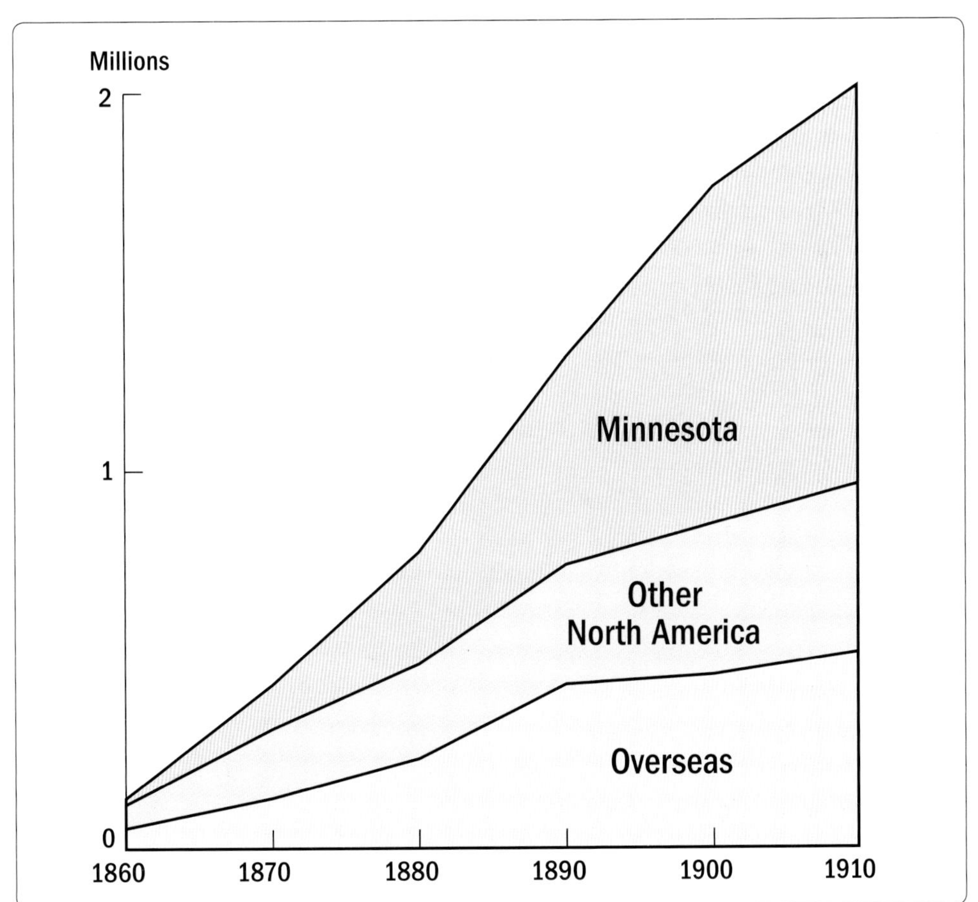

Until 1900 more than half of the people of Minnesota had migrated to the state from other areas.

Net Migrants by Place of Birth, by Decade

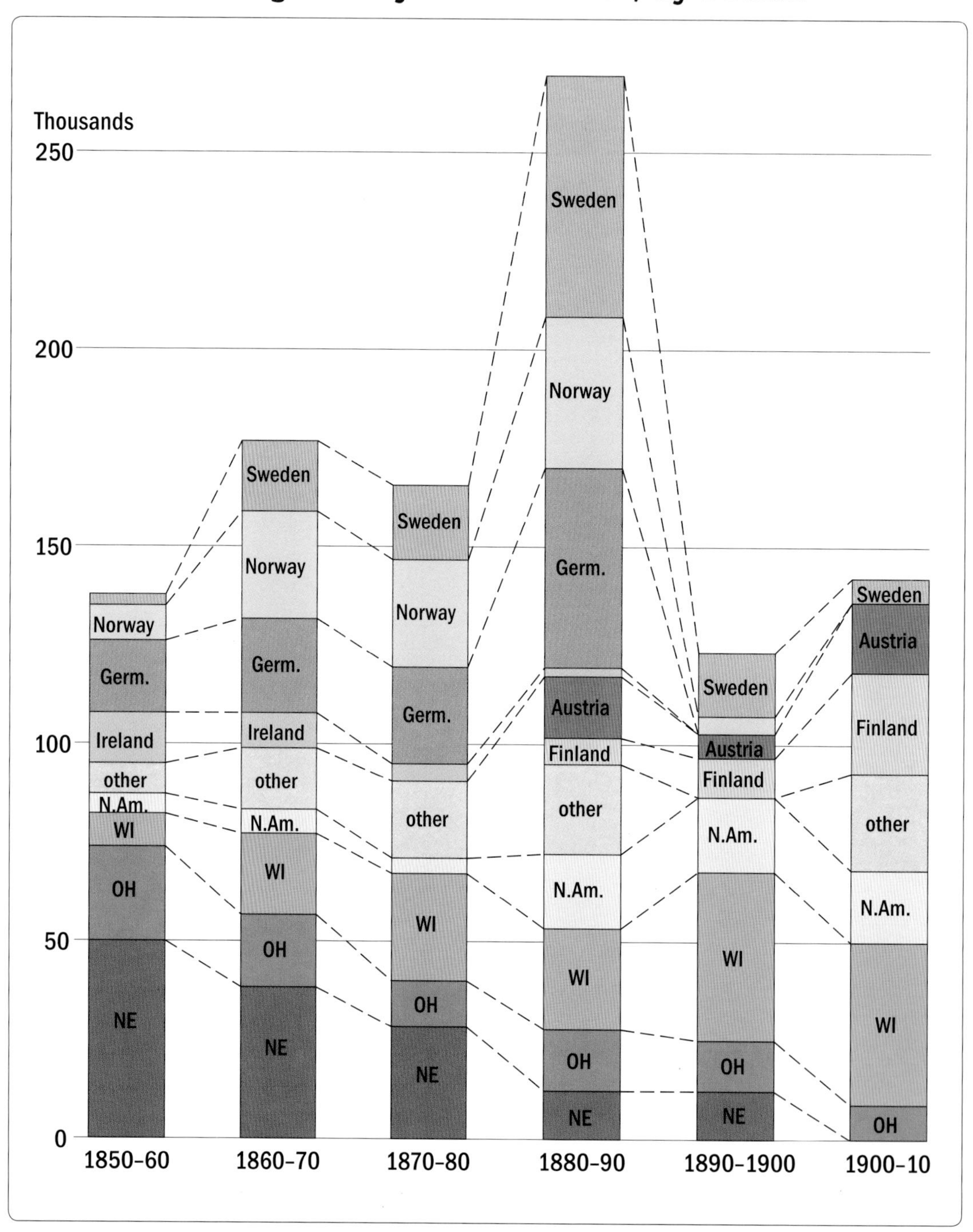

Most of the net migrants to Minnesota have come from three major areas in eastern North America or six major source areas in Europe.

attracted by the productivity of the land rather than by its scenic beauty. They pored over the handbooks for immigrants that individuals, states, and railroad companies compiled to help them assess the quality of land in the United States.

Many of the German-born immigrants actually had lived in states to the east before moving to Minnesota and had accumulated enough capital to enable them to make a good start in their new lives. They were townspeople as well as farmers and had a wide variety of occupations, from professionals to laborers. Some of them settled in the Minnesota River valley, where a group from Chicago had founded New Ulm in 1857, and after 1860 Father Pierz recruited many German Catholic settlers to Stearns County west of St. Cloud.

Many Norwegian immigrants to Minnesota were wary of cities. When over-crowding forced them out of rural areas at home, they decided to seek their fortunes in rural Minnesota instead of moving to cities in Norway. Homesteads of 160 acres in Minnesota seemed like princely estates to people who had farmed tiny hillside plots of land in Norway. The settlers wrote enthusiastic letters home and sent pre-paid tickets to friends and relatives in Norway to enable them to migrate to Minnesota.

Norwegians largely preferred wooded areas. The first groups of immigrants settled at the edge of the bluff country of southeastern Minnesota and went on to settle the park region between the prairies and the boreal forest in the west central part of the state. Norwegians migrated as families, and many settled near people who had come from the same area, even the same parish, in Norway. They cherished their Norwegian heritage and their rural way of life, and the Lutheran parish church was a central institution for preserving their ethnic identity.

Swedes, like Norwegians, migrated to Minnesota because economic conditions forced them to leave home. The earliest Swedish immigrants were families who sought farms, and the largest cluster of rural Swedes was in the logged-over areas in and around Chisago County north of St. Paul. On the prairies of western Minnesota, Swedes settled near the railroad lines, and many of them earned their seed money by working for the railroads.

Minnesota, which was opened to immigrants in an era of national railroad construction, was one of the few states settled by people who came directly from Europe. Trains in Europe carried emigrants straight to the port of departure, and from New York, trains carried them straight to the frontier in Minnesota. Before railroads, travel was so slow and difficult that most parts of this country were settled by people who had been born only a few hundred miles farther east. The children born on one frontier leapfrogged to open up the next frontier to the west. Children born

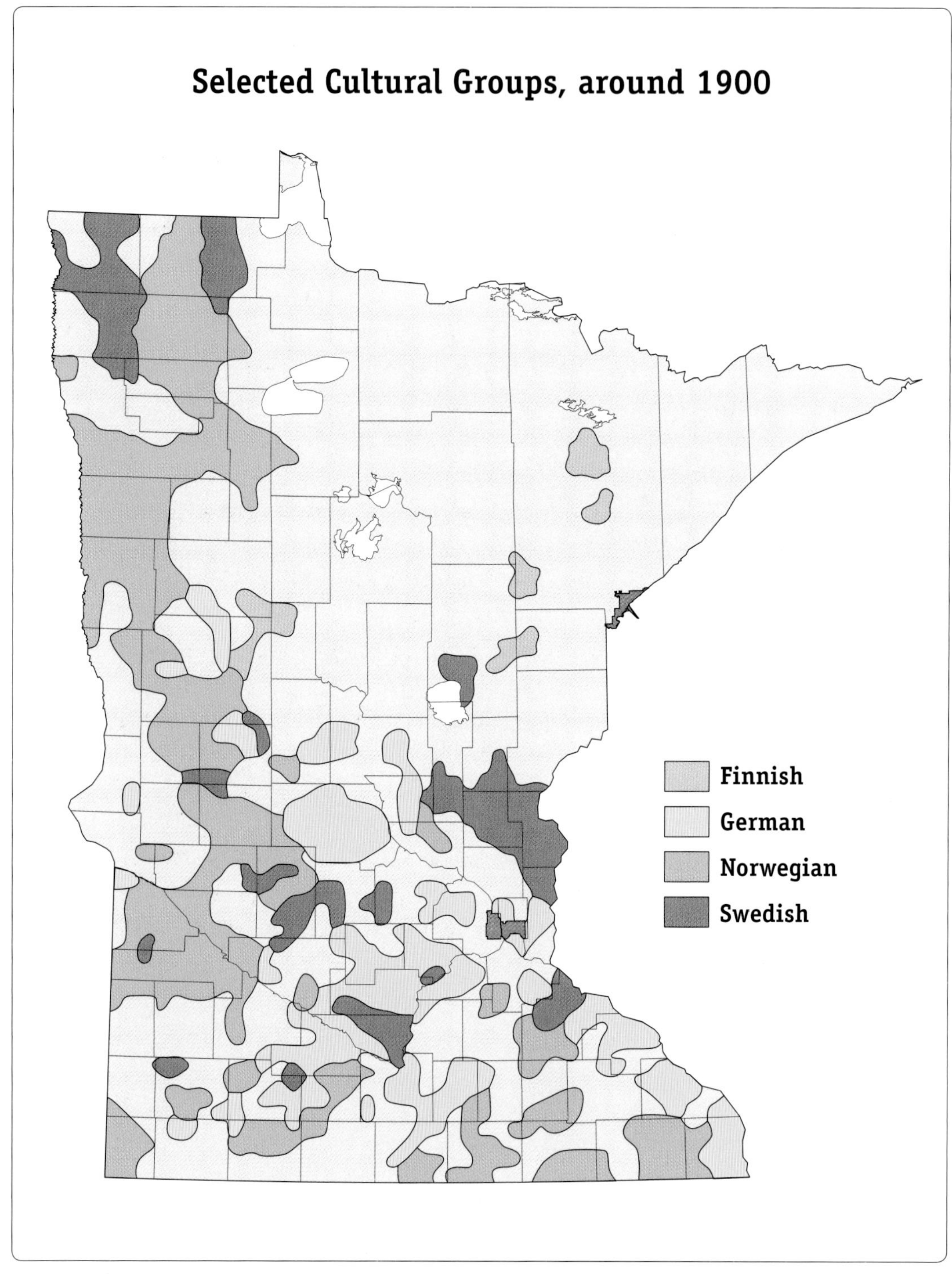

Many immigrants to Minnesota chose to live near other immigrants from their native land.

in Ohio, for example, leapfrogged to the frontier in Illinois, and children born in Indiana leapfrogged Illinois to open up the frontier in Iowa and Missouri.

More than half of all Swedish immigrants to Minnesota arrived in a single decade, the 1880s. Earlier Swedish immigrants had been families who wanted to start farms, but most of the new group were single young people seeking urban jobs. They were so numerous that the state had more Swedes than Norwegians by 1900, but they were also the tail end of large-scale immigration from Scandinavia to Minnesota.[4]

The state's last major group of immigrants from Europe before World War I consisted of Finns, Poles, southern Slavs, Italians, and people of the Austro-Hungarian and Russian empires. These immigrants were mostly unskilled young men from rural areas. Some worked on farms in summer and in the woods and sawmills in winter, and some found factory jobs, but most wound up working in the iron mines of northeastern Minnesota.

Temperance groups formed social clubs to give these young single men a healthy alternative to the saloons, gambling, and prostitution that were rampant on the Iron Range. Some of these clubs evolved into socialist organizations, many of which became active in the labor movement. Three-quarters of the participants in the first major strike on the Mesabi Range in 1907 were probably Finns, who thereby won their reputation for radicalism.

Before World War I, many people saw immigrants as a threat to the United States, so native-born residents pressured immigrants to learn to speak English and to adopt American values and the American way of life. In 1917, in one of the darker episodes in the state's history, the legislature established the Minnesota Commission on Public Safety, which had near-dictatorial powers to impose conformity on ethnic groups. Since then the state has matured and has been able to accept and sometimes embrace the diversity of its ethnic groups, including those who have arrived from Latin America, Africa, and Asia since World War II.

Kittson
Roseau
7
Lake of the Woods
Marshall
Beltrami
Koochiching
Pennington
Red Lake
Polk
Clearwater
8
Itasca
6
Norman
Mahnomen
Hubbard
Cass
St. Louis
4
Clay
Becker
10
Wadena
Aitkin
Carlton
Crow Wing
Wilkin
Otter Tail
9
Pine
Todd
Mille Lacs
Kanabec
Grant
Douglas
Morrison
Benton
Traverse
Stevens
Pope
Stearns
Sherburne
Isanti
Chisago
Big Stone
Swift
Anoka
Kandiyohi
Meeker
Wright
Washington
Lac Qui Parle
Chippewa
Hennepin
Ramsey
1
Yellow Medicine
Renville
McLeod
Carver
Scott
Dakota
Sibley
Lyon
Redwood
Nicollet
Goodhue
Le Sueur
Rice
Wabasha
Brown
5
Murray
Cottonwood
Watonwan
Blue Earth
Waseca
Steele
Dodge
Olmsted
Winona
Freeborn
Rock
Nobles
Jackson
Martin
Faribault
2
3
Mower
Fillmore
Houston
1. Twin Cities
2. Albert Lea
3. Austin
4. Red River Valley
5. Winona
6. Arrowhead
7. Warroad
8. Bemidji
9. Brainerd
10. Duluth

8

From Wheat to Dairy Farming

Minnesota became the final stage for one of the great dramas of American agriculture: the westward surge of the wheat frontier and its subsequent replacement by dairy farming. During the nineteenth century, wheat production swept westward like a giant wave across the northern United States, from New England to the Hudson River valley to upstate New York to northeastern Ohio and then across Wisconsin into Minnesota. Even in southeastern Minnesota, some early settlers abandoned their farms and migrated west in search of new wheat lands after the wheat frontier had crested. In each of these areas wheat was the pioneer crop, but wheat monoculture quickly exhausted the soil, and dairy farming replaced it.

Many people who came to Minnesota in the early days expected to make a living by farming. Some had come from worn-out areas farther east in search of virgin land in the West and brought practical farm knowledge. Immigrants who came directly from Europe quickly adopted the American farming system. Farm buildings and farm practices in Minnesota show scant sign of Scandinavian or German influence, although traditions from the old country often are preserved in the home and in dietary patterns.

Wheat dominated early farming in Minnesota, first in the southeast, then in other parts of the state as the wheat frontier swept westward. Early settlers grew potatoes and vegetables for themselves and corn, oats, and hay for their horses, cattle, and hogs, but wheat was their main money crop, and for a brief time wheat reigned as the undisputed king of the countryside.[1]

Both the price of wheat and yields were good. Prices and yields fluctuated from year to year but trended upward, and farmers made good money growing wheat.[2] The invention and continuous improvement of new farm machinery, such as reapers, binders, and threshers, made growing wheat less backbreaking. In their peak wheat year most of the state's agricultural counties had more than one-fifth of their total area devoted to growing the golden grain. The wave of wheat cultivation crested in southeastern Minnesota in 1870 and thence swept westward across the state. Today it is hard to realize how totally and completely wheat dominated the life of Minnesota during its brief heyday.

In the early settlement years, farmers grew wheat as their principal cash crop, but the continuous cultivation of any crop may create problems. Each crop has its own particular suite of plant nutrients that it must take from the soil in order to grow. Continuous cultivation of the crop removes these nutrients from the soil but leaves other nutrients that the crop does not use. Cultivation of the same crop year after year also lays out a bounteous feast for the insects and diseases that prey on it, and they multiply.

Problems of selectively depleted soil, insects, and diseases inevitably reduced the profitability of wheat cultivation, and eventually farmers decided to quit growing it. Some shifted to other crops, but many loaded all their household goods onto the farm wagon, hitched up the team of oxen, and made their slow way westward in search of virgin wheat land. Women drove the wagons, and the men and boys plodded along behind droving the farm animals.

In southeastern Minnesota, as in the rest of the state, the crest of wheat cultivation was sharp and short. Exceptionally high prices triggered small increases in 1900 and in 1920, but since 1920 the crop has virtually disappeared from the eight southeastern counties. At the crest of the wheat wave, farmers in these counties grew more than 1 million acres of wheat, but in 2002 they grew less than 4,000 acres, and the crop was little more than a memory.

By 1880 the crest of the wave of wheat cultivation had swept west to the second tier of counties, which extend south from the Twin Cities to Albert Lea and Austin. The crop made a brief comeback in 1900, when yields were exceptionally good, but thereafter both yields and acreages in these counties plummeted. The eight southwestern counties also had a brief flirtation with wheat in 1900, when their acreage spiked at 800,000 acres, but since then these counties have never grown as much as 100,000 acres of the crop, and in 2002 they had dropped to only 5,800 acres.

By 1900 the crest of the wave of wheat cultivation had spread to twelve counties in the Red River Valley, the only part of the state in which wheat still is a major crop, albeit for lack of better alternatives. The wheat acreage in these counties spiked in

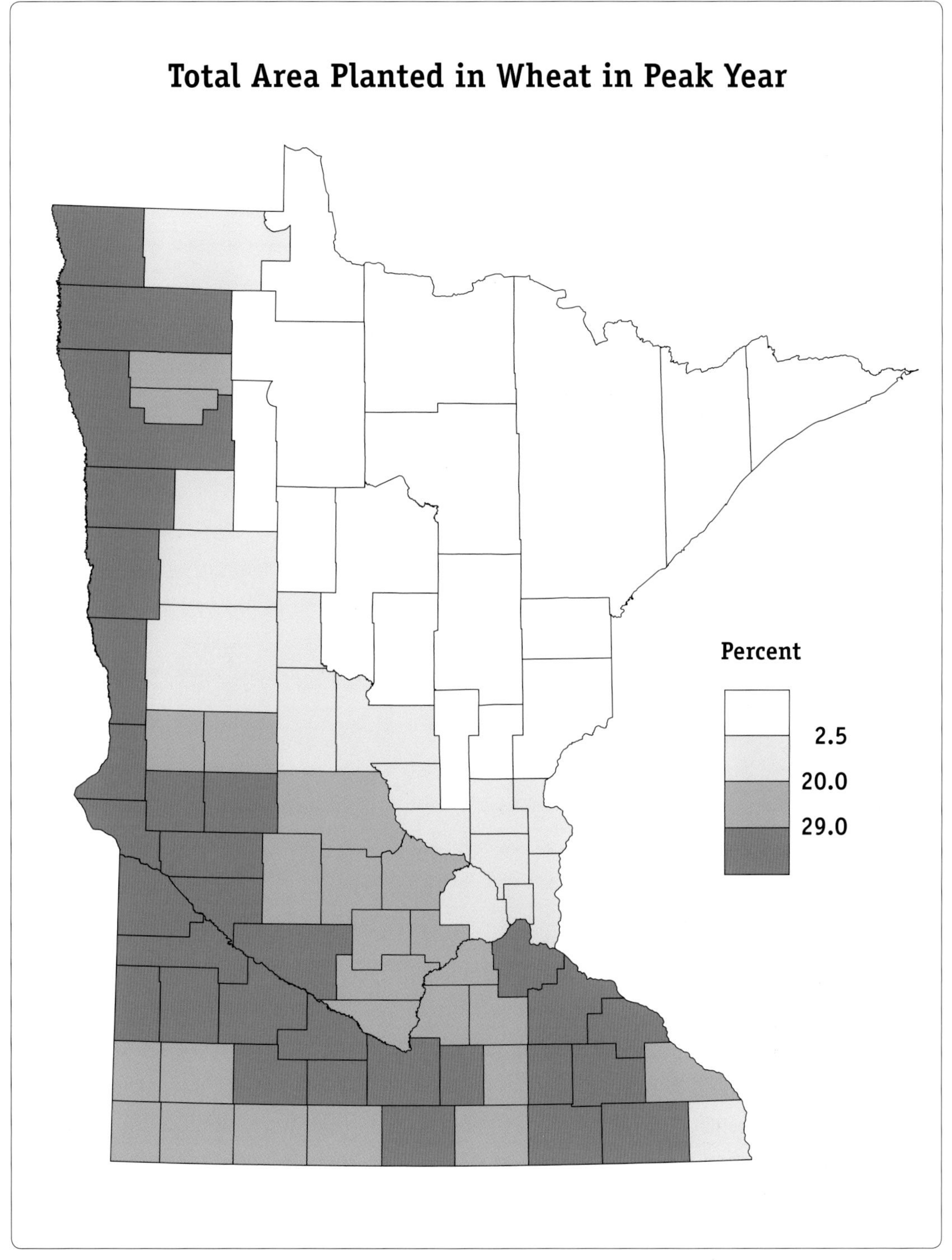

Wheat was king of southern and western Minnesota in the early days of settlement.

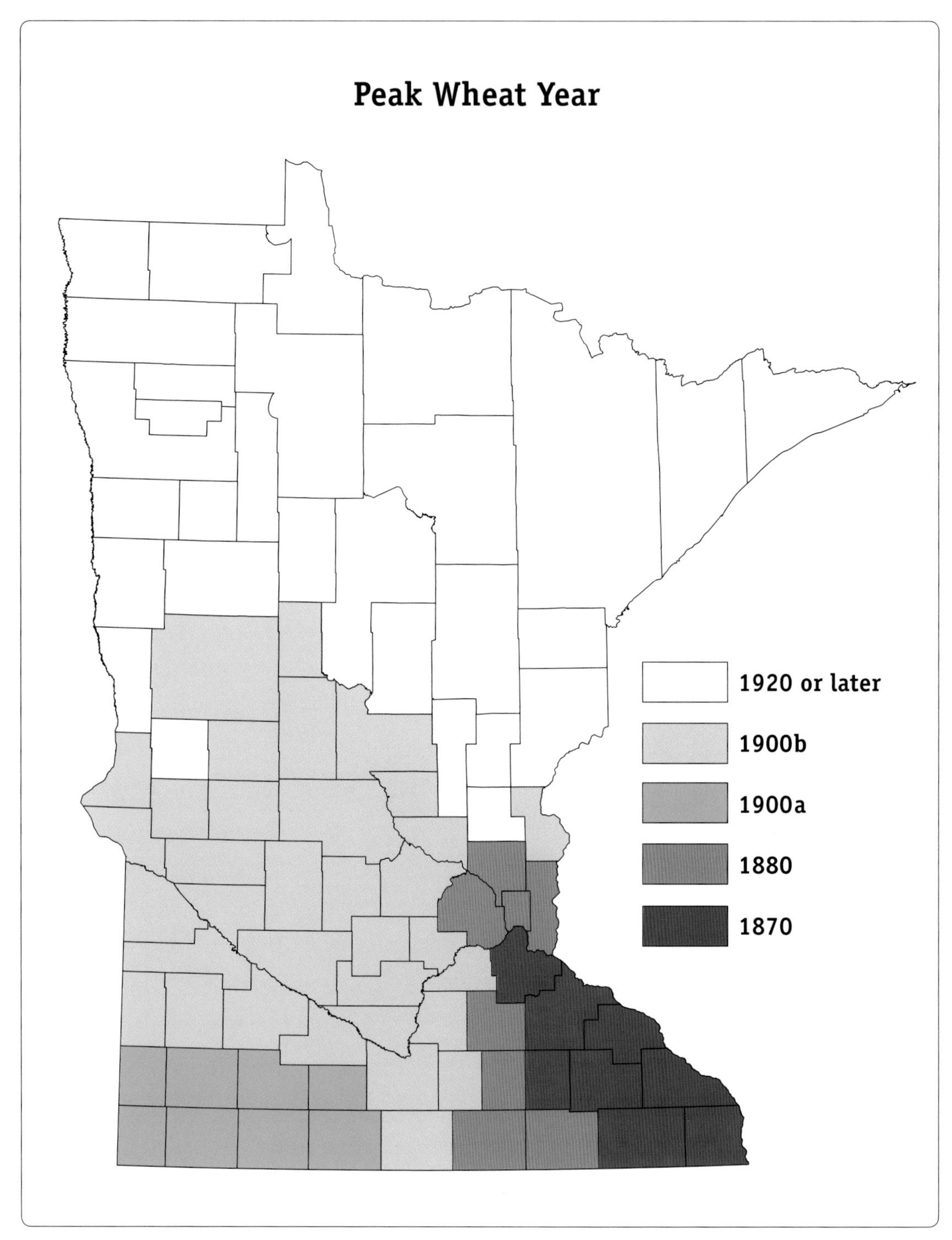

A wave of wheat cultivation has swept across Minnesota from the southeast to the northwest. The eight southwestern counties had a brief flirtation with wheat (1900a), and wheat cultivation spread toward the Red River Valley (1900b).

1900 and 1920, dropped for four decades, with a minor spike during World War II, but then resurged sharply after 1970, when wheat prices became propitious. In 2002 these counties contained 82 percent of the state's 1.9 million acres of wheat. In the rest of Minnesota the crest of the wave of wheat cultivation also spiked in 1900, but by 1925 the wave had passed. Outside the Red River Valley, wheat is only a minor crop today.

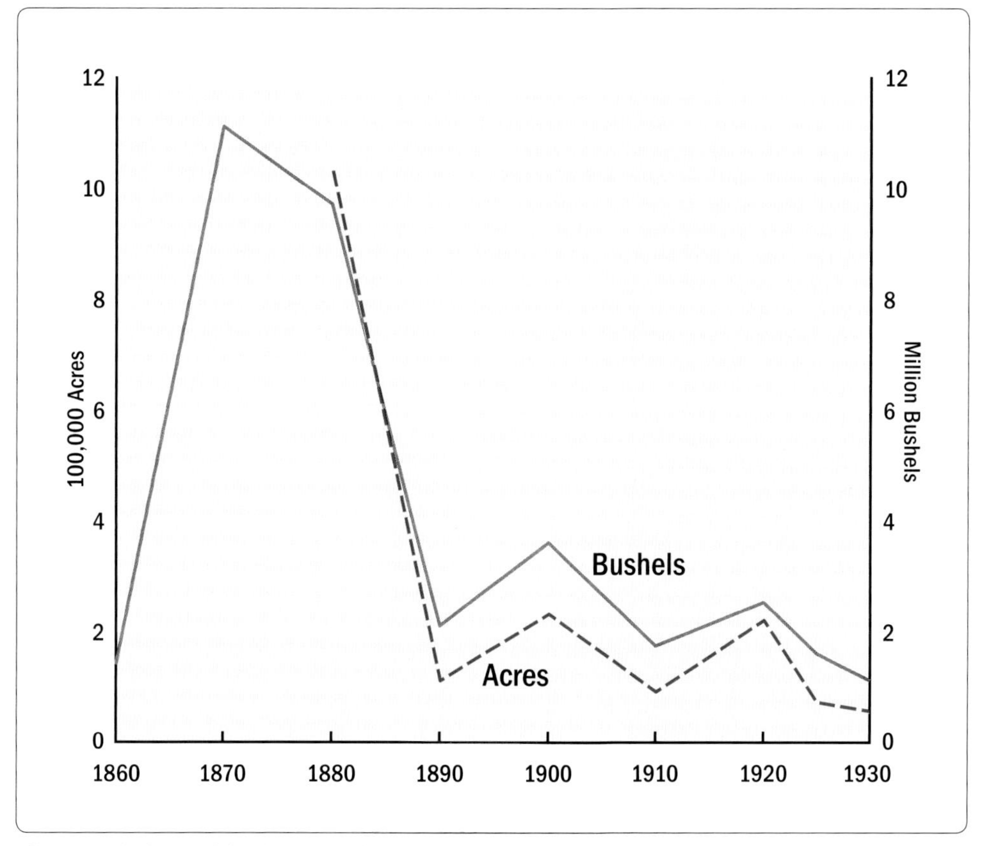

The wave of wheat cultivation crested in southeastern Minnesota in 1870.

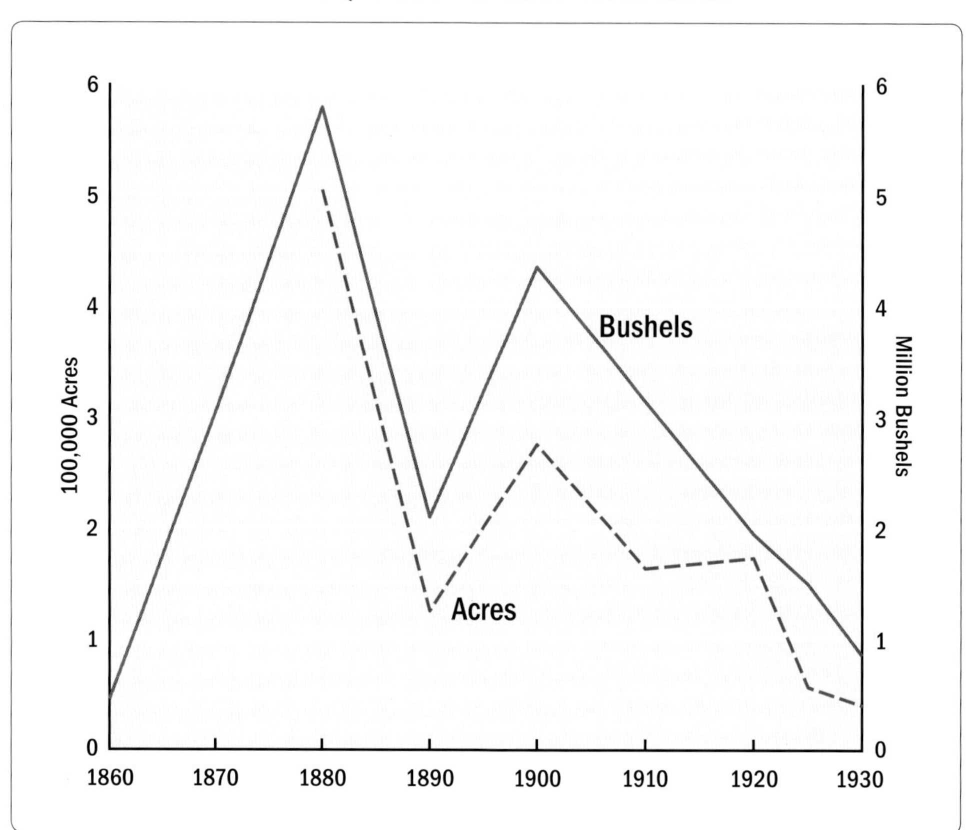

In the second tier of counties, wheat crested in 1880 and came back slightly in 1900.

WINONA WAS THE STATE'S LEADING WHEAT MARKET and shipping port when the crest of the wave was in southeastern Minnesota. Wagons loaded with wheat inundated the city after the fall harvest, and farmers begged steamboat captains to load on just a few more bags of wheat to get it out before winter ice ended the shipping season on the river. Flour mills downriver were the principal buyers of wheat, but local entrepreneurs built gristmills to grind it into flour at virtually every waterfall and stream rapids in the bluff country, and gristmills graced power sites in other parts of the state as the wheat frontier surged westward.

After 1870 Minneapolis, which had the state's largest waterpower site at the Falls of St. Anthony, began to replace Winona as the leading wheat market, and by 1890 it had become the world's leading flour milling center. Entrepreneurs such as Pillsbury, Washburn, and Crosby eagerly latched on to technological innovations such as the middlings purifier (which removed brown bran flecks from the flour)

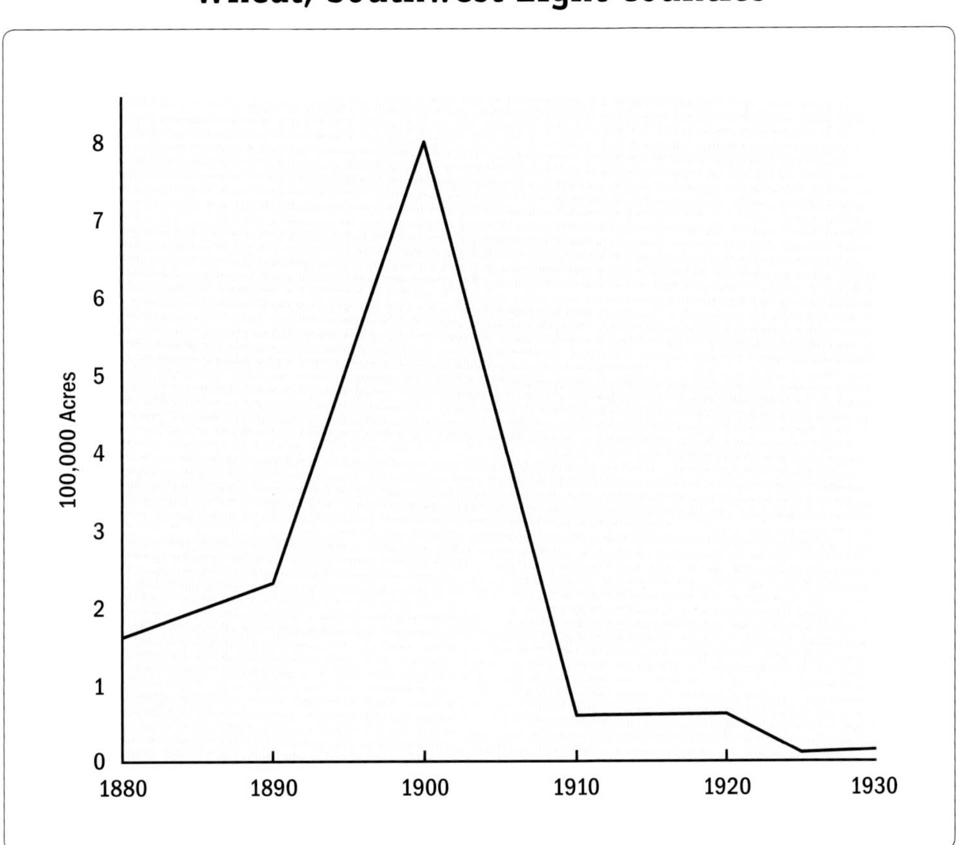

Wheat cultivation spiked dramatically in the eight southwestern counties in 1900, but it has never come back.

and replaced millstones with steel or porcelain rollers to make more flour and better flour at a lower cost. The millers generously passed on a small share of their profits to farmers by paying them a bit more for their wheat, but most farmers believed that the mill owners could and should have paid them more.

The Minneapolis milling magnates encouraged the construction of railroads to haul wheat in from the farms and to carry flour to eastern cities. At regular intervals along their lines the railroads built trackside grain elevators to which farmers could haul their wheat for sale, and the elevators stored the wheat until the flour mills at the falls needed it. The railroads emancipated wheat farmers from their dependence on the seasonality of river transportation, but the haul grew longer and longer as the crest of the wave of wheat cultivation surged ever westward, and the farmers left behind in the older wheat-growing areas were forced to diversify their operations.

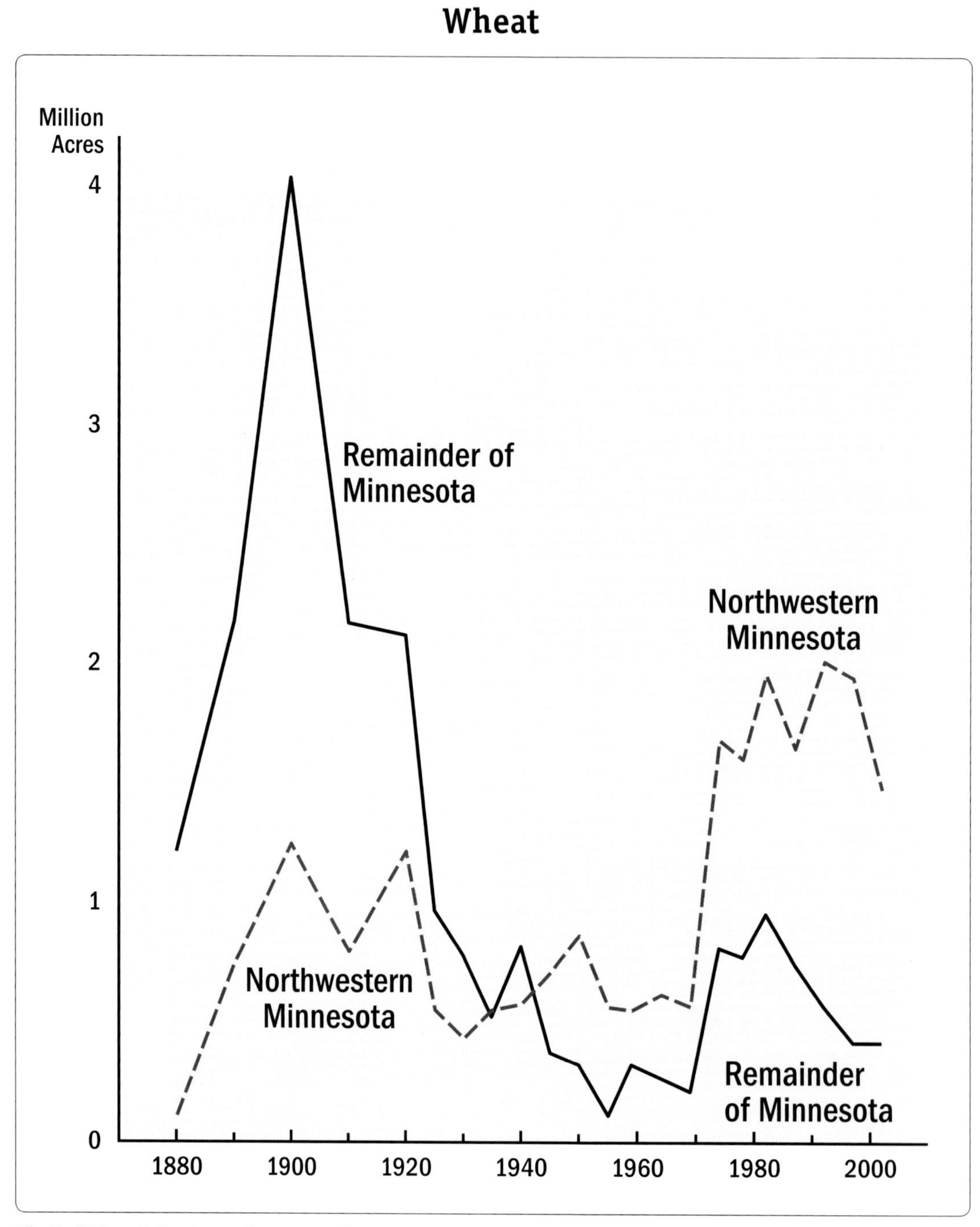

The Red River Valley in northwestern Minnesota is the only part of the state in which wheat is still a major crop.

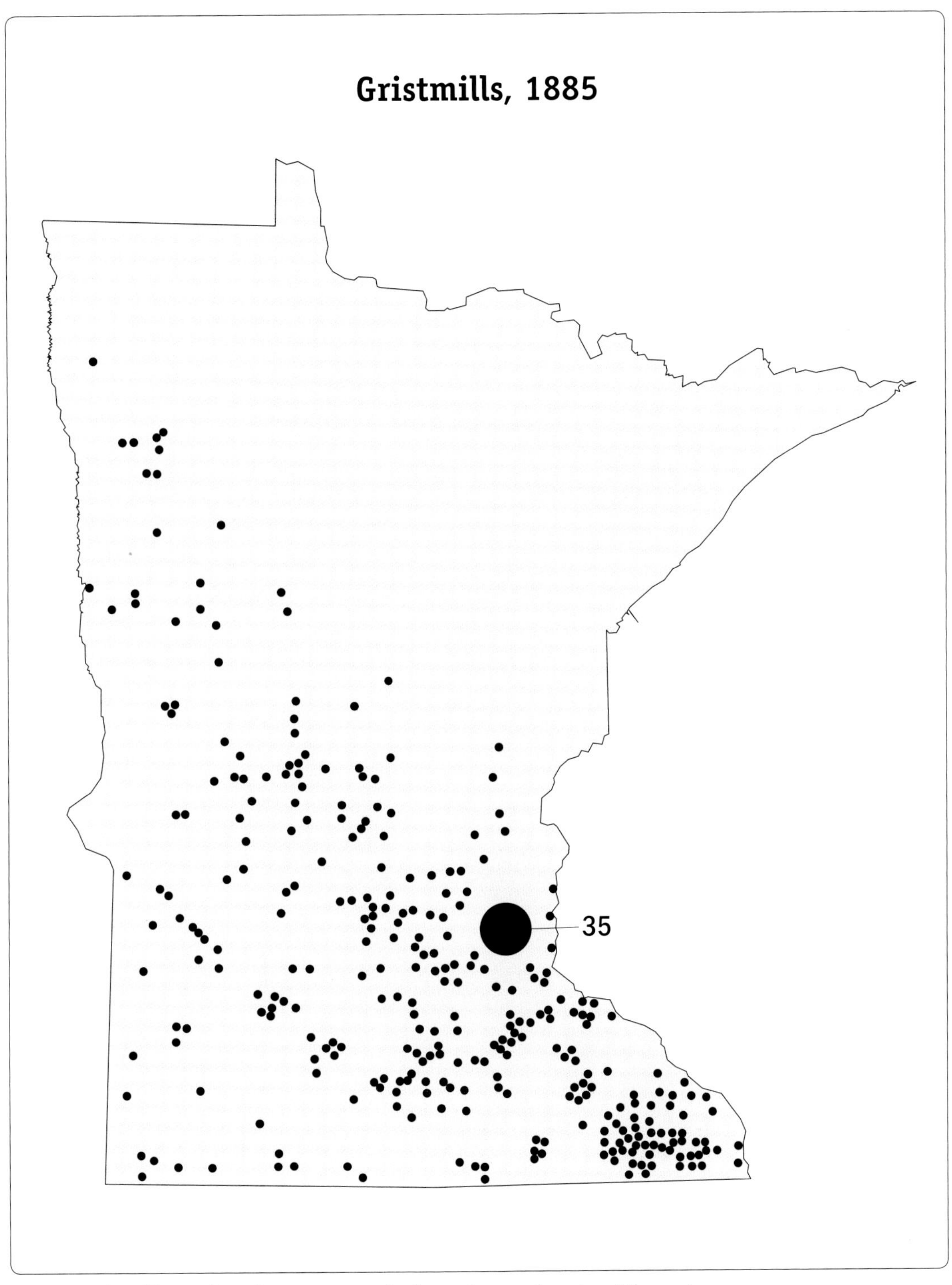

In 1885 a gristmill graced nearly every power site in southern and western Minnesota.

Knowledgeable observers had long urged farmers to reduce their dependence on wheat by growing other crops and to place greater emphasis on livestock. In most of Minnesota, farmers shifted from wheat to dairy farming based on a rotation of corn, oats, and hay. To understand how farming has evolved in the state, observers rely on the census of agriculture, the only complete, reliable, and comparable source of information, last taken in 2002.

The census defines "farmland" as all land that is owned or rented by people who live on farms. It defines a "farm" quite generously as any place on which a minimal amount of agricultural products (only $1,000 worth in censuses between 1974 and 2002) was produced and sold in the census year. In 2002 the census classified 54 percent of Minnesota as farmland, with values ranging from less than 10 percent in the Arrowhead counties of the northeast to more than 90 percent in some of the prairie counties of the central south.

Farmland acreage in Minnesota peaked in 1945 at 33 million acres and then dropped to 27.5 million acres in 2002. People who do not understand census terms have found cause for alarm in this "loss" of agricultural land because they fail to realize that a goodly part of the land classified as farmland actually is woodland of only limited agricultural value. For example, although Minnesota "lost" nearly 6 million acres of farmland between 1945 and 2002, more than 2 million acres of this lost farmland consisted of tracts of woodland that farmers sold to nonfarm buyers. A tract of woodland ceases to be farmland when a farmer sells it to a nonfarmer, even though the only thing about the land that changes is the name of the person who pays taxes on it.

In 2002 a total of 2 million of the state's 27.5 million acres of farmland, or about 7.5 percent, were woodland. More than one-quarter of the farmland is woodland in the nonagricultural counties of northeastern Minnesota, which also have the smallest percentages of farmland. A high percentage of farmland also is wooded in the bluff counties of southeastern Minnesota, but the percentage of farm woodland tails off toward the prairie counties in the southwestern part of the state, where woodlands are rare to nonexistent.

A more realistic definition of agricultural land than the census's "farmland" is the land it defines as "cropland." This includes all land on which crops were or could have been grown in the census year. The census did not start publishing data on cropland until 1925. The acreage of cropland in Minnesota peaked at 23 million acres in 1940; since then it has declined erratically at an average rate of around 25,000 acres, or only slightly more than one-tenth of 1 percent, a year. At this rate the state is in no imminent danger of running short of cropland.

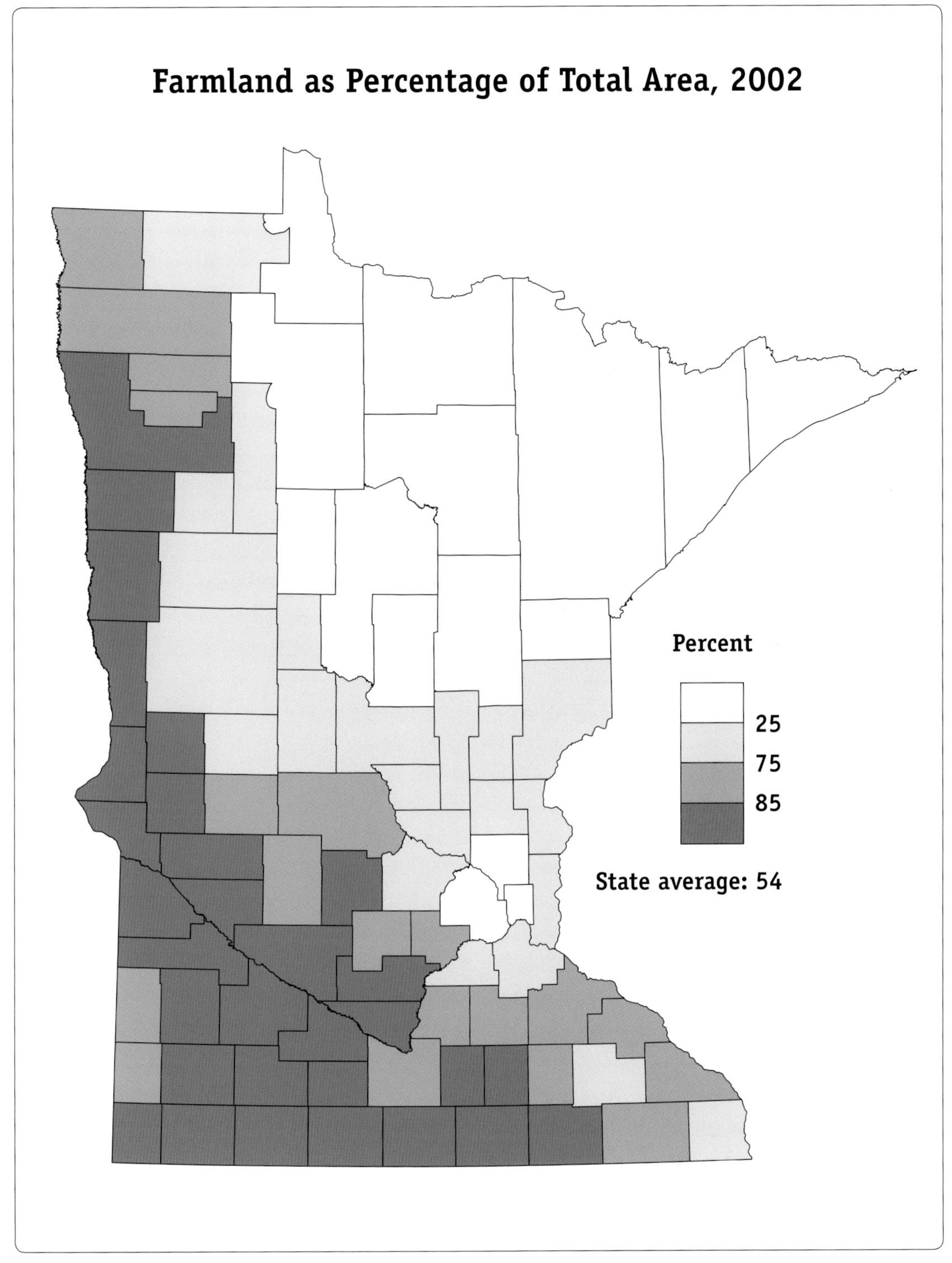

The percentage of farmland in Minnesota is greatest in the southwest and least in the northeast.

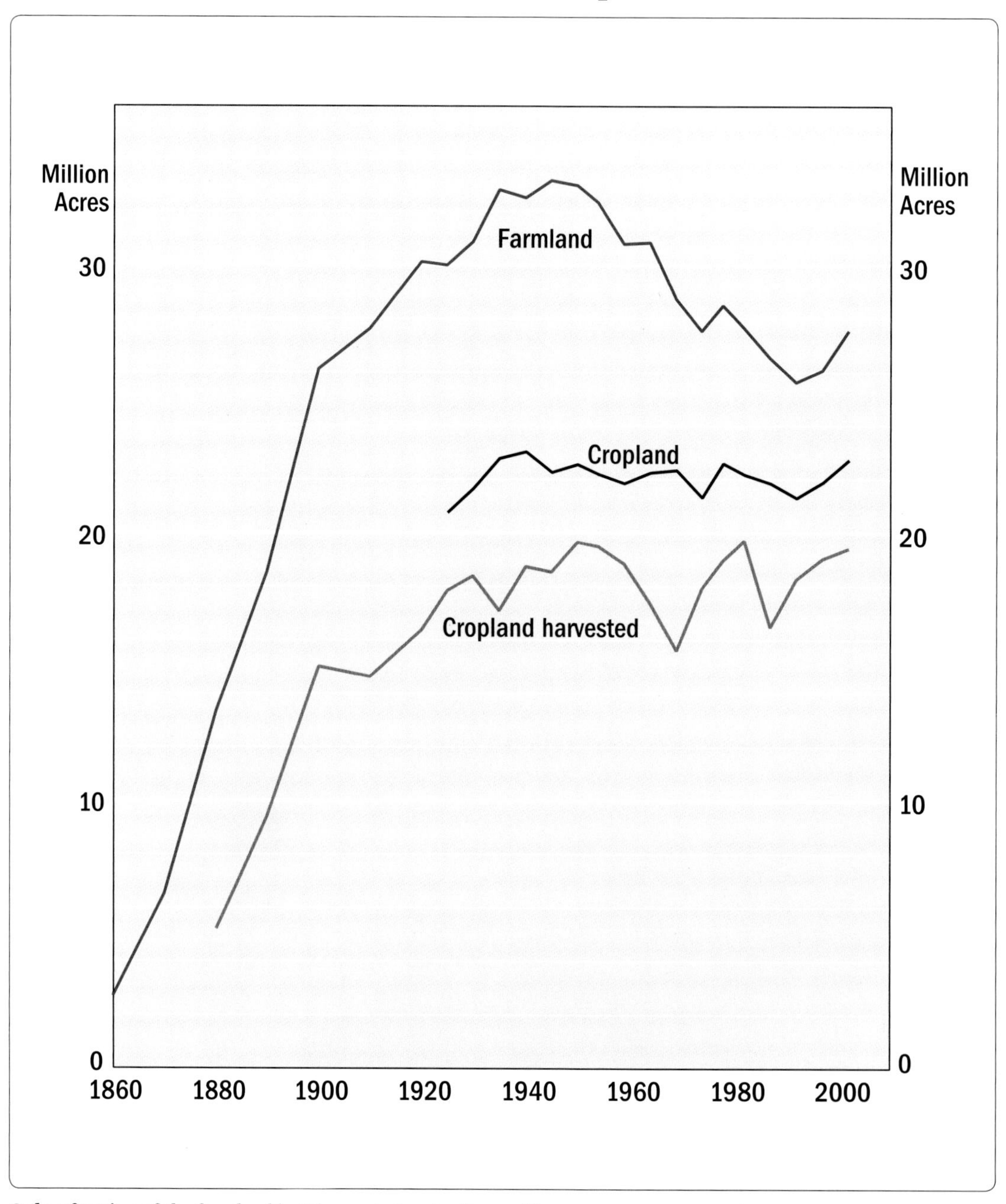

Only a fraction of the farmland in Minnesota is actually used to grow crops.

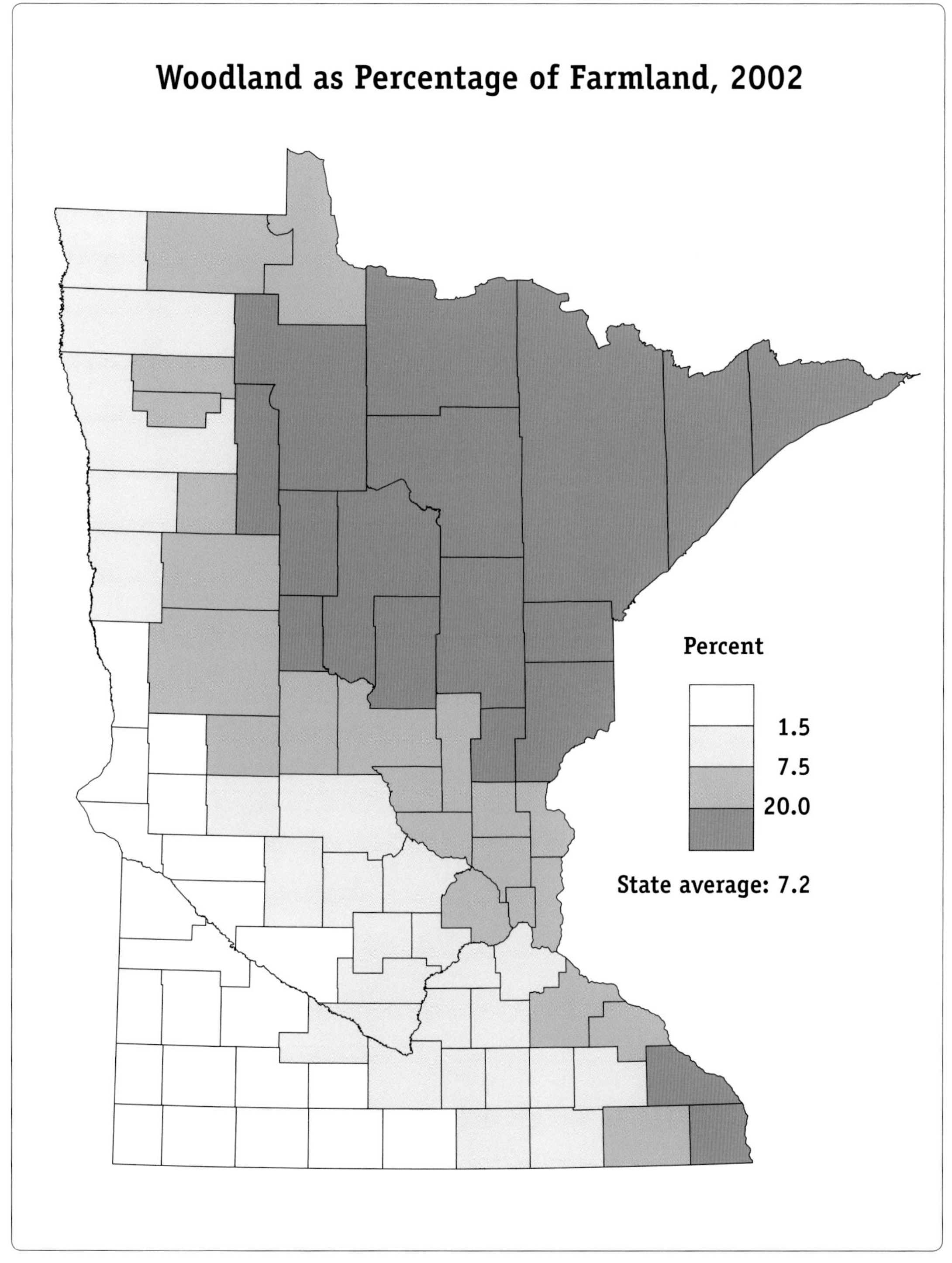

More than one-fifth of the farmland in northeastern Minnesota is actually wooded.

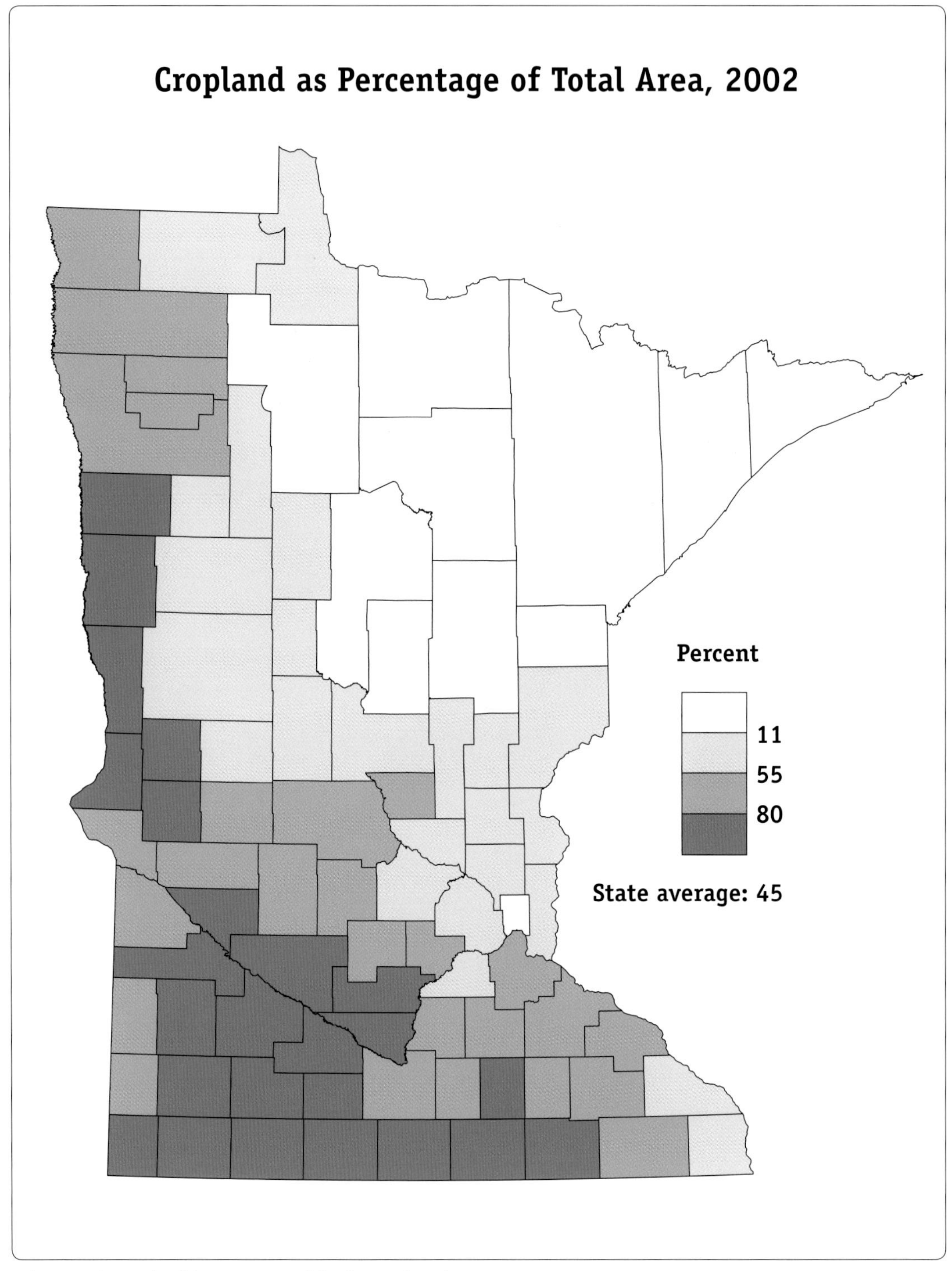

Minnesota's cropland is concentrated in the south and west.

Minnesota's cropland is concentrated in the south and west, and the percentage of cropland drops off rapidly to the northeast. In 2002 the twelve nonagricultural counties of northeastern Minnesota, which constitute 32.6 percent of the state's total land area, had only 6,000 (7.5 percent) of its 81,000 farms and only 814,000 (3.6 percent) of its 22,700,000 acres of cropland. Most of this cropland was used only for hay crops and grazing. The cropland of Minnesota lies west and south of a line from Warroad through Bemidji to Brainerd, and thence east to Duluth.

The acreage of cropland from which crops actually are harvested fluctuates from year to year. Drought, flood, storms, diseases, insects, and other destructive forces of nature can undo the best-laid plans of any farmer, but complete crop failure is only a part, and usually quite a small part, of the explanation of changes in the acreage of cropland harvested. Good farmers rotate their crops. Many crop rotations require that in some year of the rotation each piece of ground should lie fallow, be used only for pasture, or grow cover crops that are not harvested, and some cropland lies idle each year because the farmer suffers poor health or because the family disagrees about how it should be used.

Despite these fluctuations, a crop's percentage of the state's total harvested acreage is the best measure of its relative importance over time. For example, wheat was the state's dominant crop until 1900, and it remained the leading crop until corn supplanted it in 1925. After four decades in the doldrums, wheat has made a remarkable comeback in the Red River Valley since 1970, but it is still a distant third to corn and soybeans.

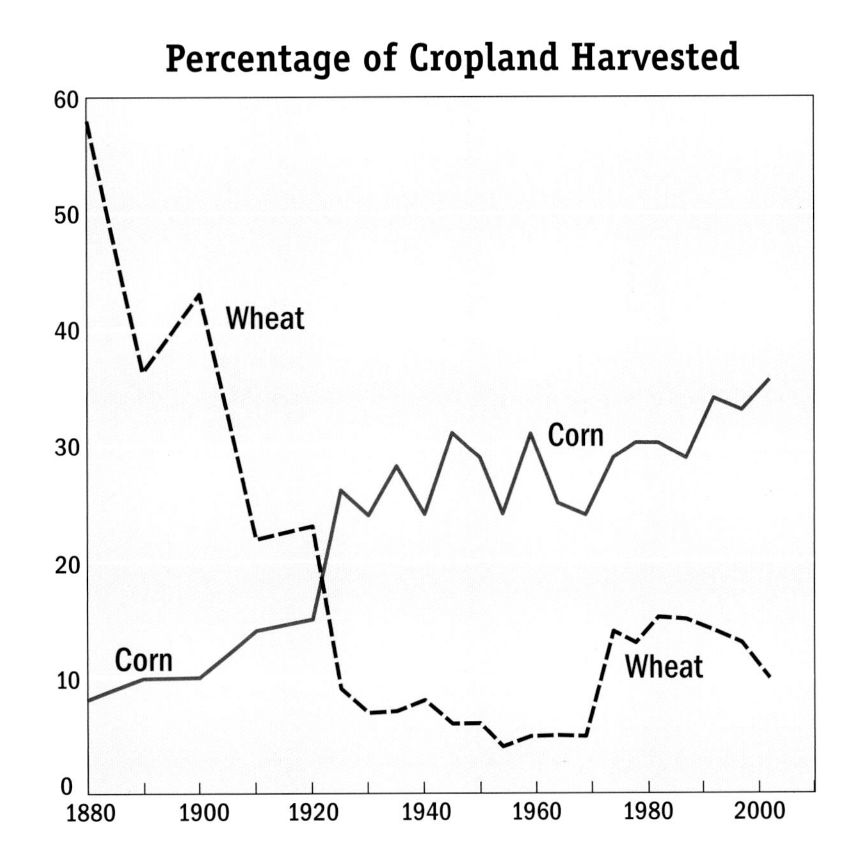

In 1925 corn replaced wheat as the leading crop in Minnesota.

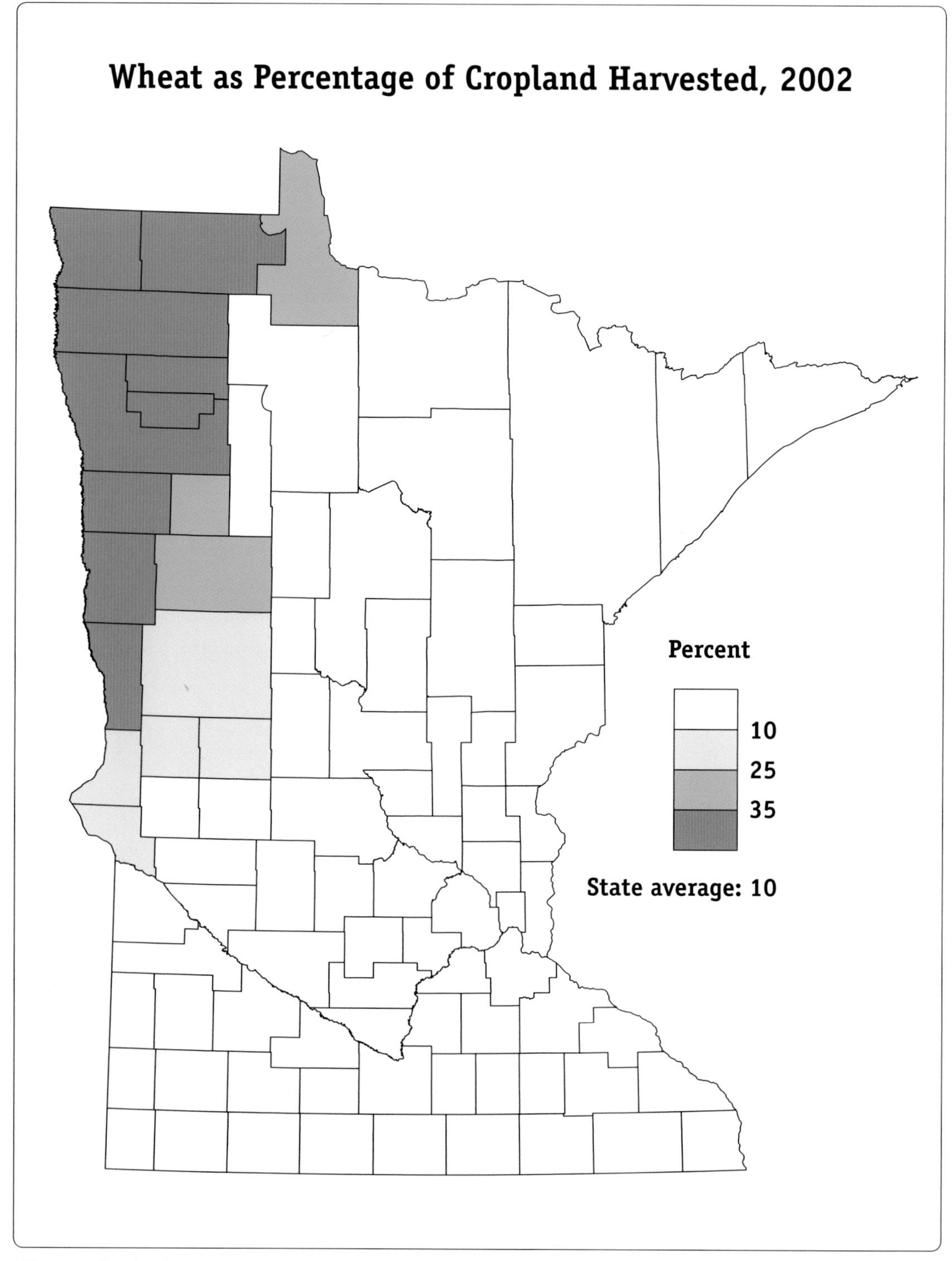

Wheat production is concentrated in the Red River Valley.

SINCE 1925, CORN HAS BEEN MINNESOTA'S LEADING CROP. The acreage of corn has increased erratically until now it is grown on just over one-third of the state's harvested cropland. In the early years corn was an easy crop to plant and harvest on land that had been poorly cleared of trees, but it was a risky crop because Minnesota is at the northern limit of corn cultivation. Corn is a subtropical plant that needs hot summers and a long growing season, and an early frost can ruin the crop before the kernels have time to ripen into grain. Even though corn is the world's finest feed for livestock, some Minnesota farmers once preferred to grow barley because corn was so unreliable.

Early snow can make corn a risky crop in parts of Minnesota.

Corn did not become an important crop in Minnesota until farmers had learned to harvest it for silage. They could cut the entire plant while the kernels were still green, chop it into pieces no larger than a person's little finger, and blow it into towering cylindrical silos, where they stored it for winter feed. Plant breeders have now developed short-season varieties of corn that can be ripened for grain in Minnesota, and today, most of the state's corn crop is harvested for grain, but corn silage remains an important feed on dairy farms.

Corn was the linchpin of the remarkably successful mixed crop-and-livestock farming system that made the Corn Belt of the American agricultural heartland one of the wonders of the world. Corn Belt farmers grew crops to feed hogs and cattle. They grew corn, oats, and hay in a regular three-year crop rotation (although this

A silage harvester chops unripe corn into pieces no larger than your little finger.

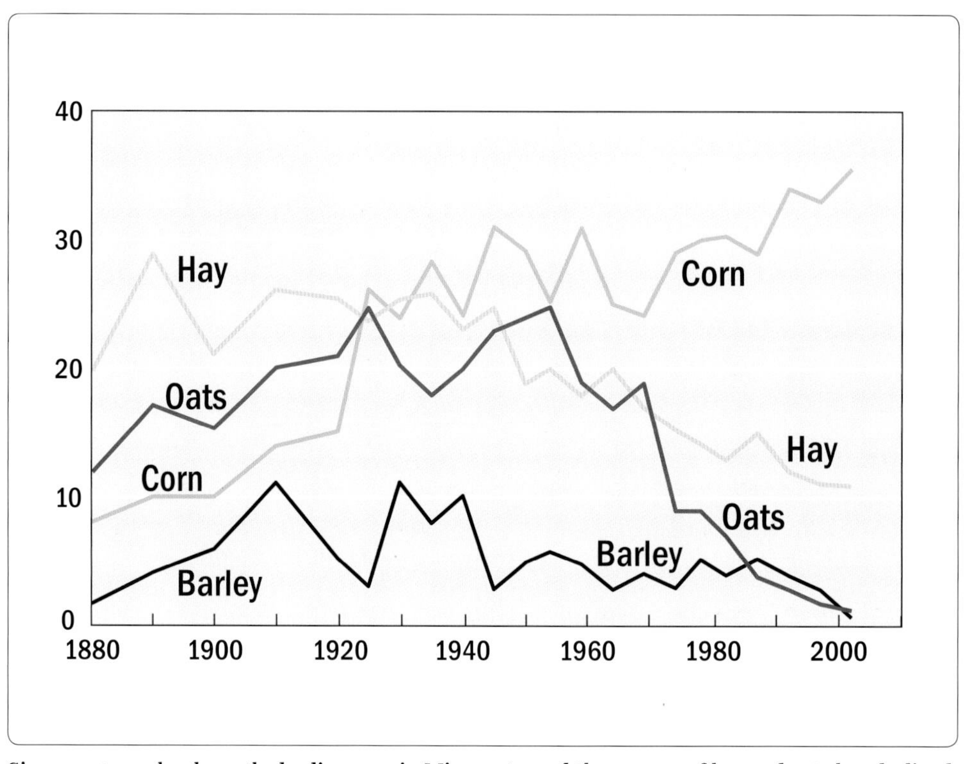

Since 1925 corn has been the leading crop in Minnesota, and the acreage of hay and oats has declined.

model was frequently modified on individual farms and fields). On good, level land they might grow corn for two years or more, but on steeper land liable to erosion they might leave a field in hay for several years. The dominance of the three-year rotation of corn, oats, and hay in Minnesota is manifest by the fact that farmers grew roughly the same acreages of these three crops from 1925 to 1950.

Oats were the second crop in the standard rotation. The seasonal labor requirements of planting and harvesting oats did not compete with those of corn, and oats were a valuable feed for workhorses until farmers replaced them with tractors. The steady decline of oats after 1950 is closely tied to the increasing use of tractors, and today oats are little more than a curiosity crop.

Oats were a valuable "nurse" crop for the hay crop that farmers grew in the third year of the rotation. Establishing a good stand of hay (clover and timothy in the early days, alfalfa more recently) on bare ground was not easy, so farmers planted their hay seeds when they sowed their oats. The rapidly growing oats protected the delicate young hay plants. The hay plants had become established by the time the oats were ripe enough to harvest, and the field was ready to be mowed for hay in the following year.

The acreage of hay in Minnesota has declined less than the acreage of oats because alfalfa, the principal hay crop, protects the soil from erosion, enriches the soil with nitrogen, and is an excellent feed for livestock. In morainic areas and in bluff country, much of the land is so steep that farmers must grow cover crops rather than row crops to slow down rainfall runoff and to protect the soil from erosion. Alfalfa enriches the soil because it is a legume. The roots of alfalfa plants have nodules in which bacteria are able to "fix," or extract, nitrogen from the air. These nodules remain in the soil when the farmer plows under the hay field at the end of the year, endowing the soil with nitrogen, a nutrient needed in abundance by the corn crop that begins the next cycle of the three-year crop rotation. Alfalfa is also an excellent feed for livestock because it is so rich in nitrogen.

The Corn Belt's mixed crop-and-livestock farming system was based on feeding grain corn to hogs and beef cattle to put meat on them, but in Minnesota this system had to be modified because the growing season is so short. Waiting for corn to ripen into grain was too risky, and farmers cut most of their corn crop for silage. Corn silage does not put good solid flesh on hogs and cattle, but it is the ideal basic ration for dairy cows when it is combined with alfalfa hay, so Minnesota farmers became milk producers rather than meat producers. They were further encouraged to become dairy farmers because the young glacial topography of the state has many short, steep slopes where farmers must protect the soil against erosion by growing hay crops and pasture rather than cultivated row crops.

Mowing alfalfa for hay is a rite of spring and summer on many Minnesota farms.

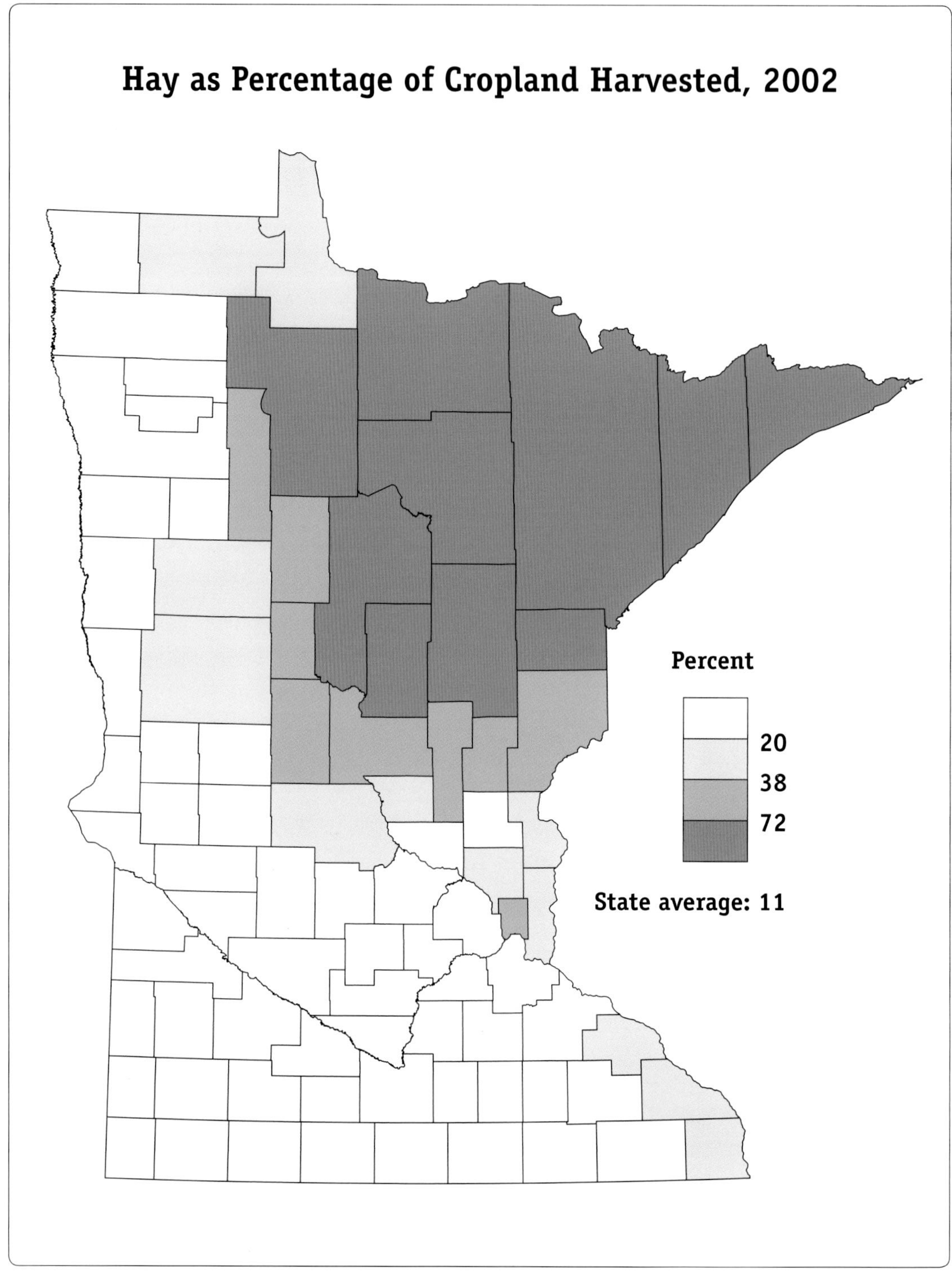

Hay is the dominant crop in northeastern Minnesota.

Even in the days when wheat was king, most Minnesota farmers had kept a dairy cow or two to produce milk for the family. They bred the cows to "come fresh" (to give birth to their calves and start producing milk) in the spring, when the lush growth of new grass provided plenty of feed. In spring and early summer the cows produced more milk than the family could drink, so the farm wife skimmed the cream from the surplus milk and churned it into butter. She took butter and eggs from the farm poultry flock to the local store, where she bartered them for coffee, tea, sugar, and other delicacies the farm could not produce.

Butter and eggs became an important source of income for many farmers when they stopped growing wheat and started keeping more cows. In 1870 the average Minnesota farm had only two or three dairy cows, enough to provide milk for the family, but by 1900 the number had doubled to slightly more than five, and some of them obviously were producing milk for sale. By 1920 a commercial dairy farming region had developed in the state's hardwood belt.

The hallmark of a dairy farm was a large barn with a cylindrical concrete silo towering over one end. The barn's spacious wooden hayloft sat above a masonry ground floor whose numerous windows let in light and air to the stanchions where the cows were tied and milked twice a day. The upper limit was ten to twenty cows when they were milked by hand, forty to fifty after milking machines became common. Most dairy farms were small, only 80 to 120 acres. The farmsteads were perched close together because a small acreage could produce all the corn silage and alfalfa hay a farmer needed to feed all the cows he could milk by hand.

Farm women first made butter and cheese in the farmhouse kitchen. They let milk sit until the cream had risen to the top. Then they skimmed off the cream and put it in a churn, saving the skim milk to feed the farm hogs. They worked the churn until their arms and shoulders ached, and globules of butter began to coagulate. They scooped out the globules with wooden paddles and packed them in tubs or crocks to take to the store. The quality of the butter varied greatly, and some was so rancid that it could be used only for grease.

Making cheese required more skill but less milk than making butter—ten pounds of milk per pound of cheese and twenty pounds per pound of butter. Cheese makers added rennet, the lining membrane from a calf's stomach, to milk to yield solid curds and watery whey. They cured and aged the curds and fed the whey to hogs. Because butter fetched a higher price than cheese, however, it could be profitable even with the cost of being shipped greater distances to market. Most Minnesota farmers were only too happy to leave cheesemaking to the "cheeseheads" of Wisconsin, who were closer to Chicago and the large urban markets of the East.

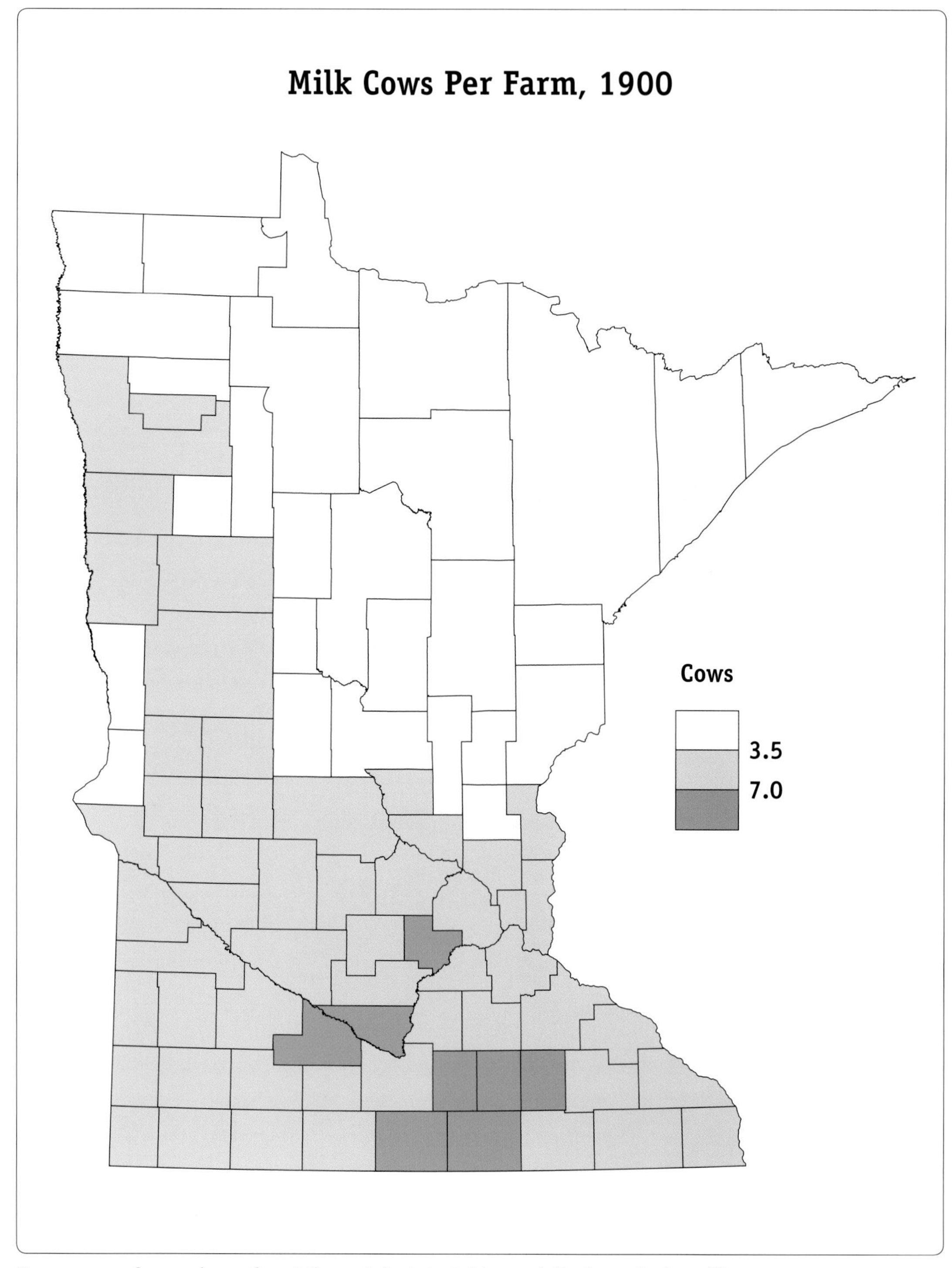

By 1900 some farmers in southern Minnesota had started to specialize in producing milk.

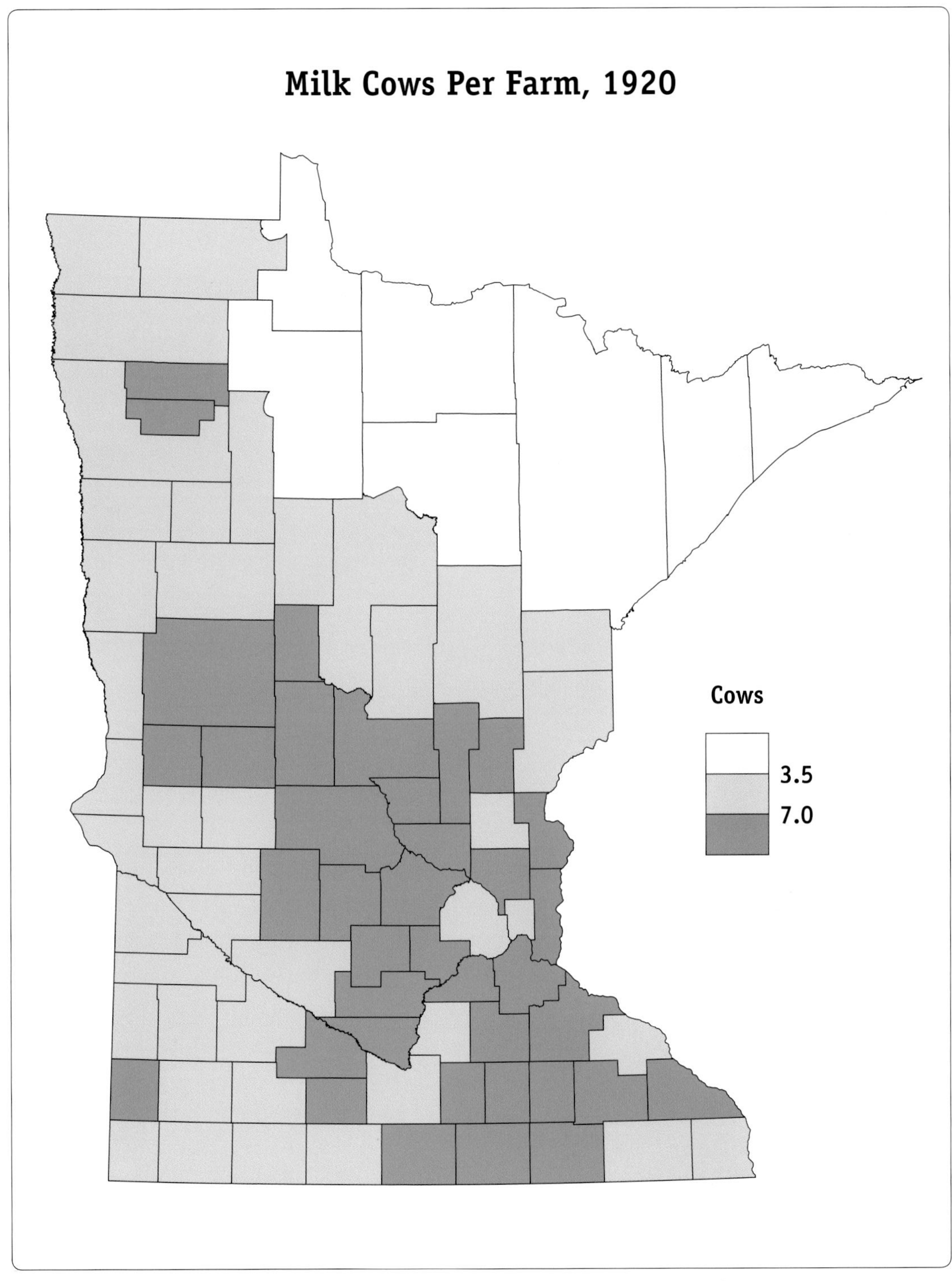

By 1920 a commercial dairy farming region had developed in the moraine and hardwood belt.

A standard forty-cow dairy barn had a silo for corn at one end and a huge loft for hay above the masonry ground floor that housed the cows.

By 1880 dairy farming was already replacing wheat farming in the original wheat counties. Dairy farms grew larger and more commercial, and farmers began to organize cooperative creameries, which moved butter making and cheese making from the farmhouse kitchen to the factory and produced a better-quality product. Farmers hauled their milk to the creamery in heavy metal cans and took home skim milk and whey for their hogs.

Both of Minnesota's two large farm organizations, the Farmers Union and the Farm Bureau, sponsored and encouraged farmers to form cooperative creameries, even though sharp ideological differences divide them otherwise. (The Farm Bureau has more larger farms, is conservative politically, is somewhat suspicious of government programs, and prizes economic success. The Farmers Union has more smaller farms, is more populist in orientation, encourages government programs to protect small farmers, and puts greater emphasis on social considerations.)

While Minnesota had only a handful of creameries in 1880, by 1900 they were almost too numerous to map. Many were so small they went broke, so the number was changing constantly, but at its peak the state probably had 600 to 700. Each served 300 to 400 dairy farmers. Every small town and every crossroads seemed to have a creamery. Many of the old buildings still stand, although they have long since been abandoned or put to other use. Most creameries switched back and forth between making cheese in the summer months, when milk production was greatest, and butter the rest of the year.

The creameries packed their butter in paper tubs that held around sixty-five pounds and sent it by rail to commission houses in New York or Philadelphia that sold it to individual grocery stores. The price of butter fluctuated daily, and most farmers assumed the commission houses were cheating them. Furthermore, Minnesota is at the tail end of the supply chain. It is farthest from the major markets on either coast, it pays the highest shipping charges, and it always gets the lowest prices, which might help to explain why some farmers in the state have such a strong populist streak.

After World War I, prices were so low that small cooperative creameries began to combine their shipments into full carloads to secure better freight rates. They started buying supplies in quantity to reduce the cost to their members. They began to sell feed, because farmers had learned that they should give their cows balanced rations. They formed a statewide association that hired technical experts to help upgrade their members and persuaded the U.S. Department of Agriculture to develop a quality grading system.

The association developed a distribution and marketing system that bypassed the commission houses and sold butter directly to individual grocery stores. It packed butter in standard one-pound cartons of four individually wrapped quarter-pound sticks that were much more aesthetic than the former method of scooping it out of paper tubs in the back of the grocery store. Most butter still is sold in quarter-pound sticks.

At one time nearly every small town in southern Minnesota had a creamery, but today they have been converted to other uses or abandoned.

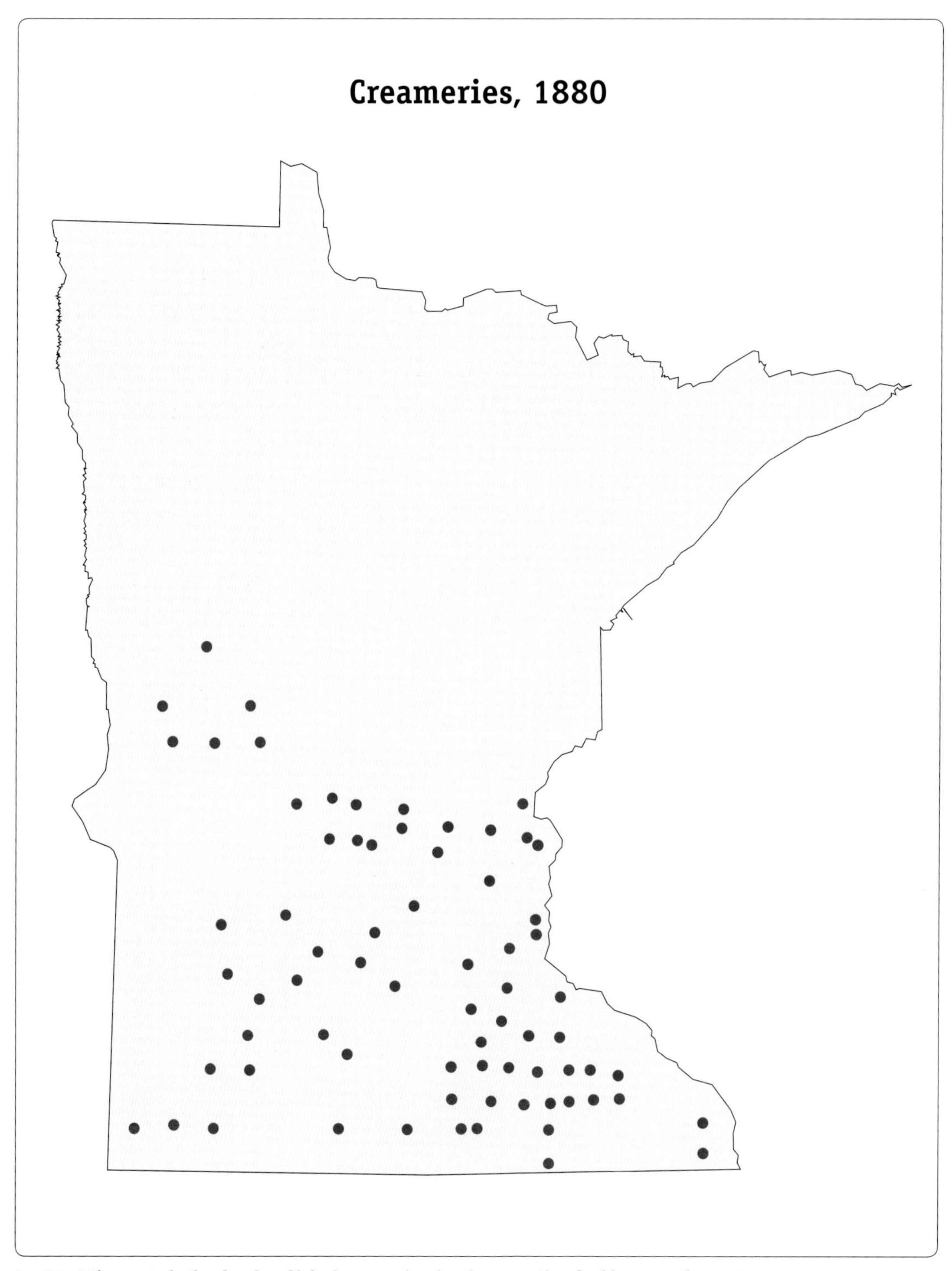

In 1880 Minnesota had only a handful of creameries, but by 1900 they had become almost too numerous to map.

The association tried to branch out by using its refrigerator cars to ship eggs, chickens, and turkeys, but it was not able to compete with efficient integrated companies that controlled complete food chains and delivered large, reliable supplies of uniform products. The co-op could not control the quality of its products because it had to accept whatever its members delivered, and it lost large sums of money trying to manage a system of many small independent producers.

Technology drove many changes in the dairy business after World War II. Margarine became a serious competitor to butter. Dairy interests fought a losing battle against it, but for many years they forced grocers to sell margarine uncolored, and customers had to mix in the color themselves by hand. The invention of continuous churning enabled one large central butter plant to replace ten to twenty smaller plants, and many small creameries closed down. Milk collection on the farm shifted from heavy metal cans that had to be handled individually to bulk tanks that were pumped into tank trucks. Minnesota farmers enlarged their herds to fill their tanks and bought milking machines to milk their larger herds, but they lagged behind other parts of the United States.

By the advent of World War II the wheat frontier that had swept across Minnesota half a century earlier was no more than a memory, and wheat was virtually a forgotten crop outside the Red River Valley in the far northwestern corner of the state. Dairying had become the predominant form of farming in the hardwood forest belt of central and eastern Minnesota, between the nonagricultural boreal forest to the northeast and the mixed crop-and-livestock farms on the prairie plains of the southwest.

A few large milk processing plants have replaced many small old-fashioned creameries.

Kittson
Roseau
Lake of the Woods
Marshall
Beltrami
Koochiching
Pennington
Red Lake
Polk
Clearwater
Lake
Cook
Itasca
Norman
Mahnomen
St. Louis
Hubbard
Cass
Clay
Becker
Wadena
Crow Wing
Aitkin
Carlton
Wilkin
Otter Tail
Pine
Todd
Grant
Douglas
Morrison
Mille Lacs
Kanabec
Traverse
Stevens
Pope
Stearns
Benton
Sherburne
Isanti
Chisago
Big Stone
Swift
Anoka
Meeker
Wright
Washington
Lac Qui Parle
Chippewa
Kandiyohi
Hennepin
Ramsey
McLeod
Carver
Yellow Medicine
Renville
Scott
Dakota
Sibley
Lyon
Redwood
Nicollet
Goodhue
Le Sueur
Rice
Wabasha
Brown
Murray
Cottonwood
Watonwan
Blue Earth
Waseca
Steele
Dodge
Olmsted
Winona
Rock
Nobles
Jackson
Martin
Faribault
Freeborn
Mower
Fillmore
Houston
1. Red River Valley
2. Mantorville
3. Perham
4. Fergus Falls
5. East Grand Forks
6. Crookston
7. Moorhead
8. Renville
9. Kandiyohi

9

The Dairy Belt and the Valley

Minnesota has three major agricultural regions. Dairy farming is concentrated in the hardwood forest belt of the center and east, where rolling glacial terrain motivates farmers to keep their land in pasture and hay crops to forestall soil erosion. The state's second major agricultural region is the flat lake-bottom plains of the Red River Valley in the northwest, where farmers have experimented with spring wheat, sugar beets, barley, soybeans, and a variety of other specialized cash crops. The state's third region is the Corn Belt on the prairie plains of the south and west (discussed in the following chapter), where farmers have shifted from mixed crop-and-livestock farming to cash-grain farming. (The northeastern third of the state is coniferous forest unsuited for farming.) While the broad outlines of these agricultural regions have changed little over the last half century, farming methods within each region have changed dramatically.

In 2002 Minnesota had 6,474 dairy farms. Ninety-eight percent of these farms were undersized and in serious trouble. Dairy farmers are bound to the cows they must milk every twelve hours, 365 days a year. Many have been unwilling or unable to invest in up-to-date technology or to develop large, modern operations that would enable them to capture economies of scale. Farmers with undersized dairy farms are understandably angry, because they work so hard and have so little to show for it. Their anger has fueled political protests, but politics cannot repeal the inexorable laws of farming economics.

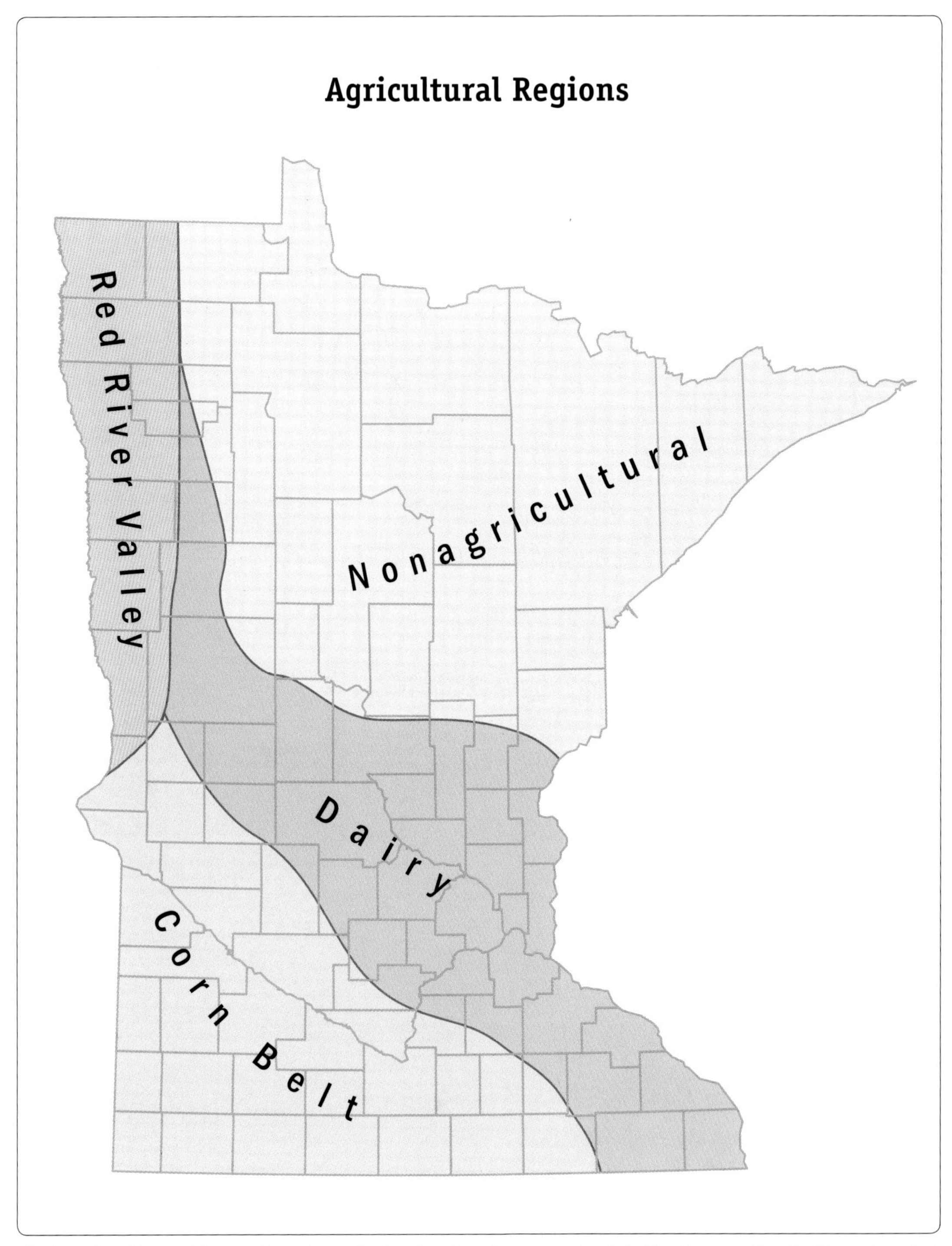

The broad outlines of the state's major agricultural regions have remained similar for more than half a century, but within each region farming has changed dramatically.

In the western United States, especially in California, the number of large, new modern dairy farms increased dramatically in the last quarter of the twentieth century. Until around 1970, Minnesota and California had almost equal shares of national sales of dairy products, but Minnesota has been slipping since 1982, while California's sales have skyrocketed. The average milking herd in California increased from 16 cows in 1950 to 559 in 2002, but in Minnesota in the same years herds grew from 10 cows to only 74.

A modern family dairy farm must milk at least 500 cows to provide an acceptable level of living for a contemporary American family.[1] Older couples who own their farms can still hang on with smaller herds by consuming their capital and accepting a lower standard of living, but they are essentially laborers who must schedule their lives around twice-a-day milkings. When they decide it is time to stop milking, the next generation may not choose to subject themselves to a lifetime of relentless toil for precious little income.

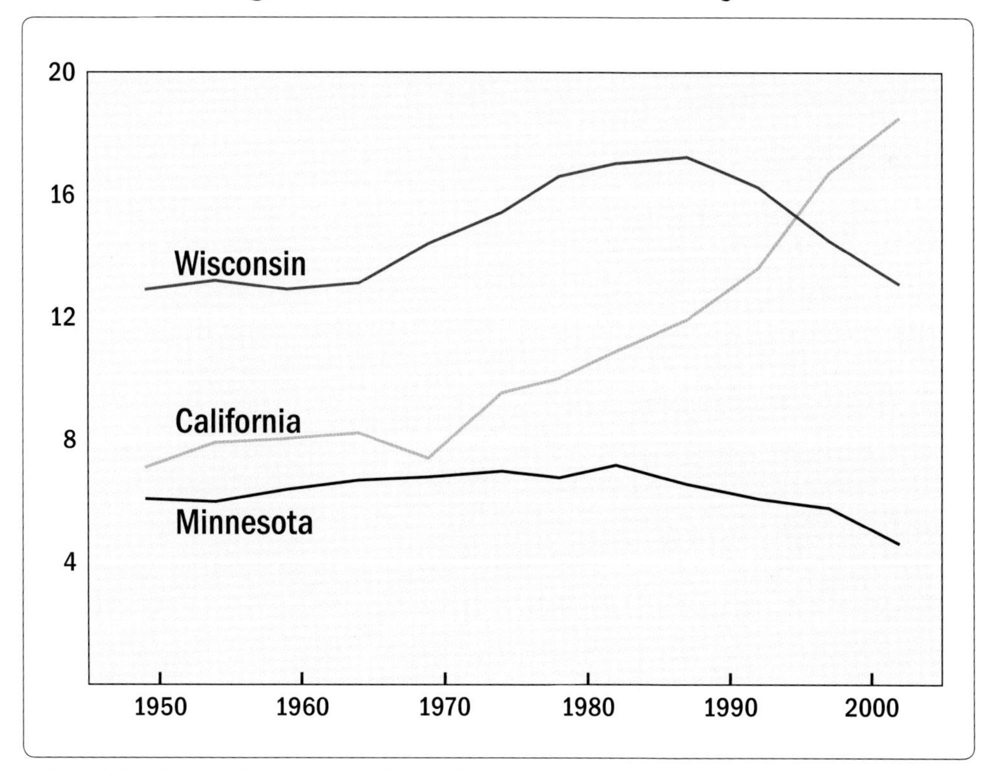

California has become the nation's leading milk-producing state, while Wisconsin and Minnesota have lagged.

Many small old-fashioned dairy barns are derelict.

Populist politicians and some small dairy farmers blame their plight on the byzantine federal system of milk pricing, a relic of the horse-and-buggy and rail-road era that tanker-truck transport has made obsolete. This system essentially is based on distance from Eau Claire, Wisconsin, and it gives a slight price advantage to producers in faraway Florida, Texas, and California. The true competitive advantage of these states, however, is their scale of production, which enables them to produce milk far more cheaply than the smaller farms of Minnesota. Many small farmers, unfortunately, cannot afford to enlarge their scale of operation in order to cut their costs.

Enlarging a dairy farm is neither cheap nor easy. A modern 500-cow dairy costs around $1.5 to $2 million for the buildings and another $1 million for the cows. No bank will lend more than 80 percent, and few farmers have the necessary down payment of $500,000. Even fewer farmers are willing to take on a staggering debt load of $2 million. They also realize that a large, modern dairy requires managerial skill of the highest order, because a large operation can lose money far faster than a small one. The dairy industry in Minnesota needs a massive infusion of external capital and sophisticated management, but state laws against corporate farms exacerbate the difficulty of securing them.

Modern dairy farms have become so large that they must be sheltered by some kind of corporate structure to ensure that they will not have to be broken up to pay estate taxes when the farmer dies. Some small-farm activists, however, confuse incorporation with size. They think all large farms are bad and assume all large farms are incorporated, even though in 2002 Minnesota had only 2,342 incorporated farms, 2,163 of which were family-held corporations for protection against excessive inheritance taxes.

The small-farm problem in Minnesota is compounded by an alien farm law that prohibits foreigners from buying farmland in the state. Overcrowded countries in Europe, especially the Netherlands, are trying to find more land for housing by encouraging dairy farmers to sell their farms and relocate. Wisconsin and other states are vying to recruit these farmers, but Minnesota laws prohibits this, even though the state needs the milk they would produce to keep its processing plants in operation. Many small processing plants in the state already have closed because they do not have an adequate supply of milk, and each plant closure means that the remaining dairy farmers have one fewer place to sell their milk.

Some politicians have the economics backward. They have argued that Minnesota farmers would receive better prices if they produced less, but they fail to understand the global economy. Cutting back on production in the state only encourages producers in other states to fill the void, accelerating the further decline of dairy farming in Minnesota.

In 2003 Jack Gherty, president and CEO of Land O'Lakes, the Minnesota-based dairy cooperative, warned that Minnesota has per-cow productivity below national averages, declining milk supplies, aging processing plants, and inadequate reinvestment on farms and in processing plants.[2] Land O'Lakes closed nine of its sixteen processing plants between 1990 and 2002, and other processing companies have had to truck milk to their plants from other states. Instead of upgrading their old plants in Minnesota, companies are investing in new plants near major milk-producing areas in the West. Dairy farmers in Minnesota may have trouble finding a place to sell their milk even if they are able to produce it.

The future of dairy farming in the state looks bleak, but a few visionary entrepreneurs are showing what the future in Minnesota could be. The Durst Brothers' dairy farm north of Mantorville is a model modern Minnesota farm. In 1978 Ron, Allen, and Ken bought their parents' 112-cow dairy. They added heifer calves to their milking herd until they had built it up to 600 cows by 1995, which gave them enough

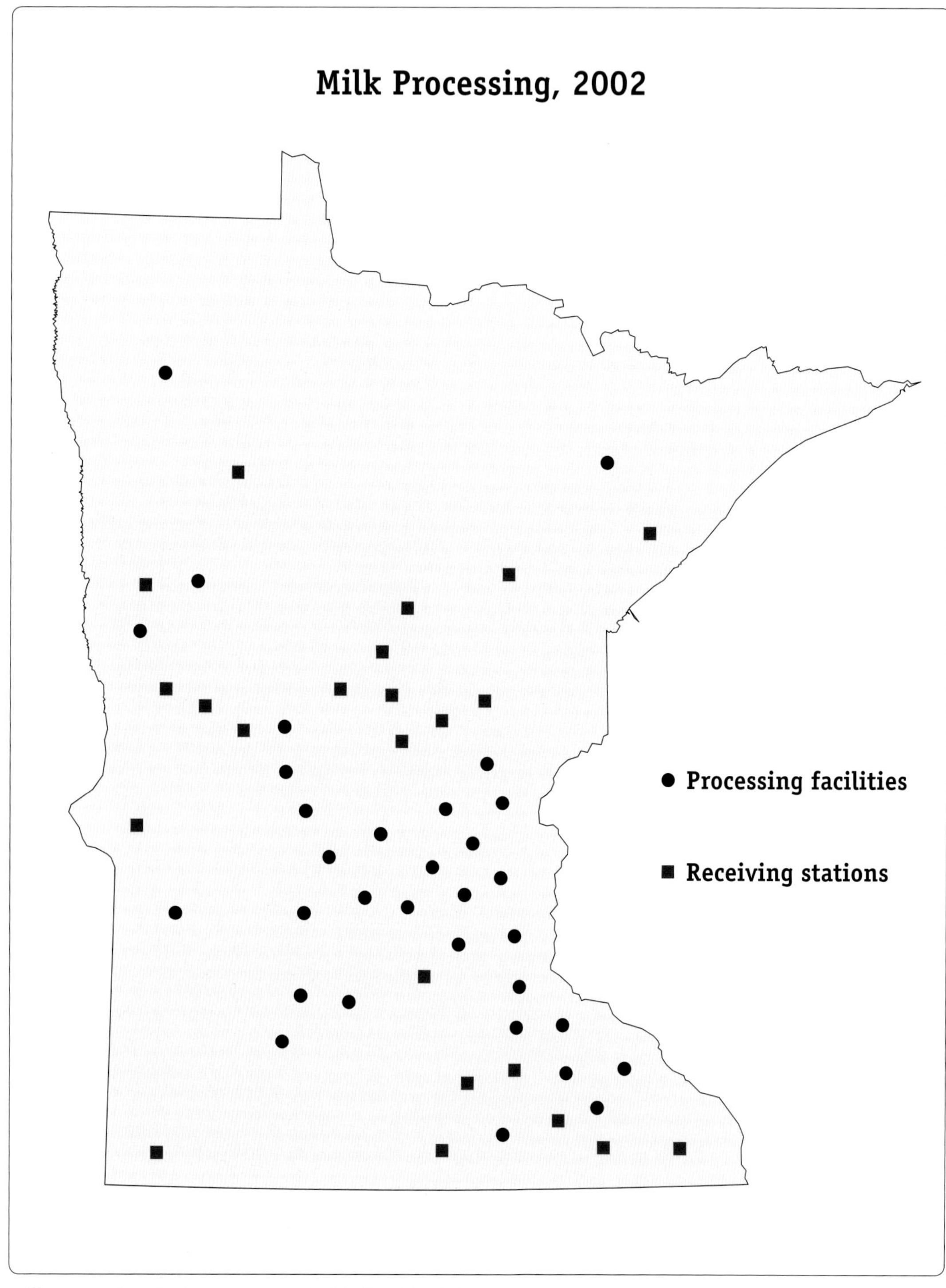

Milk-processing plants are concentrated in the dairy belt.

equity to secure the financing they needed, and the next year they opened their new 1,250-cow dairy, with space for 150 more cows. The Durst Brothers' beautifully landscaped dairy showplace is immaculately maintained. The heart of the dairy is the milking parlor, which is in the same building as the farm office and the spotless lunchroom and locker rooms for the workers. Three times a day the cows crowd into the milking parlor in groups of forty-eight. Their udders are so full that the animals are eager to be milked.

In the central pit of the milking parlor stand teams of three workers who clean and disinfect the cows' teats and attach the milking cups. Each cow wears an electronic transponder that records the amount of milk she gives at each milking. The farm average is 10 to 10.5 gallons per cow per day, much better than the statewide average of 6.5 gallons. Most of the workers are Mexicans. The Dursts employ twenty-six workers full-time and many more part-time.

The Dursts employ top-quality genetics to maintain their herd, and use their own heifers to replace the older cows. Usually they have enough surplus heifers to sell. They keep the cows in the milking herd for an average of four lactations, which is the length of time that a cow produces milk. The cow starts to produce milk when her calf is born. Three months after her first calf is born, the farmer breeds her to produce her second calf, but she continues to produce milk for her first calf. Her milk production gradually declines, and ten months after the first calf was born, the farmer stops milking the cow and "dries her off" for a rest period of two months before she has her second calf.

After they have been milked, the cows amble through covered walkways back to two 110-by-330-foot, one-story free-stall barns with curtain sides. They are comfortably bedded on carefully shaped sand to encourage them to lie down, which promotes blood flow to their udders and thus increases their milk production. (Blood flows to the cows' udders only when the cows are lying down.) The dairy also has a 104-by-185-foot special-needs barn for sick cows and those about to calve.

Three times a day, manure on the barn floors is scraped into two large basins. The first holds 5 million gallons of solid waste, and the liquid is strained into the second, which holds 8 million gallons. From these basins the manure is pumped out onto 1,185 acres of corn, 1,250 acres of alfalfa, and 450 acres of soybeans, which produce most of the feed the cattle need, although the Dursts also buy feed concentrates. Some critics complain that large, modern dairy farms harm the environment, but careful analysis has revealed that most of our air and water pollution problems have been caused by small, poorly managed farms that cannot afford to invest in the kinds of effective management the Dursts use.

On the Durst Bros. Dairy the small building in the foreground houses offices and the milking parlor, and the larger buildings house the cows.

In the milking parlor teams disinfect the cow's teats and attach the milking cups to them.

A large curtain-sided dairy barn houses seven hundred milk cows.

A special truck has unloaded feed in front of the cows in a curtain-sided dairy barn.

The hay chopper will stop automatically if it detects a piece of metal even as small as a paper clip.

The hallmark of a large modern dairy farm is huge mounds of corn silage and alfalfa hay covered by white plastic sheets weighted down by the sidewalls of old truck tires.

Ron Tobkin and his family operate the Little Pine dairy farm on the sandy outwash plain north of Perham. In 1995 they were growing 4,000 acres of irrigated cash crops—kidney beans, potatoes, and corn—but Tobkin worried about beating the sandy soil to death by not putting enough organic matter back into it. He knew they should add 1,400 acres of alfalfa to the cropping system to increase the nutrients, but then what could he do with 1,400 acres of alfalfa? He decided that the best use was to feed it to dairy cows, so the family borrowed $6.3 million to build a model dairy.

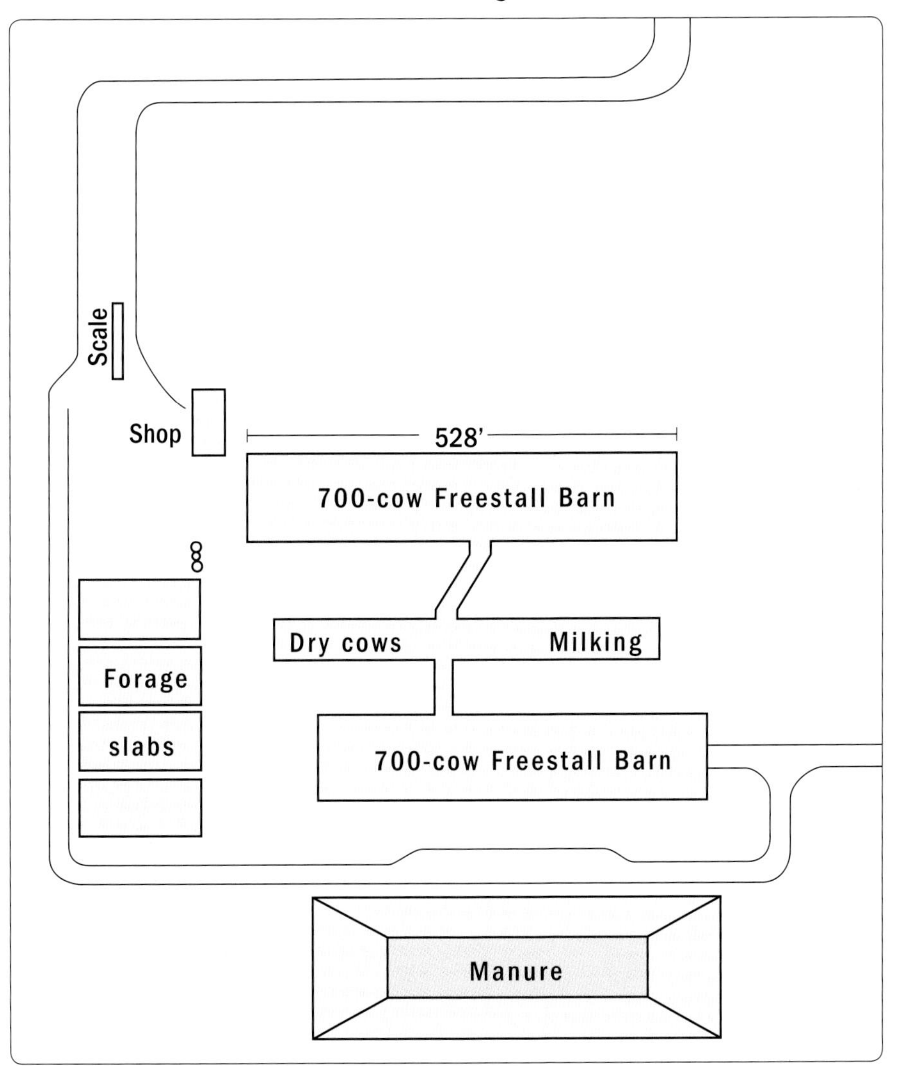

Two 700-cow curtain-sided freestall barns flank the central milking parlor at Little Pine Dairy.

Feeding a dairy cow for a year requires about an acre of alfalfa hay and an acre of corn silage, plus special feed concentrates, so Tobkin planned a 1,400-cow dairy. Two large barns housing 700 cows each flank the central 48-cow milking parlor. Teams of milkers on eight-hour shifts milk the cows three time a day, and the daily herd average is 8.5 gallons per cow. The milk is cooled and stored in a stainless steel tank until a truck comes to collect it and haul it to the processing plant.

Tobkin pumps manure from the barns into a 16 million gallon earthen lagoon, where he stores it until he can spread it on cropland. The lagoon is crusted over, and its odor is not particularly strong. Ron treats manure just like commercial fertilizer. He analyzes it to find the nutrients it contains, tests the soil in each field, and applies enough manure to supply the nutrients necessary for a normal yield of the crop he plans to grow in the field. Although he is happy to recycle the manure in this fashion, he dreams of finding a way of transforming it into a marketable product.

Tobkin has a degree in agronomy, but he is largely self-taught about dairying because university agriculture programs cannot afford to invest millions of dollars in experimental facilities just for dairy students. When Tobkin realized that others could benefit from his hard-won expertise, he formed a management services company to advise investors who want to enter the modern dairy business, and his company manages five other large, new dairy operations patterned after his own.

The Durst and Tobkin family dairy farms are success stories, but in 2006 the entire state of Minnesota had only thirty dairy operations with 1,000 cows or more.[3] Some farsighted farmers have tried to develop larger operations, but they struggle to obtain approval at the state, county, and township levels, and opponents have further frustrated their efforts. The future of small dairy farming in Minnesota looks bleak.

THE RED RIVER VALLEY of northwestern Minnesota is the state's second main agricultural region. It has some of the finest physical geography and poorest economic geography in the United States. The Red River and its tributaries meander sluggishly north toward Hudson Bay across a flat, featureless plain that was formed on the bottom of glacial Lake Agassiz. This vast, shallow inland sea was dammed on the north by the mile-thick ice of the continental glacier, and the water drained away when the ice melted.

The sediments carried into this sea were sloshed back and forth by waves and currents to create one of the flattest plains on earth. The soils formed on these sediments are deep, stone-free, and fertile, albeit slightly alkaline. Farmers had to drain these soils before they could cultivate them. Spring floods are a routine hazard in the valley, because the northern reaches of streams are often blocked by ice while their headwaters at the southern end of the valley have thawed.

Although the soils of the valley are productive, no part of the North American continent is farther from access to export markets. People in the valley often feel that they are ignored and forgotten even by the rest of the state. The region is still an agricultural frontier, where farmers continue to grope for the most suitable crop and the best agricultural system. It was the final stand of the wheat frontier in Minnesota, and wheat remains a major crop, by default, because farmers still have found none better, despite extensive experimentation with barley, sunflower, canola, and other specialty crops.

At harvest time combine harvesters lumber across the wheat fields of the level Red River Valley.

The area's latest specialty crop is sugar beets. The valley is the nation's leading sugar beet–producing area, and its economy depends on them, but they are a highly controversial and intensely political crop. Sugar beet farmers owe their prosperity to an elaborate government sugar-control program that dates back to the 1930s, and they are aware that the policy could change virtually overnight. This program does not pay farmers directly but instead tells processing plants how much they can produce and makes loans that enable the plants to guarantee farmers a minimum price for their beets. Most of the plants are grower-owned cooperatives; each share owned in the co-op permits a farmer to produce a specified quantity of sugar beets.

The sugar-control program also imposes strict import quotas on the amount of sugar that other countries are allowed to sell in the United States. Critics argue that these restrictions encourage other countries to limit their imports of other products from the United States. They say that tropical countries can produce cane sugar more cheaply than the United States can produce beet sugar, and these countries badly need our dollars. They add that these restrictions keep the price of sugar in the United States two or three times higher than the world price.

Sugar beet producers counter that this "world price" of sugar is a myth, because other countries subsidize sugar heavily and dump their surplus on the world market at less than the cost of producing it. The producers add that government programs also protect U.S. consumers from wild fluctuations in this "world price," which has ranged as high as sixty cents a pound in 1974 and as low as three cents a pound in 1985. The producers continue to grow sugar beets, but they keep a wary eye on what is happening in Washington, D.C.

Larry Morris has a 1,400-acre sugar beet farm sixteen miles west of Fergus Falls near the southern end of the Red River Valley. He owns 500 acres and rents the rest, because beets are so prone to disease that he can grow them on the same ground only every third year. In 2002 his rent ran sixty-five to seventy dollars an acre. His farthest rented land was twenty miles away. He wasted many hours going back and forth, but nothing closer was available, and he rationalized that at least he reduced his risk of damage from hail, wind, rain, or any other kinds of natural disaster by having his land widely scattered.

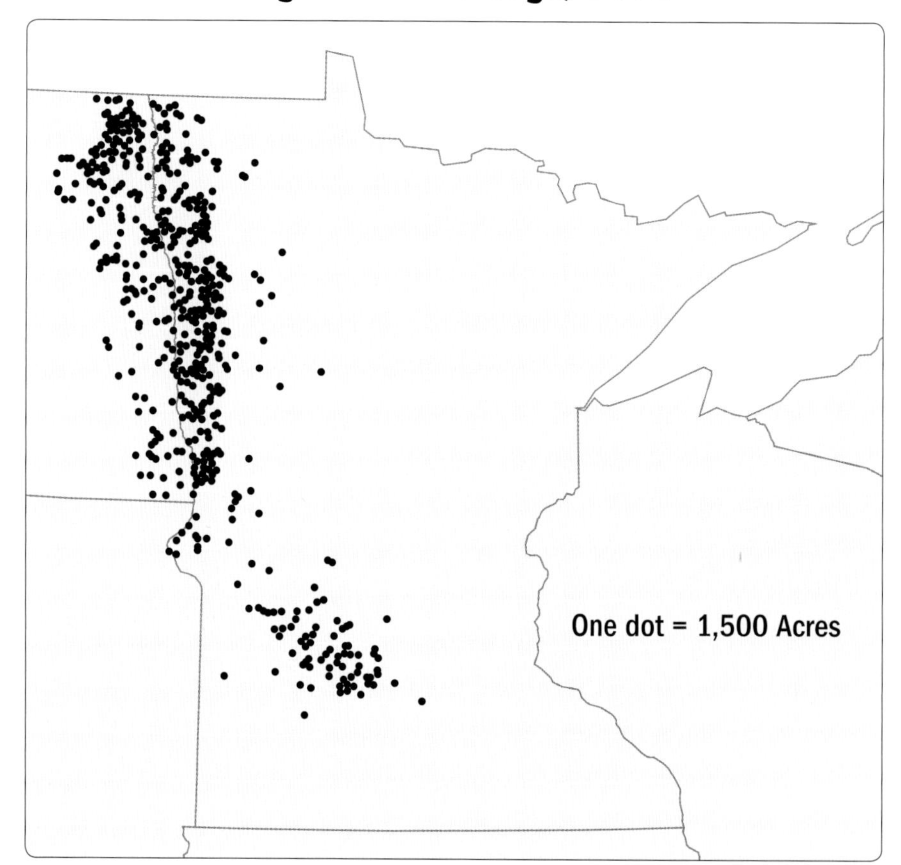

Sugar beets are a valuable crop in the Red River Valley and in west central Minnesota.

In 2002 Morris grew 420 acres of sugar beets, 420 of soybeans, 360 of wheat, and 200 of corn. He normally rotates sugar beets, soybeans, and wheat. He may plant corn rather than soybeans, because soybeans have more disease problems but withstand drought better. Some of his neighbors plant sunflowers rather than soybeans when the price is good. He may plant corn rather than wheat if wheat prices have gone south, but it is tough to prepare a proper seedbed for beets after corn. After harvest he plows wheat straw into the soil to keep it from blowing, which can be a serious problem in this perpetually windy area. Some of his neighbors plant winter cover crops.

He plants sugar beets around April 15 and hopes for a yield of twenty tons per acre. The harvest in October is three weeks of frenzied activity because farmers must pull their beets when the temperature of the soil is below 55°F but above freezing. First the topping machine cuts off and shreds the tops, which Larry works into the soil because they have nitrogen value. Then the lifter pulls the roots out of the ground, bounces them around to knock off some of the dirt, and loads them into the truck that drives right alongside it while it is lifting.

The trucks haul the beets to pilers that remove the tops, weeds, and dirt, and heap the roots in piles ten feet high and as long as a football field. Sugar beets are so bulky in relation to their value that they can be hauled economically only for short distances, so growers like to be within thirty miles of a processing plant. The valley has three plants in North Dakota (Drayton, Hillsboro, and Wahpeton) and three in Minnesota (East Grand Forks, Crookston, and Moorhead). A plant in Renville serves growers in Renville, Chippewa, and Kandiyohi counties. The plants squeeze out the juice, boil it down to make sugar, and pelletize the pulp for sale as livestock feed.

Sugar beet pilers stack the roots until time to haul them to the processing plant.

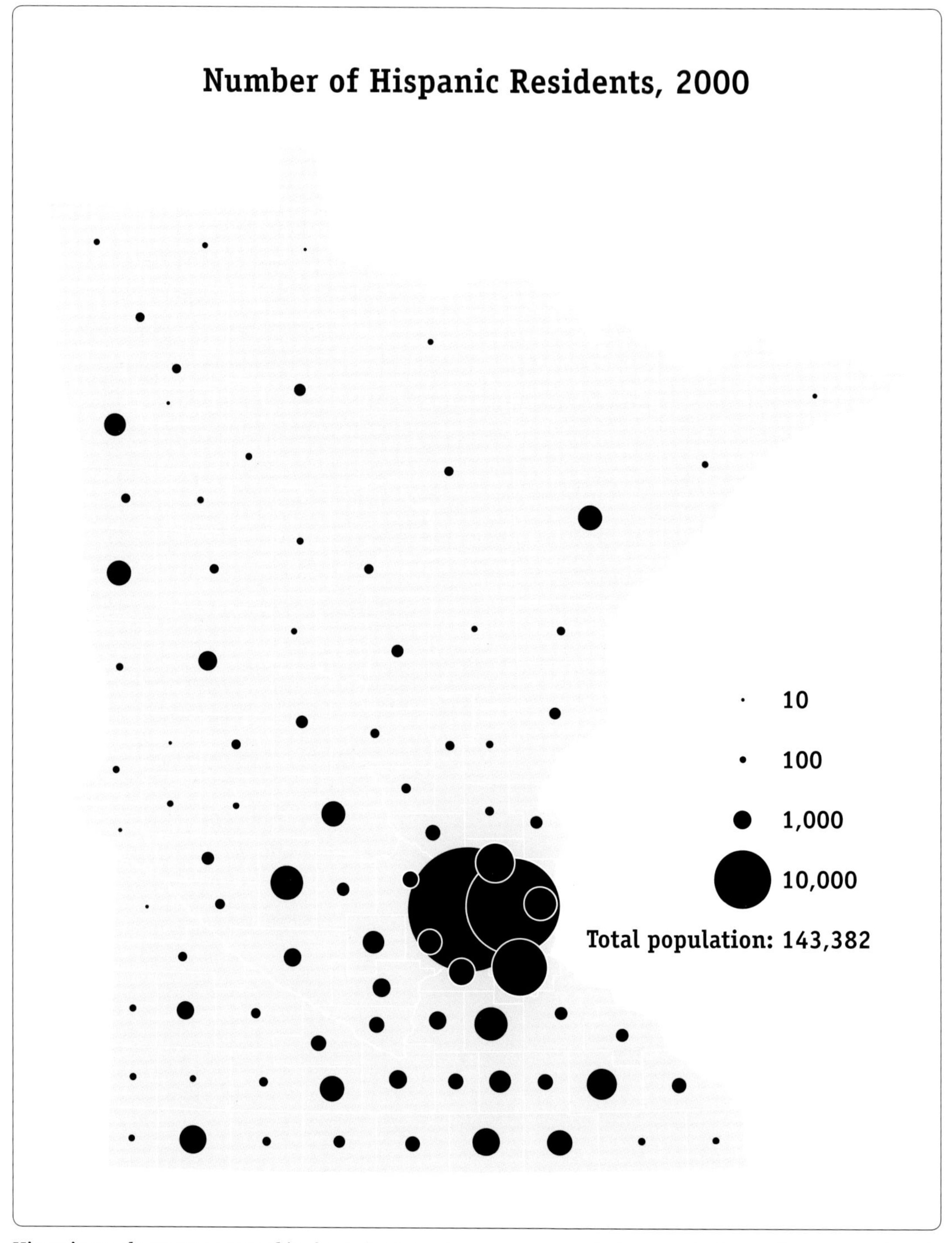

Hispanic people are concentrated in the Twin Cities, but they are essential workers in the rural economy of nearly every county in Minnesota.

Morris owns shares in the Wahpeton co-op. Each share gives him the right to grow a specified acreage of beets. The price of a share, which varies with the price of sugar, was around $2,000 in 2002. The co-op allows each grower to deliver a specified number of truckloads of beets during each twelve-hour shift at harvest time, when 7,000 beet trucks are racing along valley roads. Larry keeps seven trucks running. Most of his drivers return year after year, because the beet harvest is a fun time for them, and some of them even schedule their vacations around it.

Morris hires labor only at harvest time, because he does all his farmwork with machines, but twenty years ago he hired crews of Hispanic workers to thin and weed his sugar beet fields with hand hoes. The beet fields of the valley have attracted Hispanic workers since the early 1920s. Some overwintered in St. Paul, where they formed the nucleus of the city's Hispanic community.

Railroads, canneries, and meat-packing plants, as well as beet growers, recruited Hispanic workers aggressively when labor was scarce during World War II, and many employers still recruit them to do menial and less desirable jobs. Some meat-packing companies would not be able to operate without them. Most Minnesota counties have some Hispanic workers. Data from the most recent census of population, taken in 2000, unquestionably underestimates their numbers, because some are in this country illegally and try to avoid being enumerated.

Hispanics have become an increasingly important presence in Minnesota agriculture. The traditional small family farm usually had a hired hand or two at some stage of the demographic cycle, before sons were old enough to be much help or after a father had grown too old. Family farms have had to hire more workers as they have grown larger and more specialized. Few Americans are willing to do the hard physical work necessary to run a farm, and farmers have hired immigrant workers. Minnesotans must understand, appreciate, and embrace these changes, whether or not they like them, just as their forebears a century and a half ago had to hire Germans, Swedes, and Norwegians.

Kittson
Roseau
Lake of the Woods
Marshall
Beltrami
Koochiching
Pennington
Red Lake
Polk
Clearwater
Itasca
Norman
Mahnomen
St. Louis
Cass
Clay
Becker
Wadena
Aitkin
Carlton
Wilkin
Otter Tail
Crow Wing
Pine
Todd
Grant
Douglas
Morrison
Mille Lacs
Kanabec
Traverse
Stevens
Pope
Benton
Stearns
Big Stone
Sherburne
Isanti
Chisago
Swift
Anoka
Chippewa
Kandiyohi
Meeker
Wright
Hennepin
Ramsey
Washington
Lac Qui Parle
Yellow Medicine
Renville
McLeod
Carver
Scott
Dakota
Sibley
Lyon
Redwood
Nicollet
Rice
Goodhue
Le Sueur
Wabasha
Brown
Cottonwood
Watonwan
Blue Earth
Waseca
Steele
Dodge
Olmsted
Winona
Rock
Nobles
Jackson
Martin
Faribault
Freeborn
Mower
Fillmore
Houston
1
2
3
4
5
6
7
8
9
1. Slayton
2. Willmar
3. Wadena
4. Austin
5. Faribault
6. Litchfield
7. Sleepy Eye
8. Fairmont
9. Pipestone

10

The Corn Belt

After the wheat frontier had surged westward, farmers on the prairie plains of southwestern Minnesota diversified into mixed crop-and-livestock farming. They did many different things, but none of them particularly well. They grew corn, oats, and hay in regular rotation. They sold some of their crops but fed most of them to cattle, hogs, or their own work animals. Some kept dairy cows, but they derived much of their income from sales of fat beef cattle and fat hogs.

During the 1970s, farmers in the Corn Belt began to realize they had to become more specialized and to concentrate on doing what their computers told them they could do best. They could not produce fat cattle and hogs as cheaply as large, modern livestock operations that were being developed in other parts of the country, but they lived in the world's finest corn-growing area, and they could sell their corn to these new operations instead of feeding it to their own animals. They became specialized cash-crop farmers.

Soybeans helped enormously. As late as 1950, they were little more than a curiosity on most farms, but they were an excellent cash crop with the same agronomic requirements as corn, and they could be grown with the same equipment. Soybean acreage in Minnesota increased steadily after 1950, and by 2002 it was slightly ahead of corn, because farmers in the Corn Belt had developed a new cash-grain cropping system in which beans replaced the oats and hay of the traditional three-year rotation. Some farmers in southern Minnesota also have lucrative contracts with vegetable canning companies to grow specialized cash crops of sweet corn and green peas.

After harvest, Corn Belt farmers shipped in lean beef cattle from western ranches and fed them on corn to market weight in their feedlots.

Unit trains haul loads of corn and soybeans from massive trackside grain elevators.

Fences are disappearing from the Corn Belt landscape, because cash-grain farmers have no livestock.

The Corn Belt is becoming a vast field of soybeans and corn.

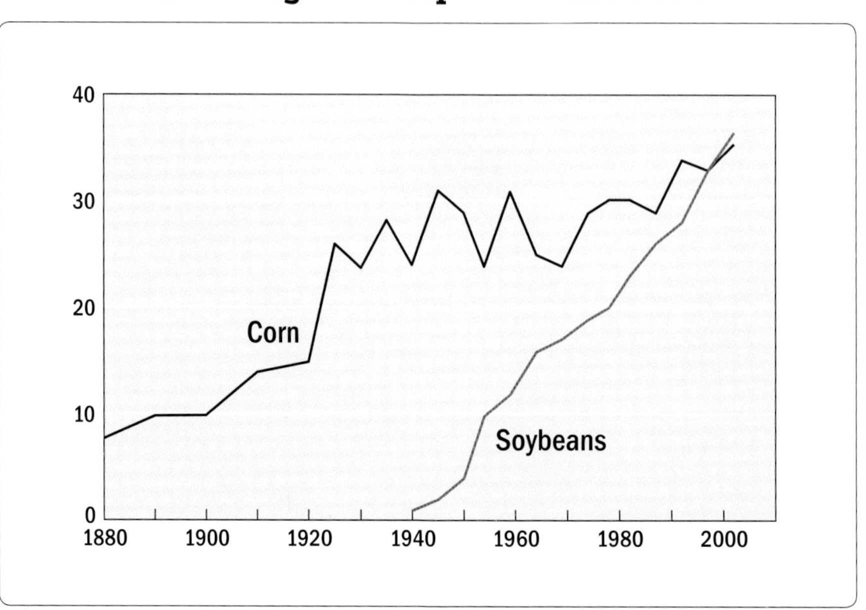

In 1940 soybean acreage was negligible, but today soybeans compete with corn for status as Minnesota's leading crop.

Farmers' profit margins per bushel of corn or soybeans have been so razor-thin that cash-grain farmers have had to increase their volume of production by producing more bushels per acre and by farming more acres.[1] Plant breeders have given them better varieties of seeds that produce more per acre, and farmers have used more chemical fertilizer to tap the full genetic potential of the new seeds. They have relied on chemical insecticides rather than physical cultivation to control weeds, insects, and diseases on larger acreages of land. Some people are concerned that continuous cultivation of the same two crops year after year will result in serious deterioration of the soil, but thus far there are few if any signs that this has occurred.

Larger acreages have necessitated bigger and better tractors and other farm machinery. The combine harvester, which lumbers across the land at harvest time, has transformed the rural landscape. The combine shells corn and beans when it picks them, and unloads the loose grain through a side spout into a trailing cart that hauls it back to the farmstead. Traditional wooden corn cribs with open slatted sides cannot hold loose grain, and farmers have replaced them with corrugated metal bins. Nests of grain bins have slender elevator legs capped with distributor heads and spiderlike networks of tubes that can transfer grain from one bin to another at the push of a button.

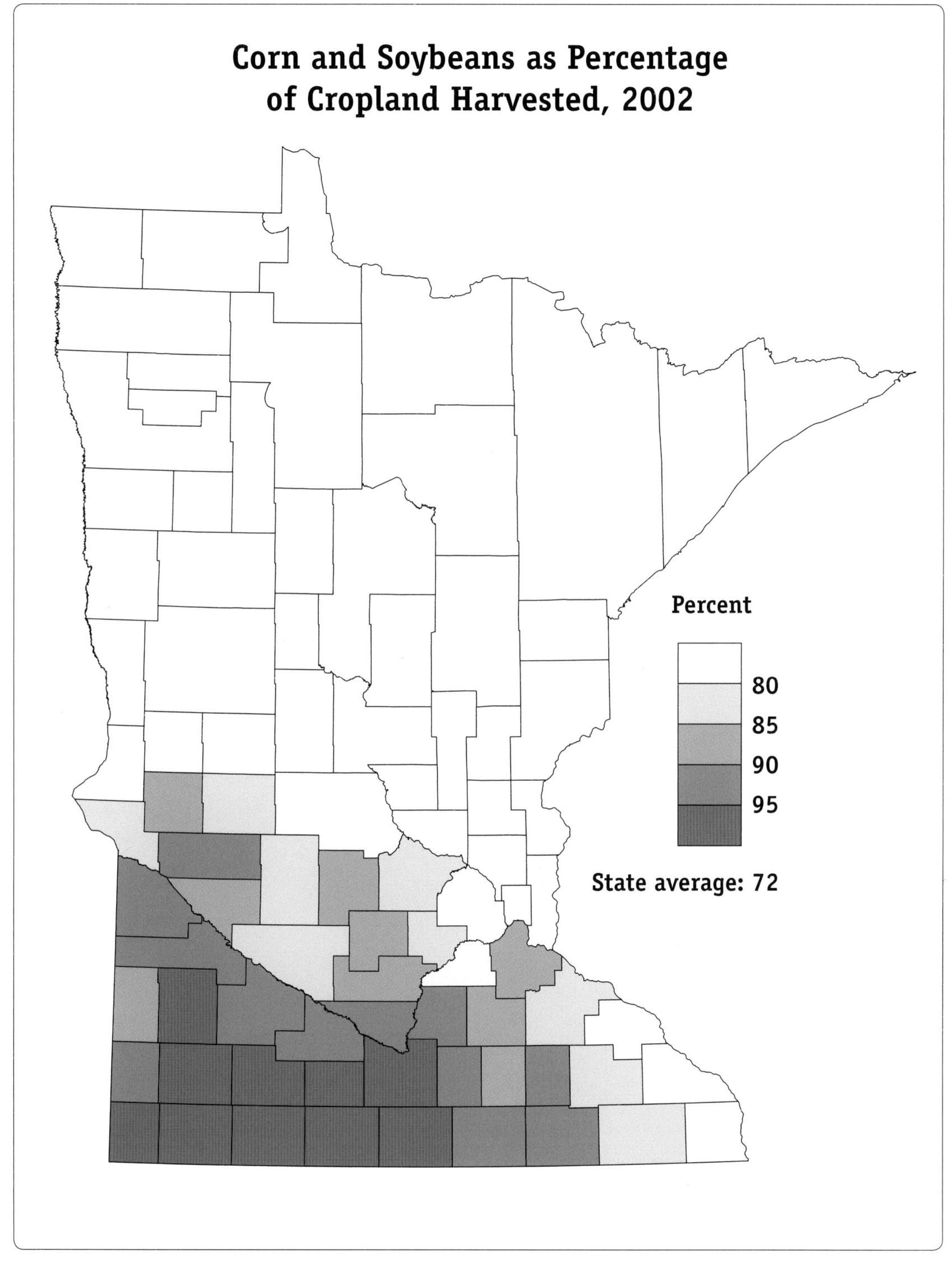

Corn and soybeans are the overwhelmingly dominant crops in southern Minnesota.

Tanks of anhydrous ammonia near a grain elevator provide nitrogen-rich fertilizer for Corn Belt soils.

Modern combine harvesters cost farmers $250,000 or more.

A combine harvester shells corn in the field, and after the corn has been harvested the farmer shifts to a reel-type head to harvest soybeans.

Combine harvesters, which shell grain as they harvest it, have forced farmers to replace their traditional wooden corn cribs with corrugated metal grain bins.

A central elevator leg lifts loose corn and soybeans to a distributor head from which it is directed through spidery tubes to individual grain bins.

A MODERN FAMILY CASH-GRAIN FARM needs around 1,600 acres of cropland to provide an acceptable level of living for a contemporary American family.[2] Most cash-grain farmers have acquired the land they need by renting it from neighbors who have stopped farming rather than by buying it, both because farmland has become too expensive to buy and because even those people who are no longer farming still are reluctant to sell land that has been in the family for several generations.

The census classifies a farmer who owns part of his land and rents the rest as a part-owner farmer.[3] The average size of part-owner farms in Minnesota more than doubled between 1950 and 2002, primarily because the farmers were renting more land. The average acreage they own has not increased much more than the average acreage owned by full-owner farmers, which indicates that part-owner farmers are renting rather than buying. Most full-owner farmers, who own all the land they farm, are older Minnesotans thinking more about retiring out of farming than about expanding their operations. The average age of Minnesota farmers is slowly inching up, from 46.9 years in 1950 to 52.9 in 2002; more than half of this increase has occurred in the last decade.

Average Size of Farm

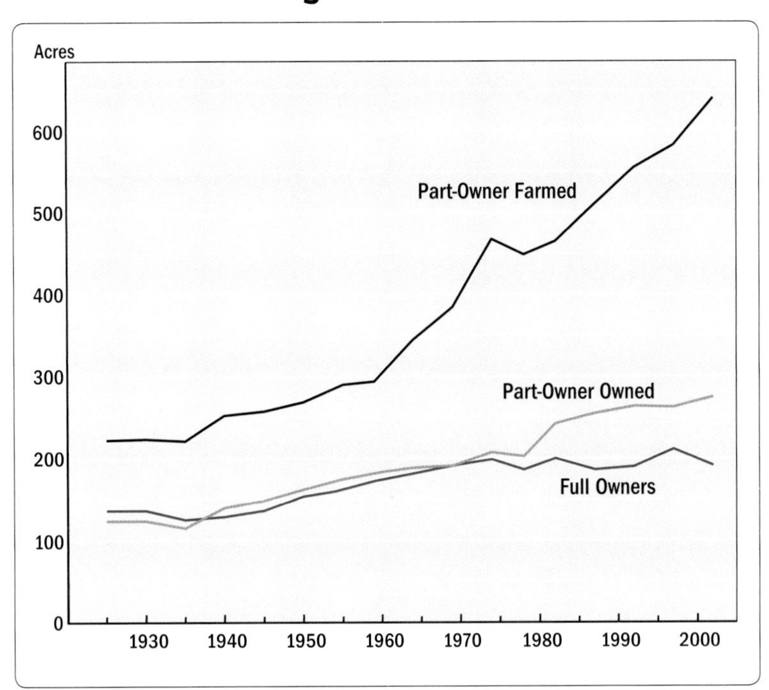

The difference between the acreage owned by part-owner farmers and the total acreage they farm is the acreage they rent.

Average Age of Minnesota Farmers

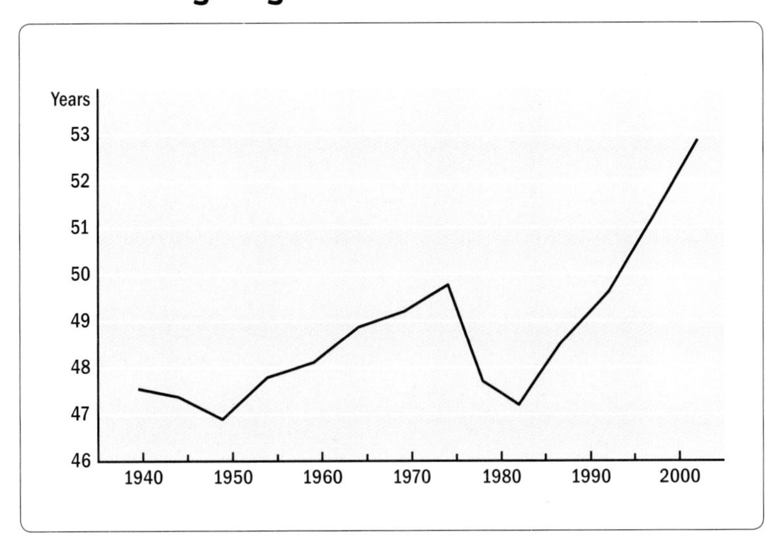

The average age of Minnesota farmers has been increasing steadily since 1982.

A family cash-grain farm of 1,600 acres is ten times the size of an original 160-acre homestead, which means that nine of every ten old farmsteads in cash-grain areas are redundant to the needs of contemporary agriculture, and the countryside has large numbers of farmsteads in various stages of abandonment. In 1950, for example, one township in Cottonwood County had 134 farmsteads, or nearly four per square mile.[4] In 1996, eighteen had been completely demolished or were marked only by old windbreaks, and twenty-four had only abandoned buildings. Forty-eight still were homes to elderly couples who were renting their land to neighbors, and they would be abandoned when the inhabitants died or moved into a nursing home or in with family. Nine were residences of part-time farmers who supplemented their farm income with off-farm jobs, and only thirty-five, or less than one per square mile, were still full-time farms.

In 2002 a full-time family farm in Minnesota had to sell at least $500,000 worth of farm products to provide an acceptable level of living. Only 3,625 (4.5 percent) of the state's 80,839 farms were in this category, but they produced 49 percent of its farm products. Only 18,542 (22.9 percent) of the state's farms had sales of $100,000 or more, but they produced 87.9 percent of its farm products. Although the census of agriculture says that the state has 80,000 farms, in fact fewer than 20,000 are viable full-time farms.

Many old farmsteads are redundant to the needs of contemporary agriculture as farms have grown larger.

The other 60,000 are undersized, part-time, or niche farms whose operators must have off-farm jobs to supplement their farm income. In 1950, 12.1 percent of Minnesota's farmers worked off their farms for 100 days or more, but in 2002 that figure had ballooned to 46.5 percent, and on most of the rest at least one member of the family had an off-farm job. Small farms, organic farms, and farms operated by women are contributing more to agricultural production than they once did, but their contribution is still small, although growing.

Each abandoned farmstead is the death of a dream.

Even the viable full-time farms in Minnesota must face the fact that they produce only commodities, such as corn and soybeans, and in order to remain competitive they must find a way to add value to these commodities by processing them. Soybean farmers, for example, are nervous about new soybean farms being carved out of the Amazon jungle in Brazil that can produce soybeans more cheaply than U.S. farmers can. Some farmer groups have founded cooperatives to produce value-added products, but many co-ops have floundered because their members lack the cutthroat business and managerial skills to make them profitable.

Minnesota's corn farmers have developed a new market for their commodity by talking the legislature into subsidizing seventeen new plants that convert corn into ethanol and by requiring that ethanol be added to all gasoline sold in the state, even though these plants are economically and environmentally suspect.[5] Calculations of

Sales of Farms Classified by Value of Sales in 2002

Value of Sales	Number of Farms	Sales (in $1,000)
Less than $100,000	62,297	1,039,624
$100,000 to $499,999	14,917	3,331,675
$500,000 or more	3,625	4,204,328
Total	**80,839**	**8,575,627**

Source: *2002 Census of Agriculture.*

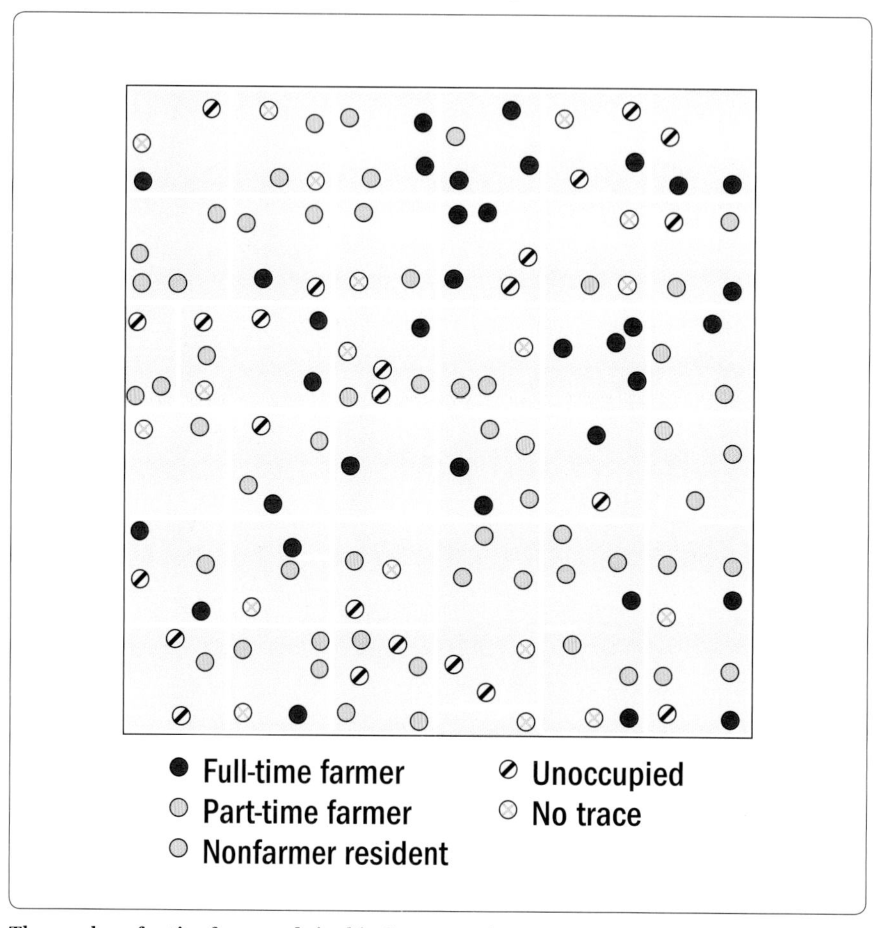

The number of active farmsteads in this Cottonwood County township dropped from nearly four per square mile in 1950 to less than one per square mile in 1996.

the costs and benefits of ethanol production are so complex that reasonable people can differ widely in their opinions about its prospects. Some argue that making ethanol requires more energy than it delivers, but others find a net benefit, although few claim that it is more than a modest one.

Producing a gallon of ethanol requires more than two-fifths of a gallon of fuel, such as natural gas, and more fuel is needed to manufacture fertilizer for the corn, to operate the farm machinery, and to truck the ethanol to market.[6] Engineers are experimenting with a number of methods to make ethanol more efficiently, but a shortage of water might also be a problem in southwestern Minnesota. Ethanol production requires prodigious amounts of steam, and the existing plants are already straining the available water resources. Plans for proposed ethanol plants have had to be shelved because not enough water is available.[7]

Doug Magnus and his father, Clarence, grow 1,400 acres of corn and soybeans near Slayton. Doug started farming in 1976 by renting a 160-acre farm from Clarence, and together they have built up the farm to its present size. They own about half the land and rent the rest. The competition for land to rent is intense. Everyone seems to need more land, and larger farmers can afford to pay higher rents because they can spread the cost over a larger acreage.

The brand marks on the flanks of these beef cattle in a Corn Belt feedlot show that they were reared on the Lazy Two ranch.

Percentage of Farmers Who Worked Off-Farm 100 Days or More

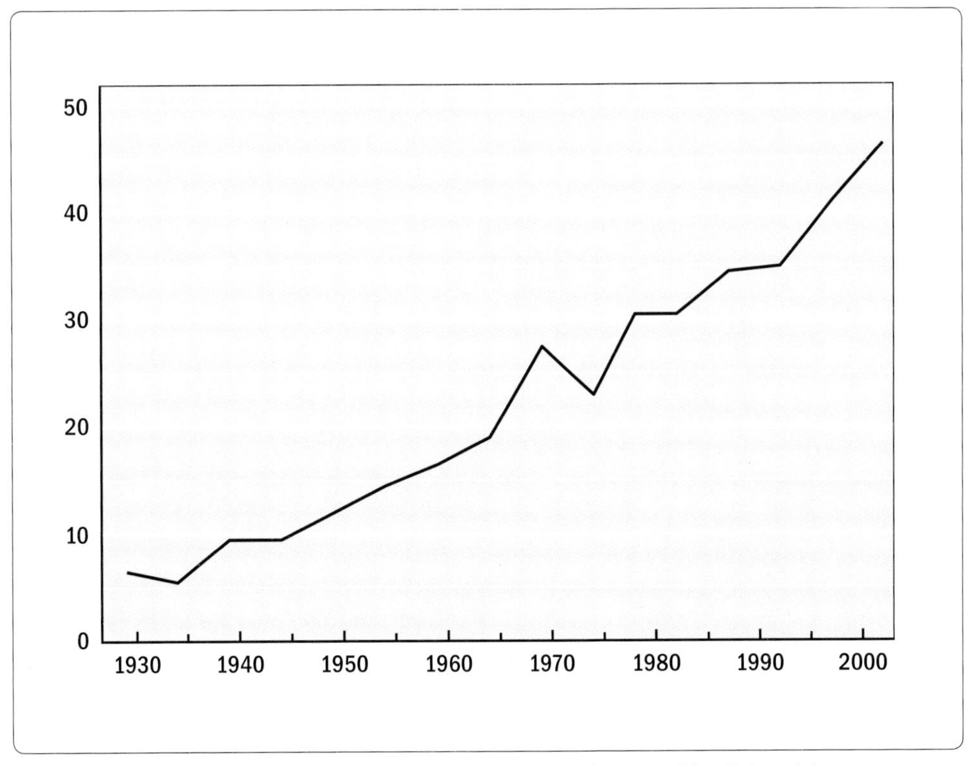

Half of the farmers in Minnesota must supplement their farm income with off-farm jobs.

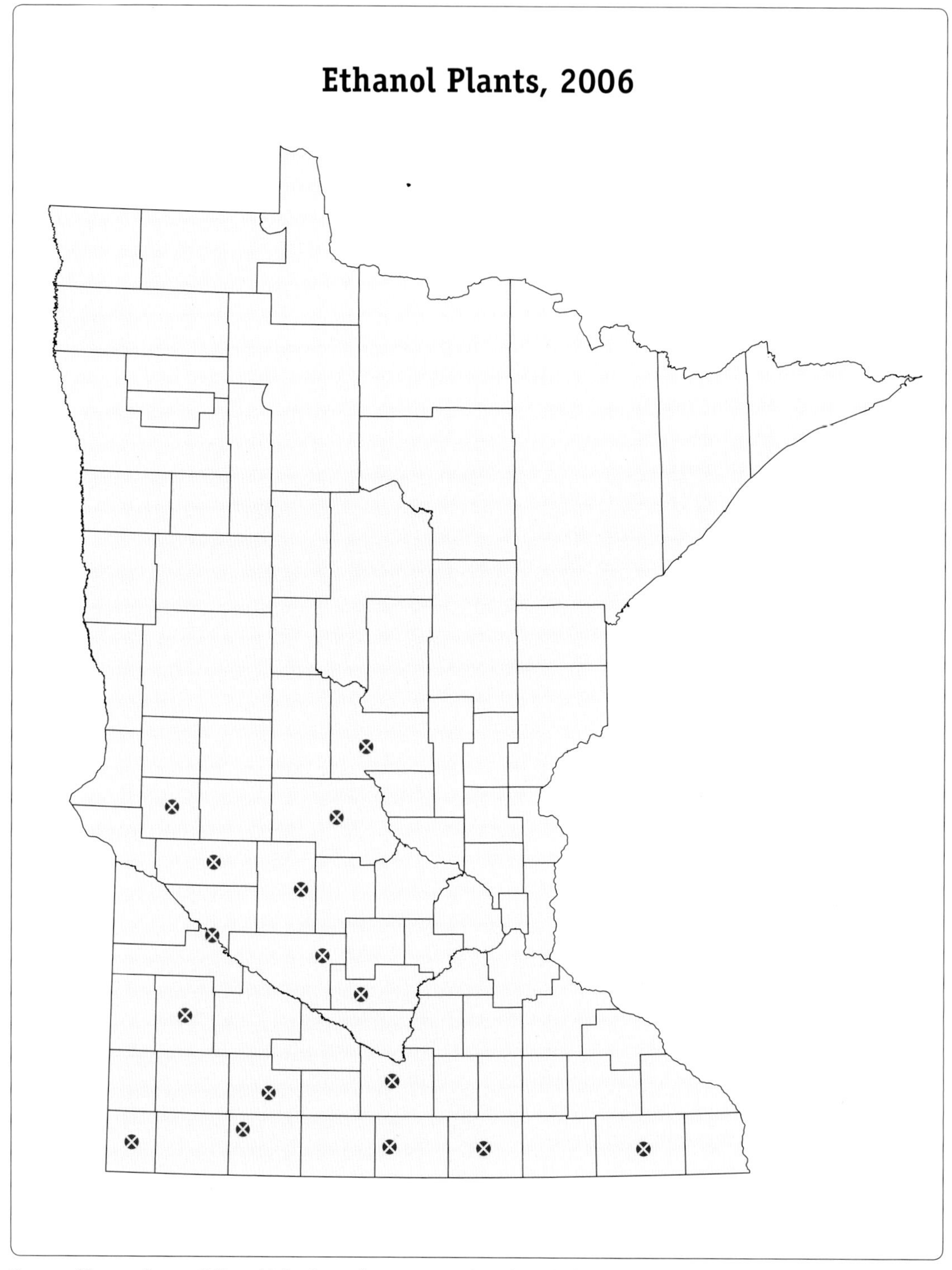

Reasonable people can differ widely about the prospects for ethanol plants.

The price of soybeans runs two to two-and-one-half times the price of corn. Magnus plants more beans if it is higher, and more corn if it is lower, but he also keeps a wary eye on soybean production in Brazil. He fed hogs under contract when he started farming because he needed the guaranteed income. He has a feedlot that can hold 500 head of beef cattle, but it is usually empty. He buys lean ranch cattle and feeds them on his corn if the price looks favorable, because he likes having them on the farm, but he admitted that the cattle business is not for the faint of heart.

In 1983 Doug's wife, Brenda, did all the plowing, chopped silage, hauled grain, chopped stalks, drove the disk cultivator, and drove the truck that hauled grain to the elevator. Brains have largely replaced brawn as a prerequisite in modern agriculture, and on many family farms the wife has taken the place of the hired man of old.

Even though most farmers in the Corn Belt have shifted from mixed crop-and-livestock farming to cash-grain farming, enough have become specialized livestock producers to make Minnesota the nation's leading turkey state, third hog state, and eleventh egg state. These producers enjoy being in the nation's principal area of feed crop production, and their advantage could increase if continuous cropping of corn and soybeans leads to soil deterioration that forces cash-grain farmers to grow more alfalfa.

Earl B. Olson of Willmar is the entrepreneur who put Minnesota on the national turkey map by turning U.S. 71 north from Willmar to Wadena into the turkey trail. He started with 300 birds in 1941 and gradually built up to 30,000 in 1949, when he opened his first turkey processing plant. He raised some turkeys on his own farm and contracted with local farmers to feed some for him.

The traditional turkey business was highly seasonal because people bought whole birds only at Thanksgiving and Christmas, but Olson had to keep his processing plant going year-round to make it profitable, so he began to produce a variety of turkey products. Turkey meat is easy to process into many different products, and now most of us eat it at least once a week

Each of these turkey barns houses ten thousand birds or more.

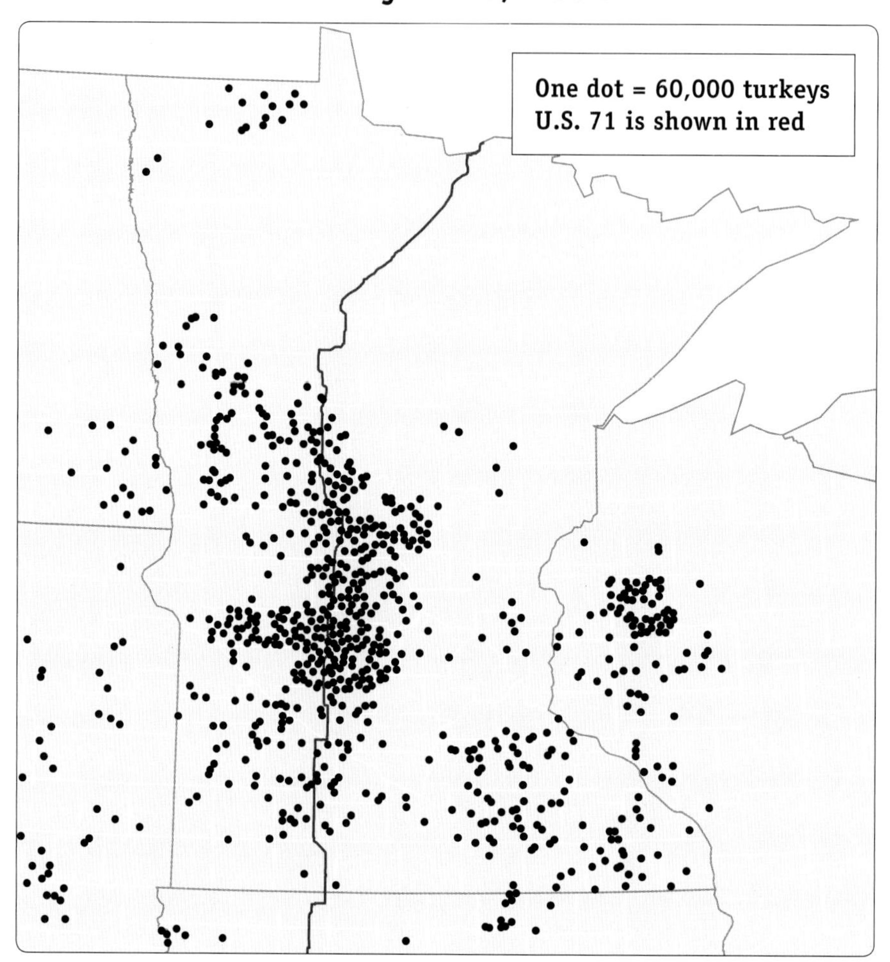

U.S. Highway 71 in western Minnesota is the Turkey Trail.

although we may not realize it. Olson advertised and marketed his products aggressively under the brand name Jennie-O, because he knew that consumers are willing to pay premium prices for branded products they know they can trust.

Initially Olson had raised turkeys on free range, but in 1968 he shifted them into confinement houses where he could protect them from predators and manage them more efficiently. A standard 70-by-500-foot turkey house for 10,000 birds is a single-story structure with curtain sidewalls. Turkey farms have clusters of three to fifty parallel houses. In 1998 Jennie-O produced one-third of its birds on twenty-five company farms and contracted with forty-five farmers to produce the rest. The birds live in a brood house for their first seven weeks and then they are moved to a

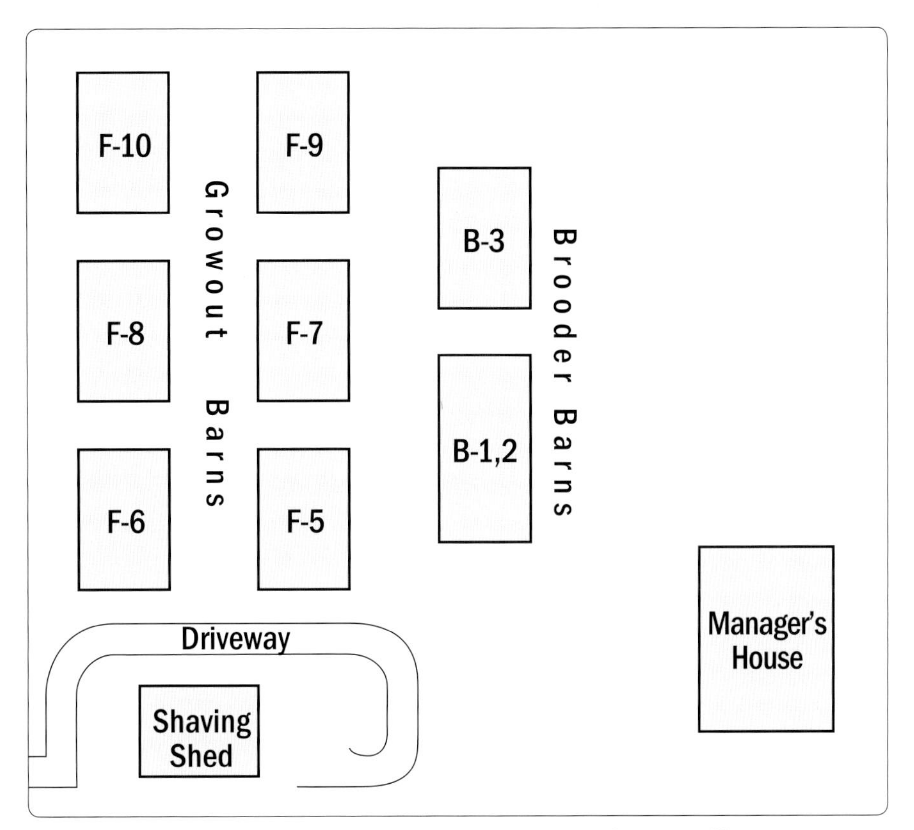

The farm manager moves turkey poults to larger houses when they grow older.

grow-out house, the hens for seven weeks until they weigh fifteen pounds, the toms for eleven weeks until they weigh thirty to thirty-five pounds.

In 1986 Olson sold his company to the Hormel Foods Corporation of Austin because he did not have the capital or the marketing and distribution system he needed, but the company kept its identity and brand name. In 2000 Jennie-O bought the Turkey Store Company, which had started in Barron, Wisconsin, but had developed a processing plant in Faribault with a feed mill, thirty company farms, and fourteen contract farms nearby. These acquisitions are part of the larger process of consolidation that is transforming livestock production in the United States.

Bob Sparboe was an entrepreneur who quietly but persistently transformed egg production from a barnyard sideline into an efficient modern system. In 1954 he started a chicken hatchery in Litchfield to sell day-old chicks to farmers who kept small farm flocks, but it was obvious that the number of farmers was dwindling rapidly, and in 1960 he shifted to commercial production of young hens and sold laying hens and eggs. He lent money to farmers who wanted to build houses for 3,600 hens and contracted with them to market their eggs.

His feed costs were so high that they were eating up his profits. In 1970 he formed his own feed production company and nutrition service. His egg sales continued to grow, and in 1974 he built one of Minnesota's first modern egg production complexes on U.S. 12 a few miles east of Litchfield. It has twelve parallel two-story 40-by-400-foot houses that hold 80,000 hens each. The houses are carefully climate-controlled to ensure that the hens are comfortable, and automatic systems bring them feed and water.

On the upper level of each house the hens are held in cages stacked five high, with five birds in each cage. They lay seven eggs every ten days. Their droppings bounce off a splatter board on the back of the cage below and fall into the pit on the ground floor, which is scraped out regularly. The droppings, which are an extraordinarily rich form of fertilizer, are later spread on cropland.

The eggs roll onto a narrow conveyor belt that carries them to the large conveyor belt at the end of the house that takes them to the packing plant. A veritable river of eggs a yard wide flows at a rate of forty eggs per second into the packing plant, where machines wash them, size them, and place them gently in cartons. Forklifts haul

Each house on the Sparboe Farm in Litchfield holds eighty thousand laying hens. Each hen lays seven eggs every ten days.

pallets stacked high with filled cartons to the vast cold storage warehouse, where they rest briefly until they are loaded onto the trucks that carry them to the grocery store. Most eggs arrive in the grocery store the day after they are laid.

Hens, like plants, become dormant in the fall when the days start to grow shorter. They stop laying and lose their appetites and some of their feathers. This natural process is called "molting." In the spring they start to lay eggs once again with renewed vigor. When the hens in a house have been laying for a year or so, the egg producer can give them a rest and induce molting by turning off the lights for longer hours each day. After eight weeks of rest the lights go back up again, and the hens go back to work. The only artificial aspect of this procedure is its timing, but it has been given the unfortunate name of "forced molting."

In 1978 Sparboe started his own trucking company to haul feed to his hen houses and eggs to grocery stores. In 1987 he bought a bank in Litchfield and reoriented it to serve the special needs of the national poultry industry. His bank also makes housing and other loans to his employees, many of whom are Latino and have trouble getting credit locally. In 1988 he began to expand in Iowa, where he built one plant that breaks 9 million eggs a week, on purpose, and sells liquid eggs by the tanker truckload to bakeries, institutions, and other large-scale users. In 2000 he had more than 3 million hens that laid more than 750 million eggs a year, and he also contracted to market 150 million more eggs for other producers.

Bob Christensen of Sleepy Eye, the leading hog producer in Minnesota and one of the top five hog producers in the United States, is an awesome success story. He grew up on a cattle and row-crop farm. In 1974, when he was thirteen, a neighbor gave him two bred sows, which he kept in an old building on the farm. By the time he finished high school, he had 200 sows and was contracting with neighbors to feed hogs for him. In 2001, Christensen Family Farms had 520 employees, 80,000 sows, and sold $300 million worth of hogs.

Christensen expanded rapidly in the 1980s. He built his first sow barn in 1977 and added one each year for ten years. Many local farmers had empty hog barns during the farm crisis of the early 1980s, because bankers were afraid to lend them money to buy hogs, and Christensen paid them to feed hogs for him. During one twenty-seven month stretch he added at least one new contract producer each month. Small farmers with only 400 to 700 acres of row crops were happy to build new hog barns and contract with him because each barn gave them manure with a fertilizer value of $4,000, and the additional income from hogs expanded the farm business enough to enable the family's children to stay on the farm.

Curtain-sided hog houses hold six hundred hogs.

From the highway, hog barns look much like turkey barns. They are long low buildings with curtain sidewalls, but hog barns have more elaborate ventilation systems. A standard 600-hog barn is divided into pens that hold twenty-five hogs each. Beneath each barn is a pit deep enough to hold all the manure the hogs can produce in fourteen months. Once a year the farmer contracts to pump out the manure and spread it on fields that will be planted with crops.

Bob Christensen is a charismatic leader who hires and inspires top-notch management talent. He uses the very best genetics and has a state-of-the-art information system. He has made a fine art of figuring out what everybody else is going to do and then doing the opposite. Until 1992 he was the only pork producer who was expanding in southern Minnesota, but when everyone else started to build hog barns, he invested in his own feed mills, sow farms, and nurseries, and he even built some of his own finishing barns. His huge grain elevator adorned with the Christensen Family Farms logo is a landmark north of U.S. 12 a few miles east of Sleepy Eye.

Veterinary clinics in Fairmont, Pipestone, and other small towns complemented the endeavors of Bob Christensen in encouraging the solid growth of the hog business in southern Minnesota during the late 1980s and early 1990s. These clinics organized hog production alliances of small farmers. Each alliance built a sow farm, which the clinic manages. Each member of the alliance agrees to receive a specified number of hogs each year, takes sole ownership of them upon delivery, and feeds and sells them. Some members have done so well that they have begun to contract with other farmers to feed some of their hogs for them.

Small hog producers worry about the possibility that the packing plants to which they sell their hogs might close or consolidate because the plants are too small to be competitive. Small producers may have no place to sell their hogs if a packing plant closes because it cannot compete with larger and more efficient plants. Packing plants themselves prefer to deal with producers who can deliver large numbers of uniform animals, because dealing with many small producers can be an expensive managerial headache.

Small farmers in the Corn Belt of southwestern Minnesota will continue to produce livestock, because this region is part of the leading feed-producing area in the United States, but most farmers will continue to concentrate on producing cash crops of corn and soybeans. The acreage planted to each crop will vary from year to year. The size of cash-grain farms has increased steadily, and the upper limit apparently is not yet in sight, although today's farms are already huge. Even larger farms will mean that ever fewer people are needed on the land, and increasing farm size will continue the depopulation of rural areas in this part of the state.

The Christensen Family Farms grain elevator and feed mill is a landmark on U.S. 12 a few miles east of Sleepy Eye.

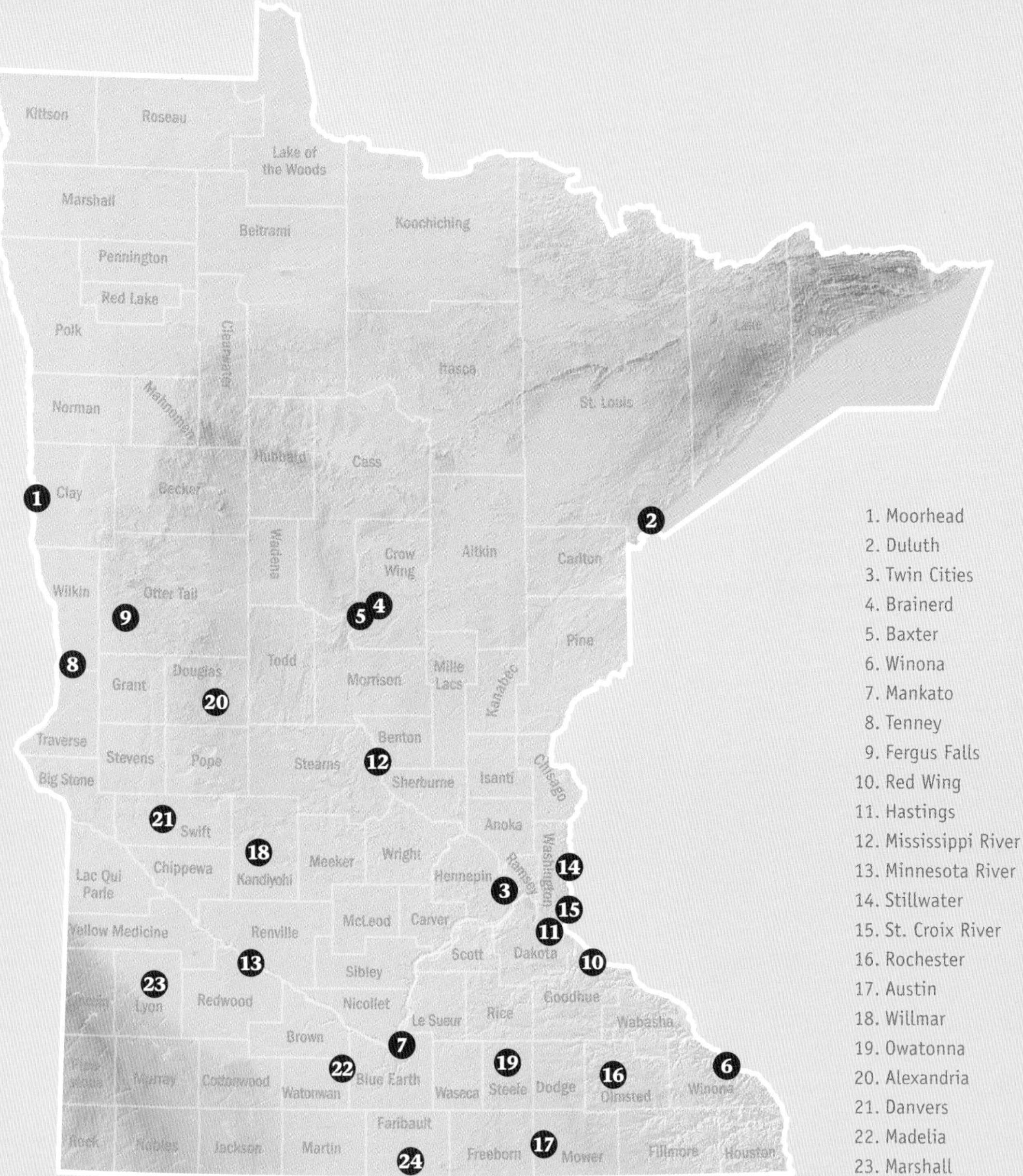

Kittson
Roseau
Lake of the Woods
Marshall
Beltrami
Koochiching
Pennington
Red Lake
Polk
Clearwater
Mahnomen
Norman
Itasca
St. Louis
Lake
Cook
Hubbard
Cass
Clay
Becker
Wadena
Crow Wing
Aitkin
Carlton
Wilkin
Otter Tail
Pine
Grant
Douglas
Todd
Morrison
Mille Lacs
Kanabec
Traverse
Stevens
Pope
Stearns
Benton
Sherburne
Isanti
Chisago
Big Stone
Swift
Anoka
Chippewa
Kandiyohi
Meeker
Wright
Hennepin
Ramsey
Washington
Lac Qui Parle
McLeod
Carver
Yellow Medicine
Renville
Scott
Dakota
Sibley
Lyon
Redwood
Nicollet
Le Sueur
Rice
Goodhue
Wabasha
Brown
Blue Earth
Waseca
Steele
Dodge
Olmsted
Winona
Murray
Cottonwood
Watonwan
Faribault
Rock
Nobles
Jackson
Martin
Freeborn
Mower
Fillmore
Houston
1. Moorhead
2. Duluth
3. Twin Cities
4. Brainerd
5. Baxter
6. Winona
7. Mankato
8. Tenney
9. Fergus Falls
10. Red Wing
11. Hastings
12. Mississippi River
13. Minnesota River
14. Stillwater
15. St. Croix River
16. Rochester
17. Austin
18. Willmar
19. Owatonna
20. Alexandria
21. Danvers
22. Madelia
23. Marshall
24. Frost

11

Small Towns

All farmers need towns. Nineteenth-century Minnesota farmers needed places where they could sell and ship their wheat, butter, eggs, and other farm products. They needed establishments where they could buy tools, machines, fabric and clothes, and other manufactured goods they could not make for themselves on their farms. They needed doctors, lawyers, bankers, and other professional people to help them. They needed schools, churches, libraries, and other institutions for themselves and their families. Farmers on treeless prairies needed coal yards and lumberyards where they could buy the fuel and building materials that their counterparts farther east had obtained from their own farm woodlots.

Country people needed access to these goods and services on a regular but not a daily basis. In the days of horse-drawn vehicles that meant it was desirable to be within six to ten miles of a town. That way they could drive their wagons or buggies into town, do their business, and drive home again the same day. Each small town served a hinterland or trade area with this small radius. Larger cities had enough people to support a greater, more specialized variety of goods and services, and their trade areas included these small towns and their trade areas.

Geographers John Borchert and William Casey classified all places in Minnesota into a hierarchy of trade centers based on the types and varieties of their businesses.[1] The order of a particular place in the trade center hierarchy is roughly, but only roughly, related to the size of its population. Southwestern Minnesota and the area north of a line from Fargo-Moorhead to Duluth have relatively few trade centers, whereas the dairy area northwest of the Twin Cities has many.

Trackside lumber and coal yards in small towns on the treeless prairie provide the building materials and fuel that farmers farther east had gotten from their own farm woodlots.

Small towns originated as collecting and retail centers for the surrounding agricultural area.

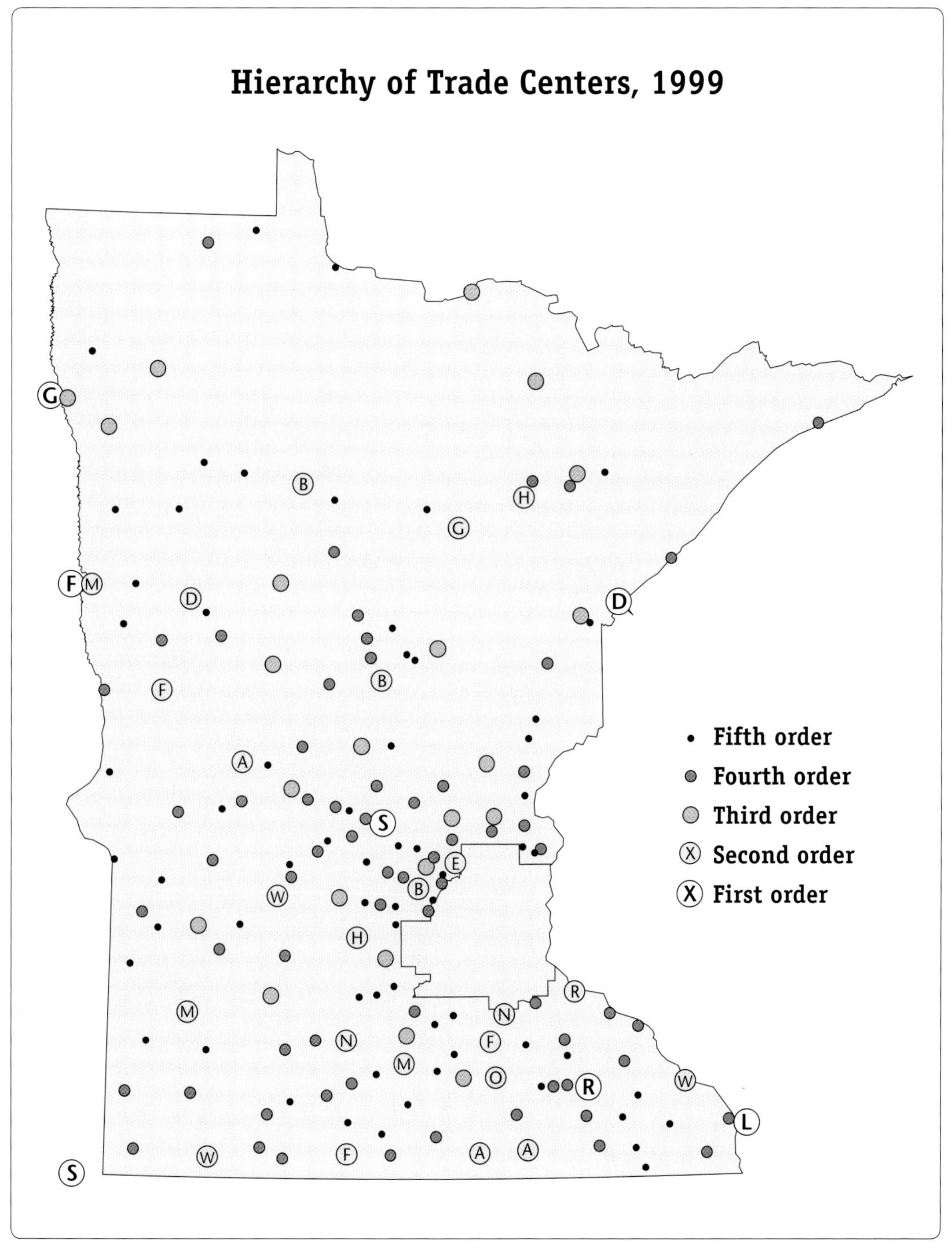

The hierarchy of trade centers is based on the types and variety of their businesses. The highest order places have the greatest number and variety, and the lowest order places have the least.

Hamlets, which are the lowest, seventh-order places in the hierarchy, typically have a filling station, a grocery store, and a café or tavern. Sixth-order places also have a bank, a hardware store, and a drugstore. Fifth-order towns add a household appliance or furniture store, a jewelry store, a clothing store, a laundromat, and specialty stores such as a shoe store, a lumberyard, a funeral parlor, a hotel or motel, and a farm and garden center.

Higher-order towns have more stores, more specialized stores, and more wholesaling. Second-order places have more than fifty wholesale businesses, and first-order places have more than one hundred. At the top of Minnesota's hierarchy is the Twin Cities metropolitan area, which is the economic capital of a region that stretches westward into the Rocky Mountains of Montana.

Small towns are the lowest-order places in the hierarchy of trade centers. Many of them are unknown to people outside their immediate vicinity. By Minnesota law, all incorporated places are officially classified as cities, but many of them seem too small to warrant that designation. Few are large enough to be considered urban by the U.S. Bureau of the Census, which requires a population of at least 2,500 people.

Contiguous clusters of separately incorporated places actually are single freestanding places, even though such places may have several names. The Twin Cities, for example, is a single built-up urban area made up of more than eighty separately incorporated places. Other examples would include places such as Brainerd and Baxter, Winona and Goodview, Norwood and Young America (which finally did decide to merge), and Mankato, North Mankato, and Skyline.

In 2000, the population of freestanding places in Minnesota ranged from 2.5 million in the Twin Cities to only 6 in Tenney, a hamlet in southern Wilkin County, twenty-five miles southwest of Fergus Falls. Tenney proudly incorporated in 1901, with a population of 185 people, and it has been losing population more or less steadily ever since. In the 1990s, however, Tenney staged a dramatic turnaround when it had one of the highest growth rates in the state, a 50 percent increase. Its population soared from 4 people to 6. Percentage growth rates for small places obviously can be deceptive.

MOST OF MINNESOTA'S EARLIEST TOWNS were on navigable rivers that connected them with markets and manufacturers in the older settled areas to the east. In 1860 these river towns included Winona, Red Wing, Hastings, St. Paul, and Minneapolis on the Mississippi; Stillwater on the St. Croix; and Mankato on the Minnesota. Some of these towns had been sawmilling centers in the lumber days, and many small towns had creameries, abattoirs, gristmills, and other small plants that processed local agricultural commodities to reduce their bulk and avoid the expense of shipping waste. Later towns developed at falls and rapids, waterpower sites that were especially well suited to the growth of manufacturing.

In 1860 a few towns such as Rochester and Austin had been incorporated at favored inland sites, and more inland and river towns had been incorporated by 1870. (Date of incorporation is a surrogate for date of town founding, or at least the time when a town became self-conscious of its identity.)

After 1870 in Minnesota railroads began to replace rivers as the principal mode of transportation, and since that date most new towns in Minnesota have been on railroad lines. By 1870 a string of towns had already been incorporated along the line from Winona to Mankato. Many towns in Minnesota were founded by the railroads, which spaced them at intervals of six to ten miles along their lines. This distance is just about the right length of track for a railroad sectional maintenance crew, so this spacing was as appropriate for the railroad as it was for horse-and-buggy travel.

By 1880 railroads had created scores of new towns along their lines as an additional source of revenue. The railroad companies had received land grants from the state and federal governments, and they sold this land to settlers to pay their costs of construction. The companies eagerly recruited settlers, who would produce goods that had to be shipped out by rail, who would need goods that had to be shipped in by rail, and who would generate passenger traffic. At regular intervals along their lines the railroads platted towns in which they could sell lots to people who wanted to start businesses. The railroad lines of Minnesota, including those that have been abandoned, are festooned with towns, like strings of lights on a Christmas tree.

Many small towns originated at sites where waterpower was available.

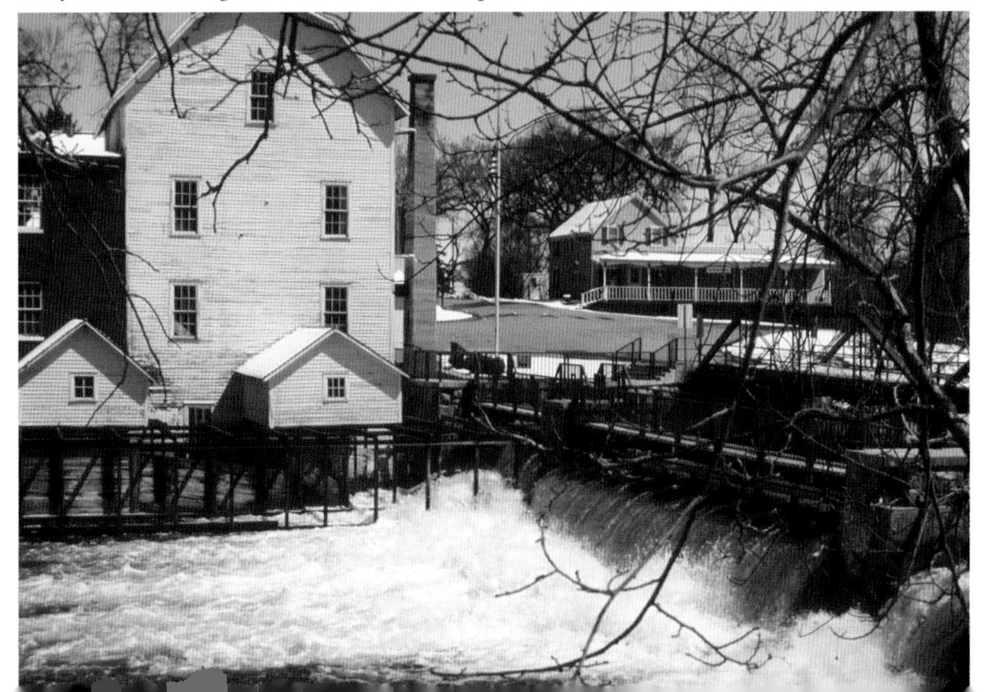

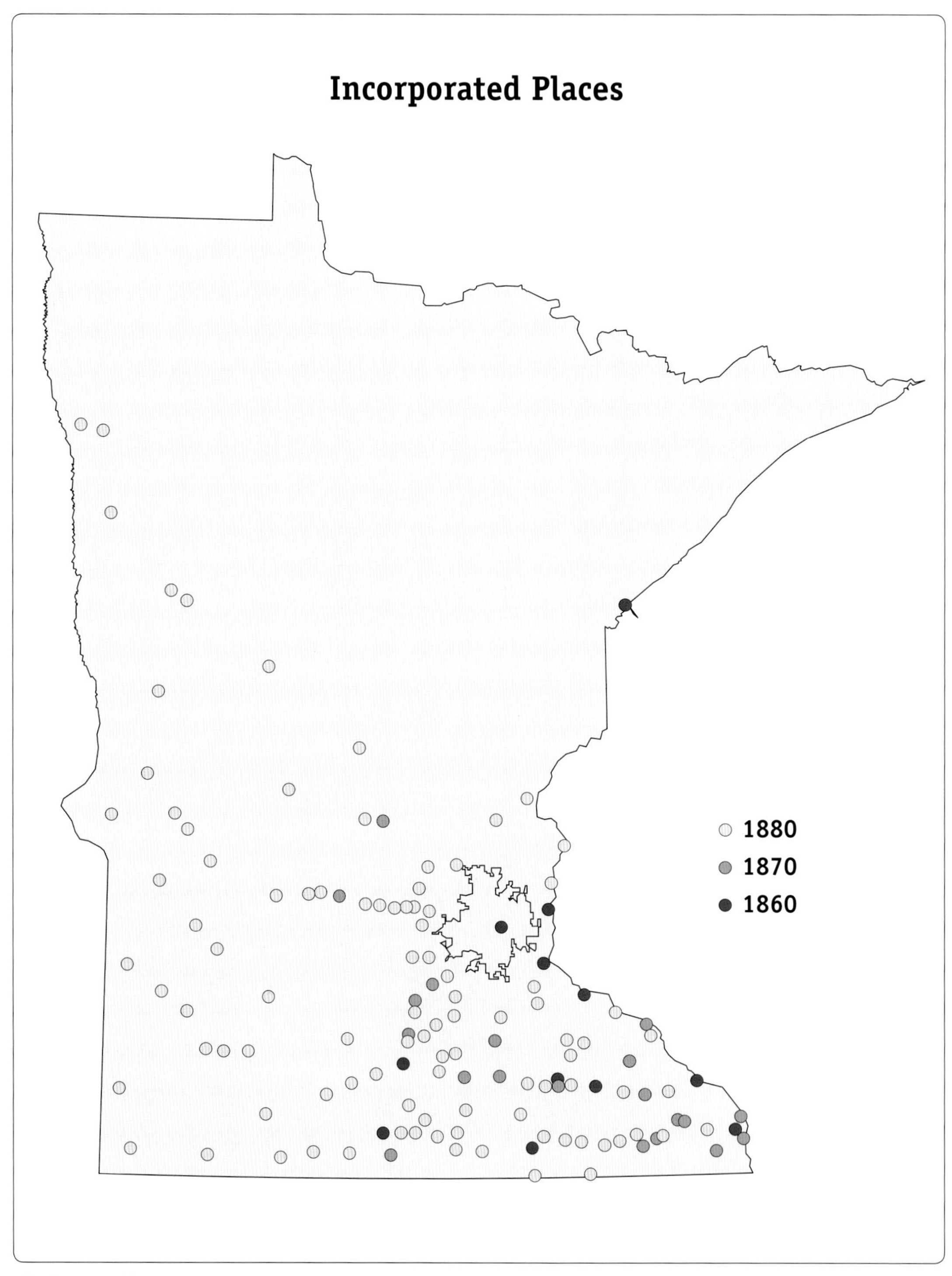

The date at which places were incorporated reflects the spread of settlement across Minnesota.

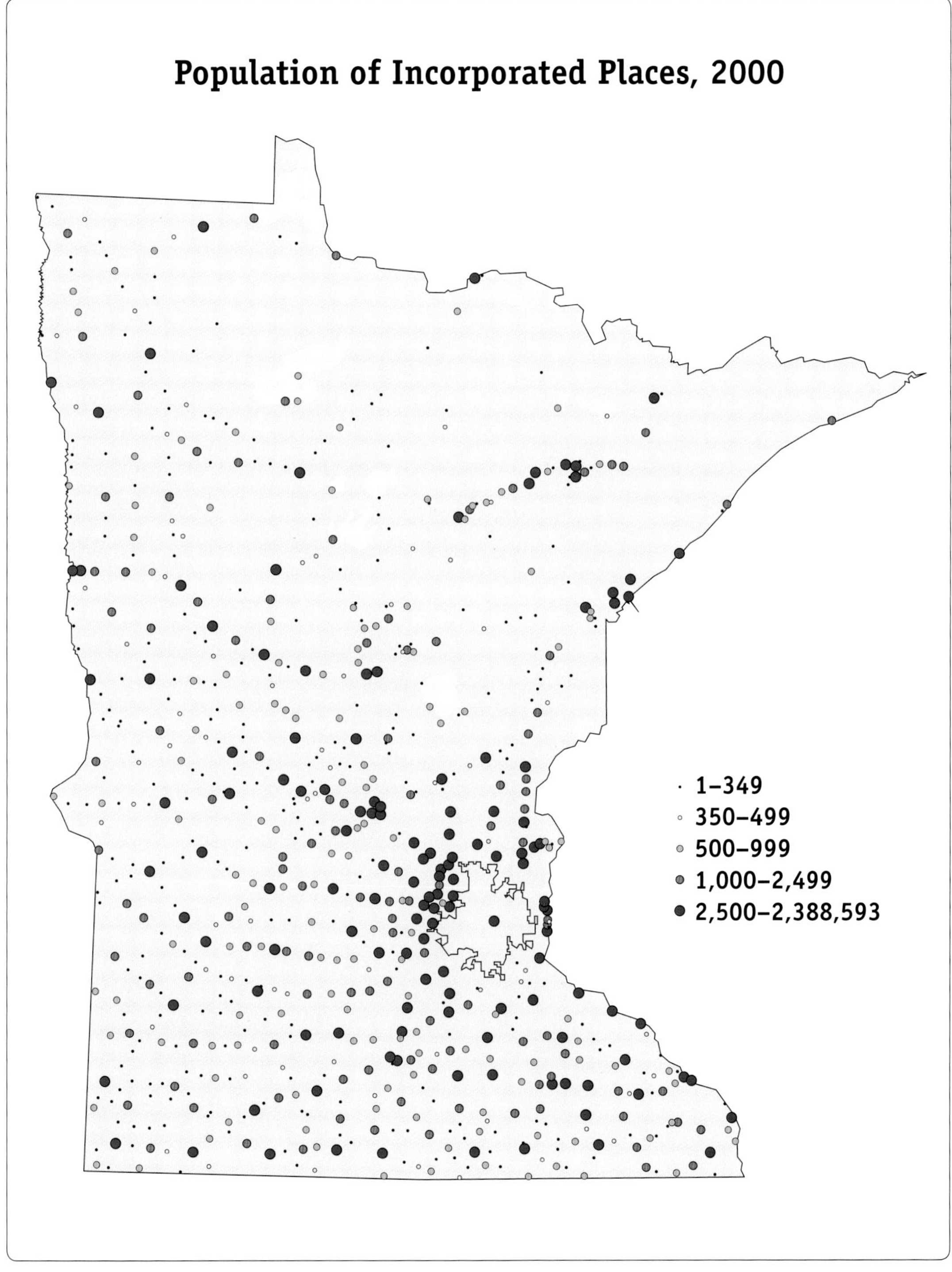

Railroad lines played a major role in creating the towns of Minnesota.

Railroad towns have grid street patterns that are oriented to the tracks.[2] Streets in towns that have outgrown the original plat may be oriented to the compass, and they bend where the two grids meet. The blocks are divided into lots, with small lots on business streets and larger lots in residential areas. In some railroad towns the principal business street parallels the tracks, but in others it runs at right angles, often leading to a major public building, such as the courthouse.[3]

Stores on the principal business street have professional offices or residences on the floors above. The railroads reserved for themselves the prime real estate along the tracks, where the hulking abandoned grain elevator still stands, but they sought an undeserved reputation for public-spirited philanthropy by donating to churches or schools the remote corner lots they knew they would have difficulty selling.

Minnesota's small-town settlement system had essentially been completed by 1900, and most towns incorporated since then have simply filled gaps in the existing pattern. The system has been remarkably robust, and the pattern in 2000 looks very much like the pattern a century earlier. Places persist once they have been incorporated. Even the smallest places, which are too small to grow, are too tough to die. Few towns have ever disincorporated, but the state does have a handful of ghost towns that have been completely depopulated and are merely names on the map. They are mainly mining camps on the iron ranges or whistle-stops on abandoned railroad lines.

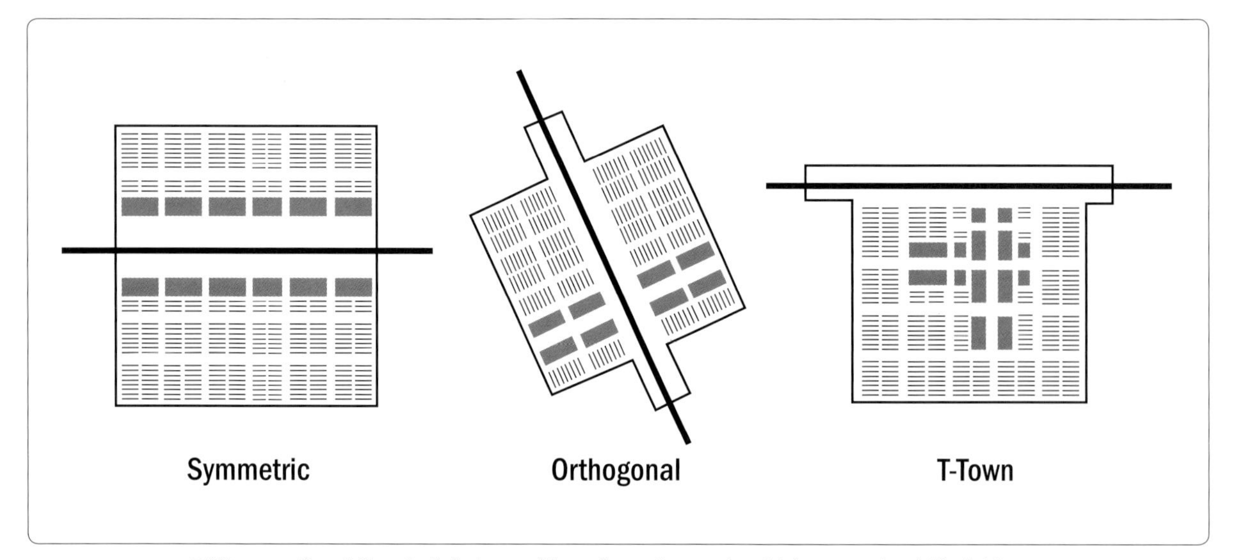

Different railroad lines had their own ideas about the way in which towns should be laid out.

In some small towns Main Street faces the railroad tracks.

Trackside grain elevators dominate Angus.

The total population of Minnesota's small town system has continued to grow, but the growth of individual places has fluctuated. It is tantalizingly unpredictable, larger at one census, smaller at the next, for no apparent reason, while neighboring places that seem similar are doing just the opposite. These fluctuations of individual places balance each other out, but their long-term trajectory has been gently upward, and the system as a whole has continued to grow, rather like a lumpy expanding balloon.

The population of most individual small towns fluctuates within a fairly narrow range, and some hardly change at all from one census to the next. These fluctuations may be exaggerated if they are measured in percentages, because their base is so small. For example, a place of fifty people that gained five people over a decade would show a 10 percent gain, although it would have added only one new person every other year.

Until 1900 most incorporated places in Minnesota grew more than 5 percent in each decade, and only a few lost, but in the twentieth century the percentage of gains

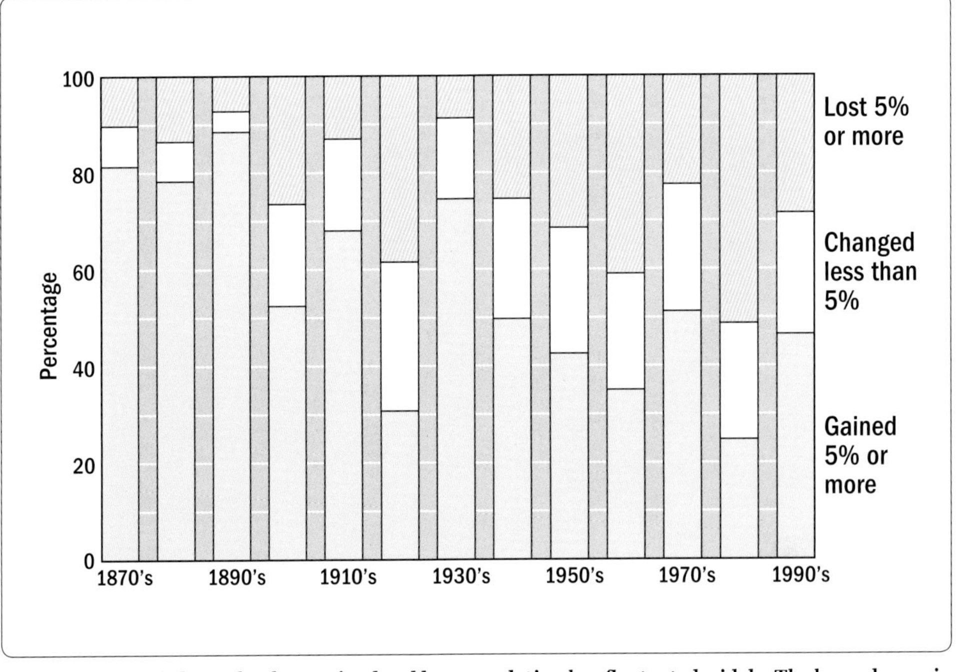

The percentage of places that have gained and lost population has fluctuated widely. The heavy losses in the 1920s reflect the impact of the automobile and the motor truck.

and losses has varied considerably from census to census. The 1920s, the first automobile decade, and the recession decade of the 1980s had the greatest percentage of losses. The percentage of places that gained in population declined steadily from the 1930s to the 1980s, except for the 1970s, and one might well wonder whether this decline will continue today, although almost half of all places gained between 1990 and 2000.

Because the urban system has an enormous amount of built-in stability, or inertia, the best statistical predictor of the population of places at any census is the size of their population at the preceding census. For example, no one expects the population of the Twin Cities to drop to zero at any time, nor does anyone expect Tenney to have a quarter of a million people in 2010.

Small incorporated places in Minnesota's small-town settlement system show impressive statistical stability in 1990 and in 2000. A few places broke from the pack and grew extravagantly, and a few places sagged severely, but most places of the same size grew at about the same pace.

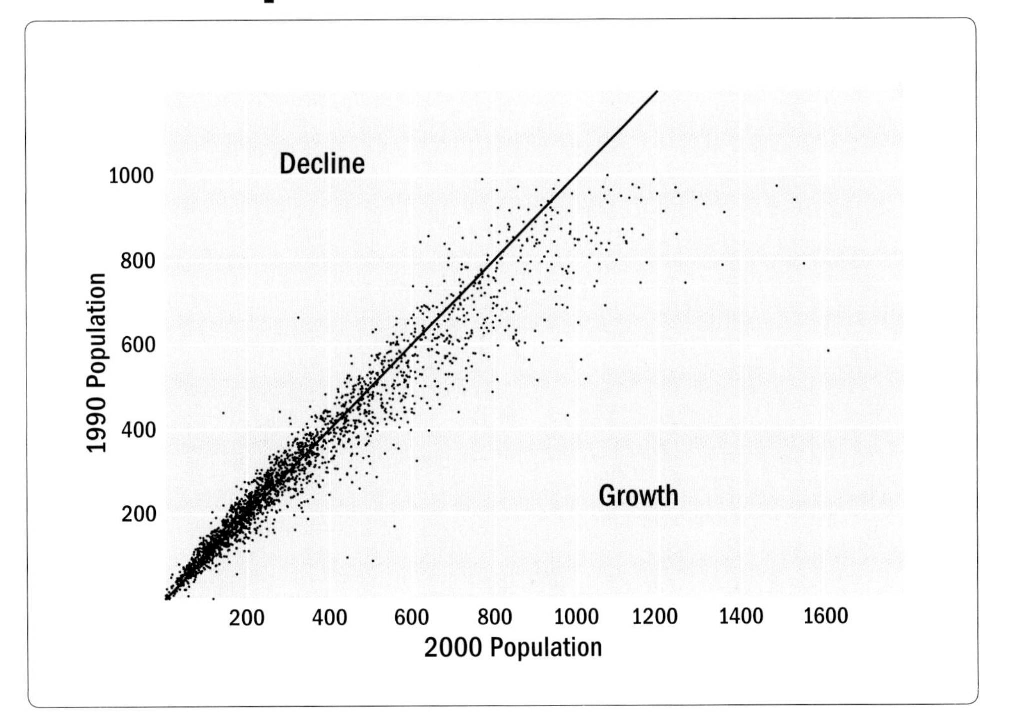

The size of small towns is the best statistical predictor of their growth.

Over the past century the population of most sizes of town grew in the first two decades, but then dipped in the 1920s, when the widespread availability of the automobile and the motor truck enabled rural people to bypass nearby small towns in search of the greater range of goods and services that were available in larger places.

The smallest towns, those with fewer than 350 people, have never recovered from the setback of the 1920s. They are disproportionately concentrated in the parts of the state that were the last to be settled. They are still hanging on, but they were founded too late to grow before they became redundant to the needs of the contemporary economy and society. Their Main Streets have a gap-toothed look, with open plots of ground separating the individual buildings, and former stores stand empty or have been converted to residential use. Once a place has attained a threshold population of around 350 people, however, it seems destined to continue growing and has a solid block of contiguous buildings, usually two-story structures of brick or masonry, on at least one side of Main Street.

Percentage of Incorporated Places That Gained and Lost Population between 1990 and 2000, by Size of Place

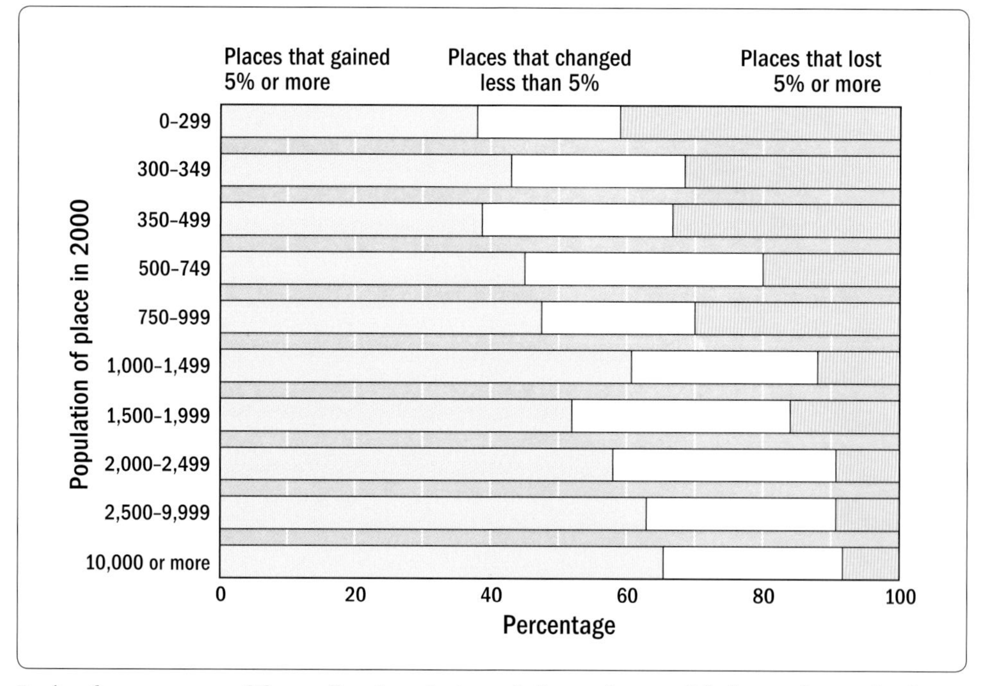

During the 1990s more of the smaller places lost population, and more of the larger places gained.

Minnesota's largest places have grown the most. The size of a town's population is related to the date at which it was incorporated, because the earliest places have had the longest time to grow. Most of the largest places in Minnesota had been incorporated by 1870 or 1880, and most of the smallest places were incorporated after 1890 or 1900.

Mean Population of Incorporated Places, 1900–2000, by Size in 1980

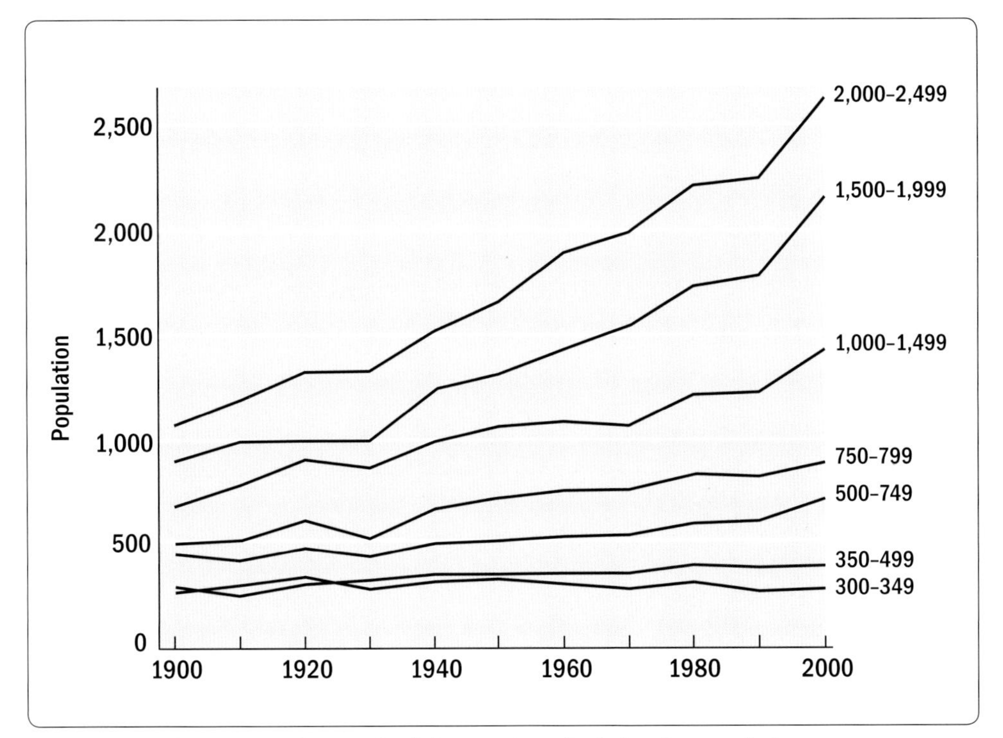

Places of more than 350 people grew slowly but consistently during the twentieth century.

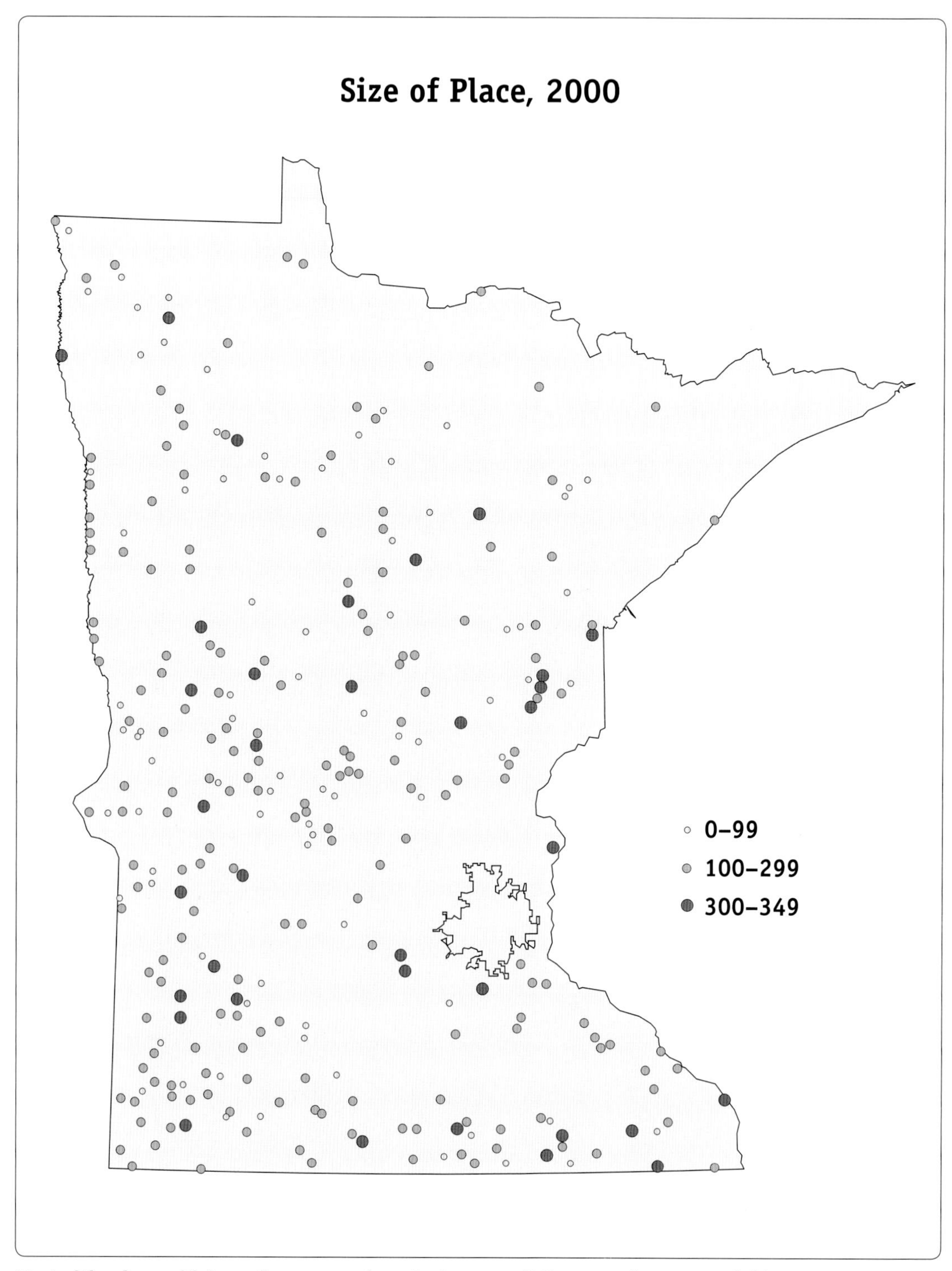

Most of the places with fewer than 350 people are in the parts of Minnesota that were settled last.

Most places of 350 people or more have a solid block of contiguous buildings on one side of Main Street.

In places of fewer than 350 people, Main Street has a gap-toothed look, with vacant lots between buildings.

Number of Freestanding Incorporated Places by Population Size and Decade of Incorporation

Population	1869 or earlier	1870–79	1880–89	1890–99	1900–09	1910 or later	Total in 2000
10,000 or more	14	14	2	3	—	—	33
2,500 to 9,999	10	33	16	17	7	5	88
1,000 to 2,499	6	14	41	18	14	10	103
750 to 999	1	4	17	13	12	6	53
500 to 749	2	13	17	25	19	10	86
250 to 499	—	8	13	41	47	30	139
6 to 249	—	5	5	43	73	100	226
Total	33	91	111	160	172	161	728

Minnesota's small-town settlement system reflects the fundamental geographic principle of *prior potior,* "the earlier one is the more influential." The basic pattern of places in the state had been established by the turn of the century and has hardly changed since then. The largest places in 1900 were still the largest places in 2000, the smallest places in 1900 were still small a century later, and two-thirds of the smallest places of today had not even been incorporated in 1900.

Half of the incorporated places in Minnesota, and 90 percent of those with a population of 2,500 or more, had already celebrated their centennial year by 2000, and more than three-quarters had been incorporated by 1910. At the other extreme, the places that have been incorporated since 1910 have remained small, with four-fifths still having fewer than 500 people in 2000.

Maps of small-town population change in the 1980s and 1990s illustrate the complex way in which individual places have grown. The 1980s were the toughest decade in history for small towns in Minnesota, and they were also the only decade in which geography predicted growth better than size.[4] Virtually all places that grew were within commuting distance of the Twin Cities and Rochester, in the lakeshore resort and retirement area north of Brainerd, or in the northern manufacturing area in Roseau County.

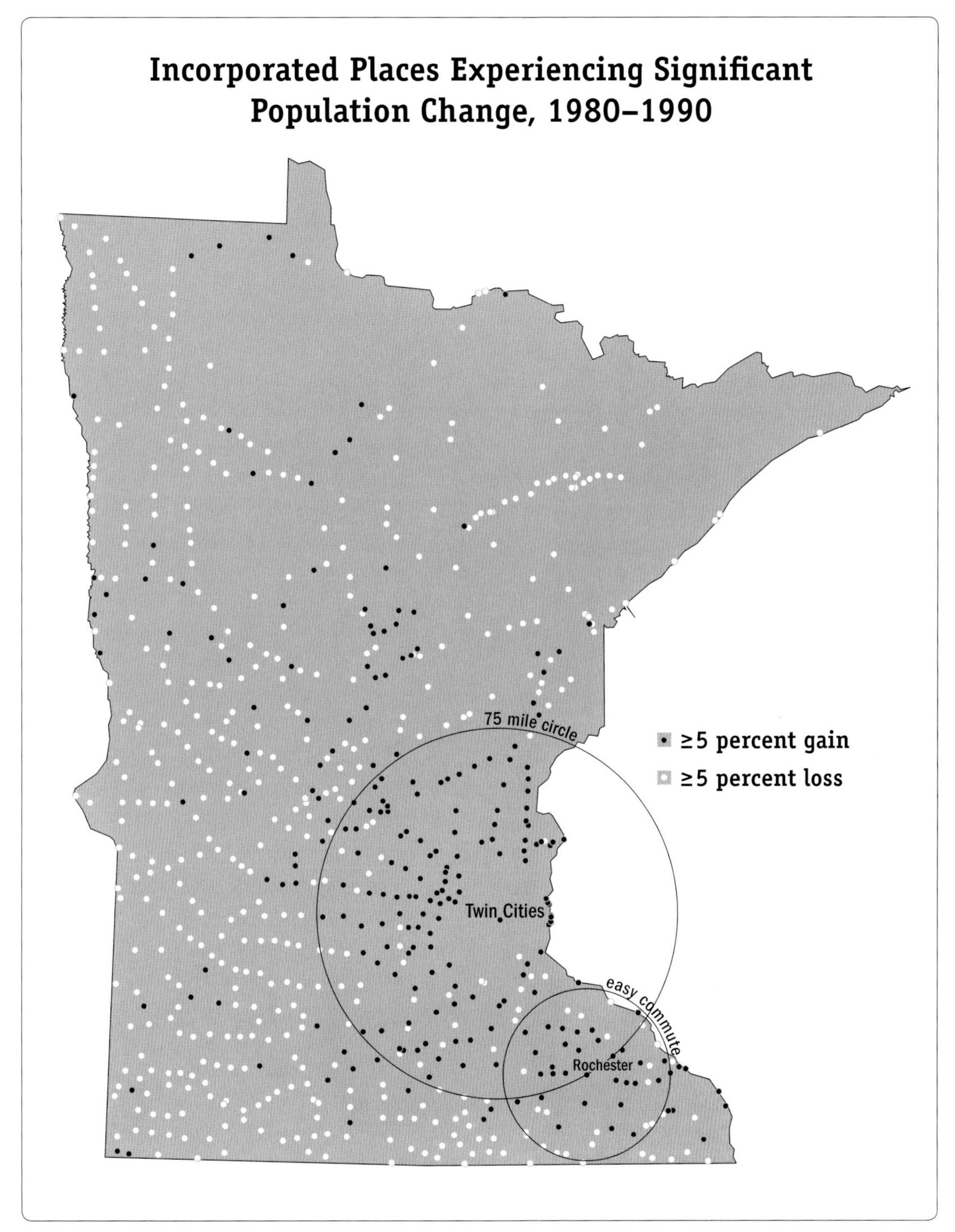

Most of the places that grew during the 1980s were within commuting distance of the Twin Cities or Rochester or in the northern lakeshore resort and retirement area.

During the 1990s, small-town growth was far more widespread, and the pattern of growth was much more complex and more typical of earlier decades.[5] Most places in metropolitan "commutersheds" and in the lakeshore resort and retirement area continued to grow. Many places also grew along major highways radiating from the Twin Cities, such as Interstate 35 north to Duluth, Interstate 94 northwest to St. Cloud, and U.S. 12 west to Willmar. U.S. 210 and 10 from Brainerd to Moorhead and U.S. 14 from Owatonna through Rochester to Winona were also corridors of growth, but many places that were not on major highways also grew. Small-town growth has no easy explanation.

Most of the small towns that lost population during the 1990s were in peripheral areas: in the northwestern corner of the state, on the Iron Range, in the southeastern corner, and southwest of the Minnesota River. These areas also had more than their share of places whose population changed only slightly during the decade. The railroad towns on the prairies of western Minnesota were too many, too small, and too late. They started too late to attain the critical size of around 350 persons that seems to ensure continued growth. Even on the prairies, however, some small towns grew during the 1990s, and like most parts of the state the prairies had a complex mixture of towns that gained, towns that lost, and towns that stagnated.

Small towns in Minnesota have grown because they have changed their function. Some have become bedroom or dormitory towns for workers who commute to jobs elsewhere, especially near the Twin Cities and Rochester, but even smaller cities have spawned dormitory towns, especially those near major highways that facilitate long-distance commuting. Other places have grown because they serve burgeoning resort and retirement areas, and these places probably will continue to flourish as the state's population ages.

Many small towns have grown because they have become minor manufacturing centers.[6] Although they originated as retail and service centers to serve agricultural areas, their prosperity is no longer tied to the agricultural areas around them. The Model T Ford killed Main Street, which has been dying a long, slow, lingering death since the 1920s. Many of Main Street's storefronts are boarded up because the retail and service function of small towns has been withering ever since the automobile enabled farmers and townspeople alike to travel greater distances to larger places in search of the goods and services they need.[7] Some people blame "big-box" stores, such as Wal-Mart, for this demise, but Main Street was killed by Henry Ford, not by Sam Walton.

Some people still cherish the romantic notion of Main Street as the heart and symbol of the small town, and even in many larger places downtowns have become

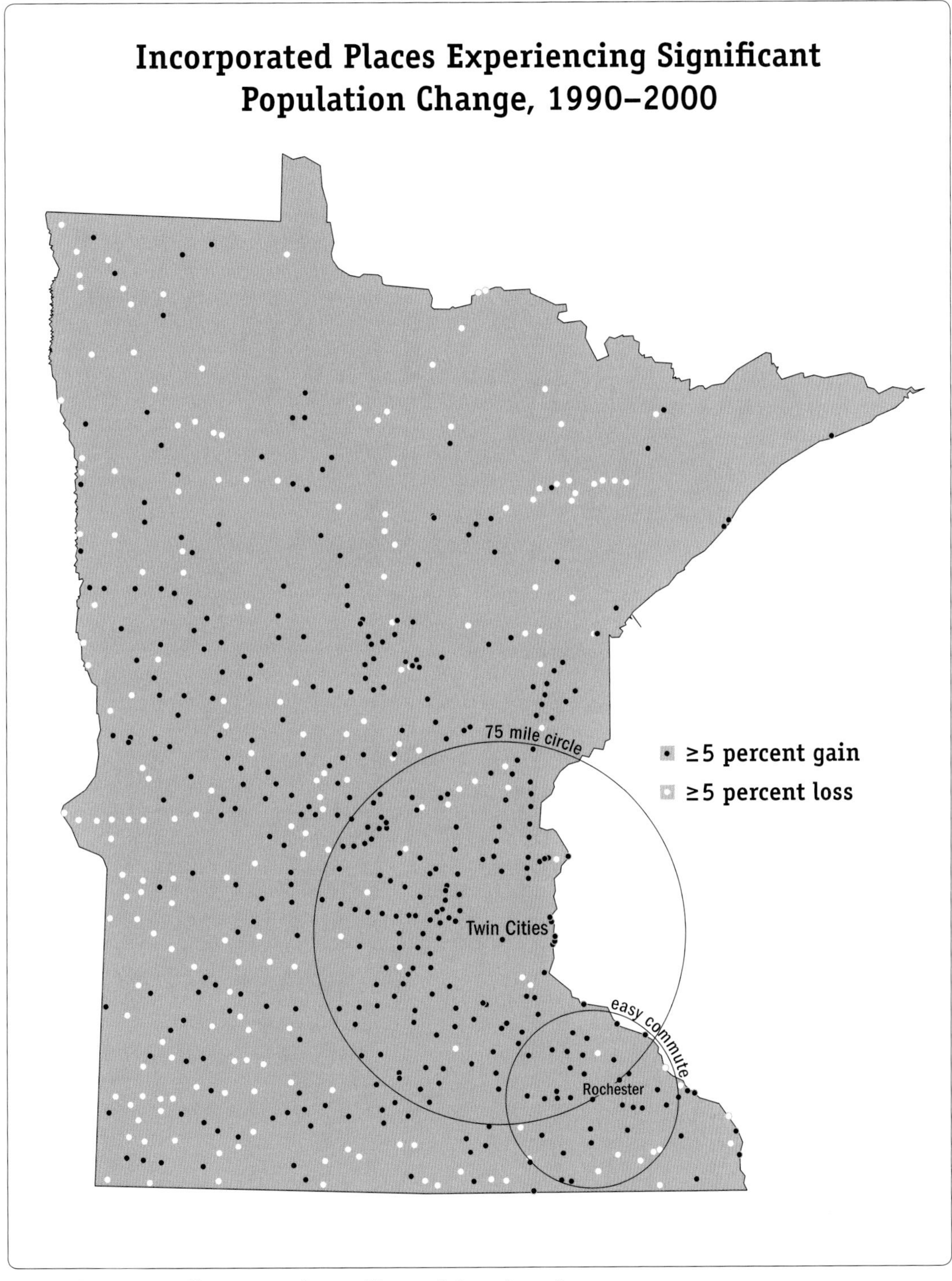

During the 1990s small-town growth was widespread throughout the state.

merely convenience shopping areas for those who work there. Most retail and other traditional downtown functions have moved out of town to the mall and to the bypass, which are more accessible and have easier parking. New stores, office buildings, shopping malls, motels, fast-food outlets, and big-box retailers are on the highway strip, and even in smaller places the new convenience stores are on the highway strip rather than on Main Street. Main Street's older buildings have been recycled for use by such low-rent businesses as video stores, fitness centers, and tanning salons, and some are even used as residences.

Since World War II, employment in manufacturing has been shifting away from the two largest cities in Minnesota. At the end of the war, it was heavily concentrated in the metropolitan area, except for paper mills in International Falls and Cloquet and the Hormel meat-packing plant in Austin. In 1950, a total of 187,000 people in the state were employed in manufacturing; 57 percent of them were in Hennepin and Ramsey counties alone, where 25 percent of the labor force was employed in manufacturing.

In 2000, manufacturing employed 419,271 people in Minnesota. The geographic distribution of manufacturing changed dramatically between 1950 and 2000. By 2000 the share of the labor force in the two Twin Cities counties employed in manufacturing had dropped from 25 percent to only 14 percent. This decrease masked the increased importance of manufacturing in the rest of the state, where the number of people employed in manufacturing increased from only 80,000 in 1950 to 297,000 in 2000. By sheer coincidence, the share of the state's labor force that was employed in manufacturing was an identical 16.3 percent in both census years, because the growth of manufacturing employment in southern and far northwestern Minnesota had neatly balanced its decline in the metropolitan area.

The greatest percentage increase in manufacturing employment between 1950 and 2000 was in a ring of overspill counties around the Twin Cities. The number of people employed in manufacturing also increased more than 600 percent near Rochester and Alexandria, and in the far northwest near Warroad (Marvin windows), Roseau (Polaris snowmobiles), and Thief River Falls (Arctic Cat snowmobiles). Employment in manufacturing increased more than 250 percent in most of the counties in southern Minnesota where manufacturing was not already a major employer.

A few counties with surprisingly high increases in manufacturing employment are a reminder that the census counts people where they live, not where they work. Some people commute considerable distances to manufacturing jobs in other counties, and some small towns that have no new factories have become dormitories for other towns that do.

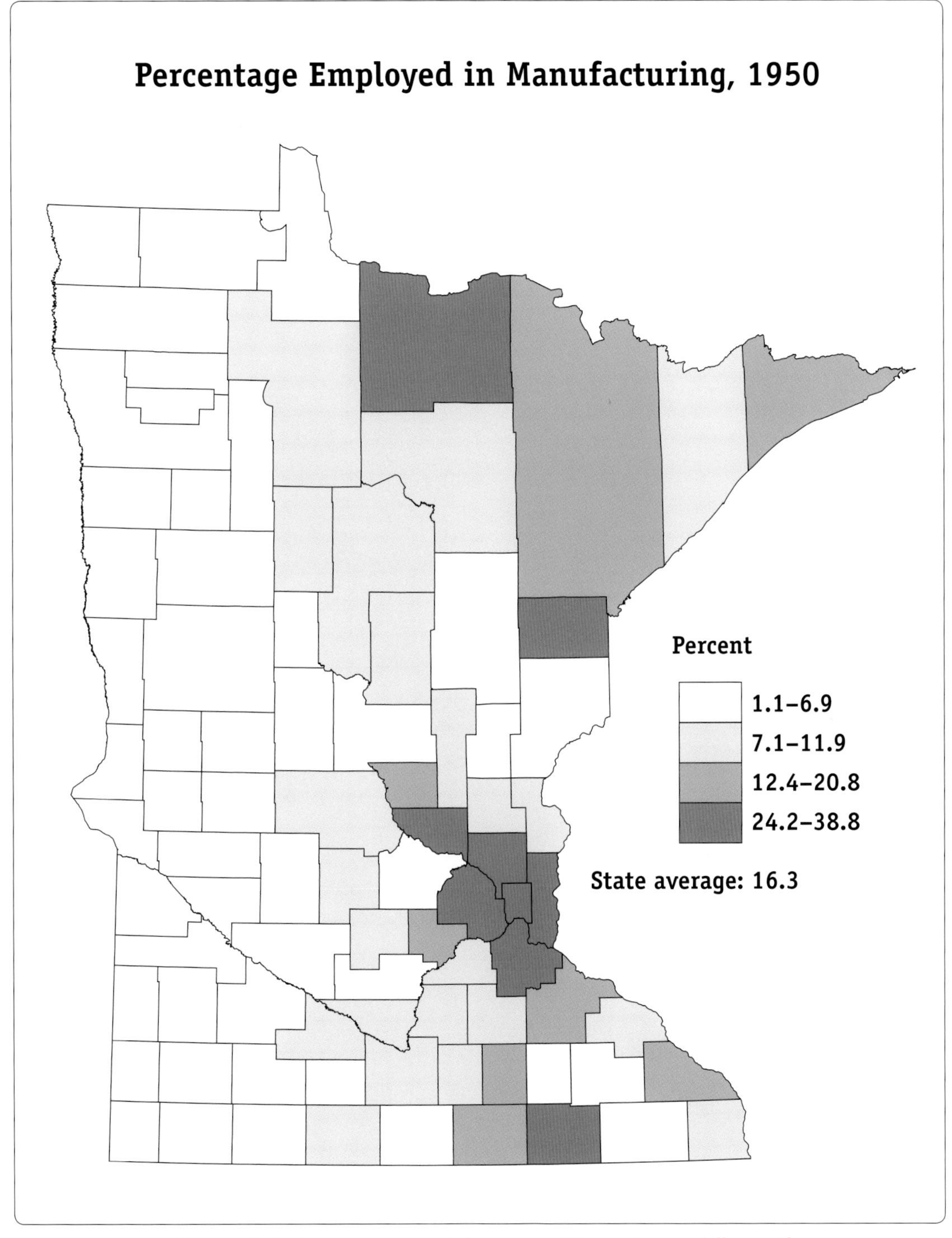

In 1950 employment in manufacturing was concentrated in eastern Minnesota, especially near the Twin Cities.

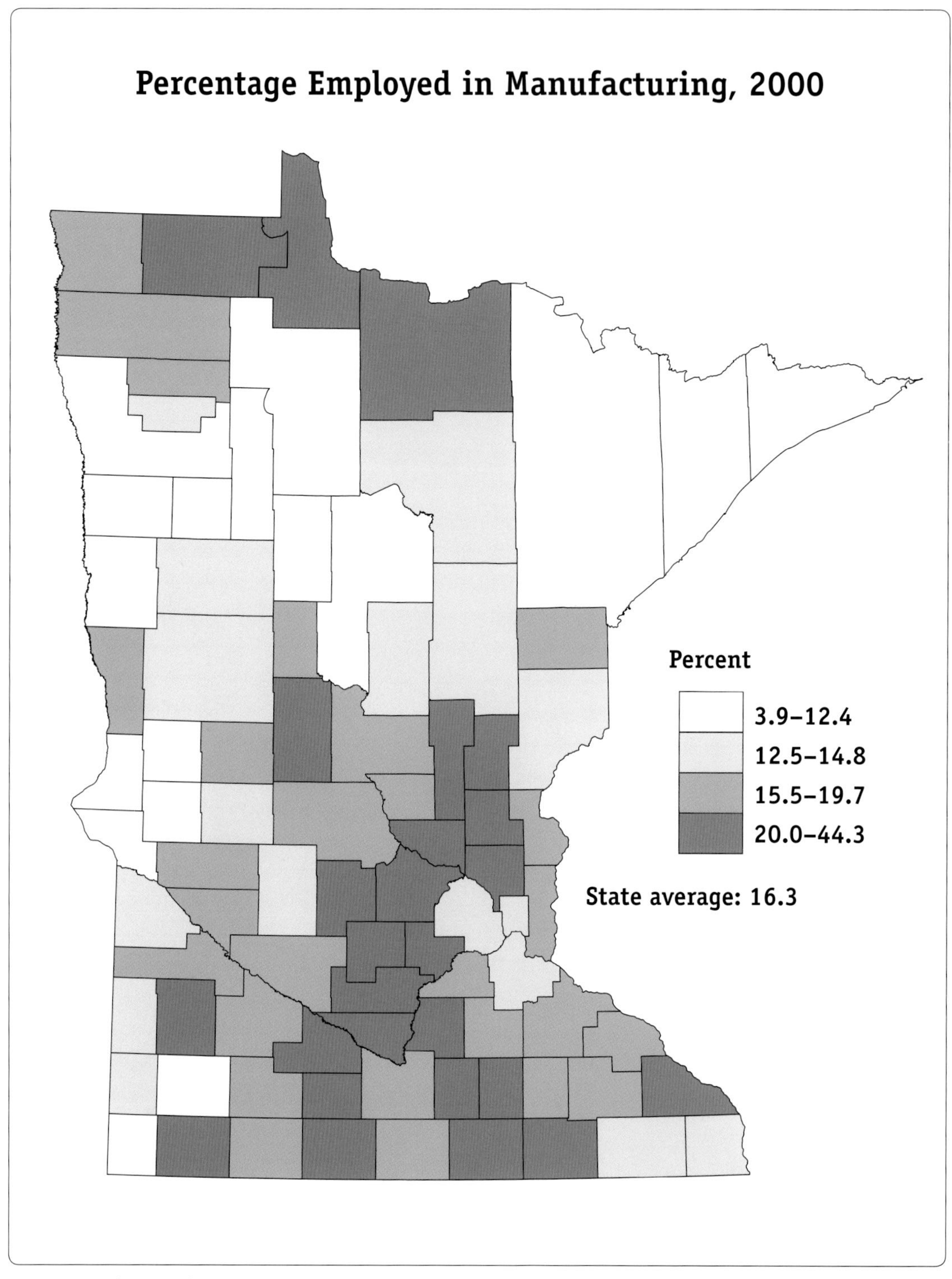

By 2000 employment in manufacturing had become more widely dispersed through Minnesota.

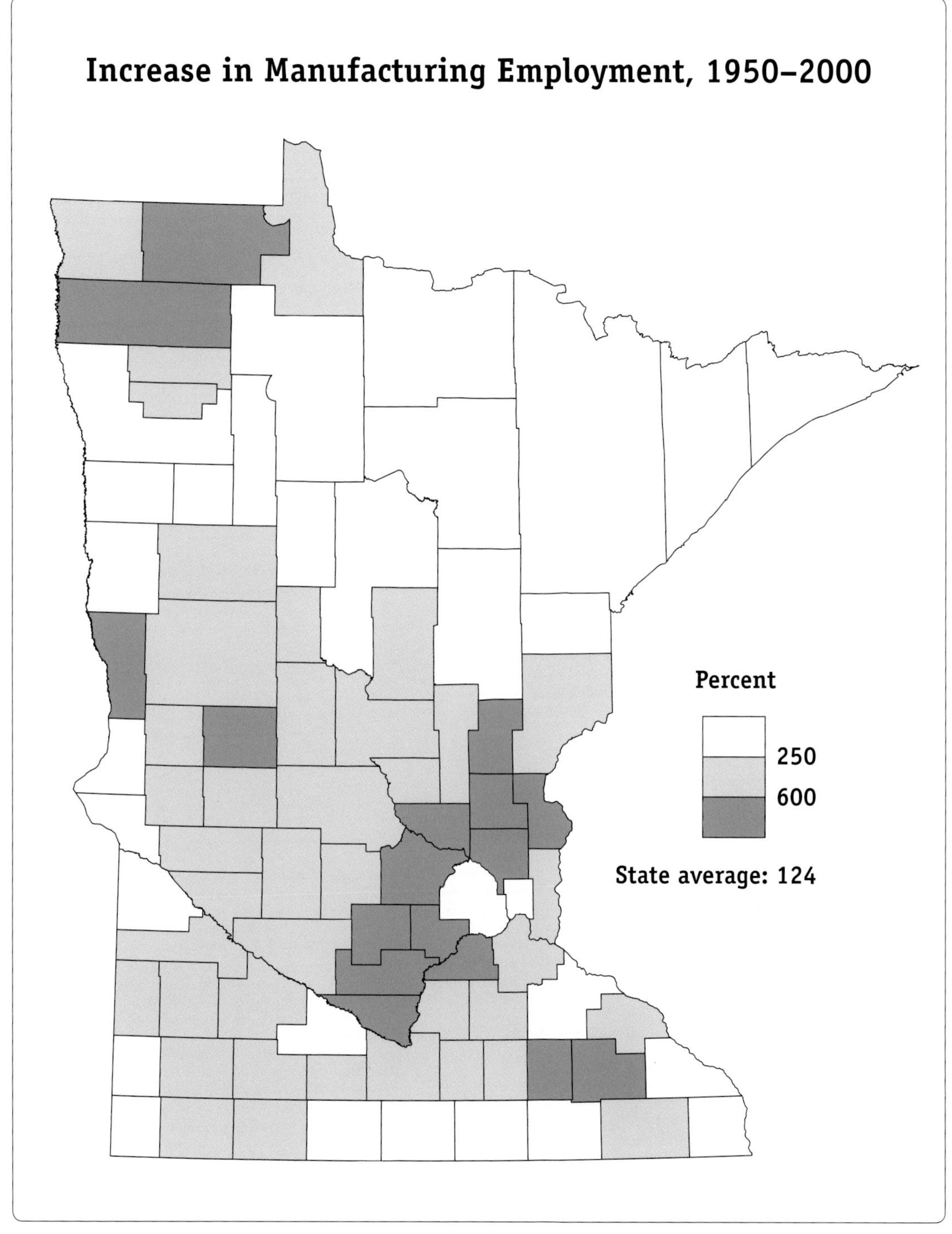

The greatest increase in manufacturing employment between 1950 and 2000 was in a ring of counties around the Twin Cities, but most of the counties of southern Minnesota enjoyed appreciable increases.

The new manufacturing plants in nonmetropolitan Minnesota produce a remarkably diverse range of products. One might expect that processing of agricultural commodities would be a major manufacturing activity in rural areas, as indeed it is, but many of the new factories have few ties to the local rural area other than the people they employ.

As one extreme example, in the late 1980s an entrepreneur set up fifteen sewing machines in an abandoned grocery store in Danvers (population 108 in 2000), which is about 150 miles west of the Twin Cities in Swift County. She hired local women to make seamless girdles for patients who had had liposuction surgery. The entrepreneur moved the operation to Mexico when the workers eventually complained that they were not being paid enough. However modest or bizarre the product, nearly anything made in a big city can also be made in a small town.

Some of the new small-town factories are homegrown, products of the fertile minds of local entrepreneurs, and some are branch plants of larger companies. In Alexandria in 2003, for example, the leading industrial employer was Douglas Machine, which Bud Thoen started in his garage in 1964. In 2003 it employed more than 600 people to make specialized packaging for companies around the world. The second-largest employer was a 3M branch plant that employed 300 people.

NO ONE KNOWS EXACTLY HOW MANY new factories have been started in small-town Minnesota, or how many are branch plants and how many are homegrown, but it is clear that some manufacturing firms have left the Twin Cities for smaller places within easy driving distance, and others have opened branch plants in such places in search of cheaper labor and cheaper space.[8] In the Twin Cities plants have to compete for a shrinking pool of increasingly expensive labor, but in smaller places they hope to have their pick of a labor force with a highly developed work ethic willing to accept lower wages.

Companies that relocate may be in trouble, however, because age-selective migration over the years has stripped rural areas and small towns in Minnesota of much of their labor force. In 2000, for example, the total employed labor force in eleven counties in northwestern Minnesota was only slightly larger than the labor force in Olmsted County, and the labor force in fourteen counties in southwestern Minnesota actually was 8 percent smaller than in Washington County alone.

Firms that locate or expand in such areas often have to recruit workers from other areas because the local labor force is inadequate. Where will the workers come from? Few of our towns and cities have managed to reproduce themselves, and their continued growth has depended on migrants from other areas. The population

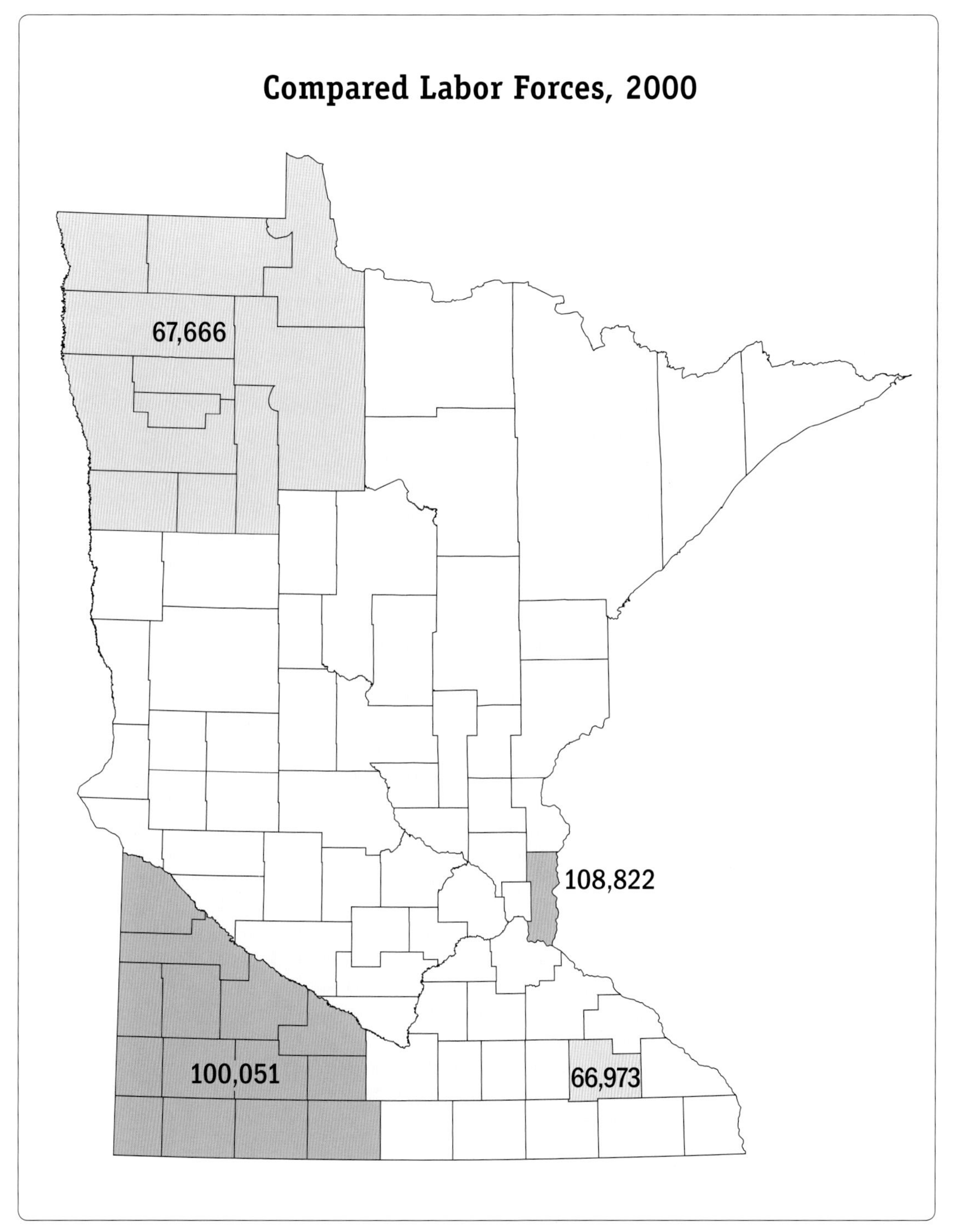

In 2000 the labor force in northwestern Minnesota was only slightly larger than the labor force in Olmsted County, and the labor force in southwestern Minnesota was smaller than the labor force in Washington County.

growth of the smallest places has been maintained by in-migration from the surrounding countryside, but the population of rural areas in western and southwestern Minnesota has been declining for half a century or more, and these areas can no longer replenish the population of the small towns that are not reproducing themselves.

Unless small town dwellers increase their birthrates, which seems unlikely, the future stability or growth of the population of most small towns in Minnesota will depend on their ability to attract new residents from distant areas, even from beyond the United States. For example, in 1990 a minimum-wage beef and chicken processing company in Madelia was forced to bus in twenty to forty new Latino workers from Texas each month to replace those who had quit, and a company in Marshall was offering on-site day care and childhood education programs in its attempt to recruit Southeast Asian workers from the Twin Cities.

The state of Minnesota has always depended on in-migrants for its growth. If continued population growth is desirable (and some people would argue that it is not), the challenge that faces not only small towns but the entire state is to attract new immigrants. These new immigrants will probably be Latinos and Asians, which may make some old-time residents uncomfortable, and Minnesota will have to recruit new residents in competition with intervening opportunities in coastal areas that are perceived as more attractive.

Some new Latino and Asian immigrants to Minnesota experience major social stress, if not downright hostility. This stress probably is most intense in small towns, where their numbers are small and daily face-to-face encounters between newcomers and natives are most numerous. Nevertheless, small towns in Minnesota will have to woo and hold immigrant workers if they hope to attract new industry, because towns no longer have large untapped reservoirs of inexpensive native-born labor. Without new workers, towns will not be able to attract the new employers they need.

Finding cheap floor space for new employers is not a problem. Most small towns have entirely too much of it in empty stores, schoolhouses, and other buildings that are no longer needed for the purposes they were originally built to serve. Some manufacturing firms have built new factories, to be sure, especially those that have established branch plants, but many, and especially those that are homegrown, initially prefer the cheap floor space that is available by recycling older buildings.

The recycling of old buildings, many on side or back streets, has disguised the importance of manufacturing in many small towns. Even in places with significant employment in manufacturing, a casual visitor may be deceived by the lack of shiny new plants and by the overgrown, weed-choked lots in the planned industrial park at the edge of town. Sometimes local people fail to recognize the change going on in their own backyard. (The postmistress in a town with a population of 251 in 2000, when asked about manufacturing activities in a converted schoolhouse, said, "Well, I wouldn't exactly call it a factory, but they are making guns and musical instruments in there.")

New industry can be a temporary Band-Aid rather than a panacea for small towns. Some firms will fail and some will leave. Too many firms have packed up and left after they have taken a town for all it had to offer. A firm that has moved once can just as easily move again, and "smokestack chasing," the mindless pursuit of new industry regardless of the cost, can milk a town and leave it high and dry.

Although encouraging homegrown industry is far better than trying to steal it from some other place, the long-term success of most small-town manufacturing remains questionable. Attracting a new industry might give a town the breathing spell it needs in which to ponder its future economic and social role. Low-skill, low-wage jobs may be better than no jobs at all, but they are not a formula for long-term prosperity. One might even question the wisdom of recruiting new industry to a place that then has to recruit new workers.

The former school building in Holt has been recycled into a taxidermy shop and woodworking plant.

Minnesota probably has too many small towns. One-third of the incorporated places in the state had fewer than 300 people in 2000. Even though they have continued to gain population slowly and erratically, they actually have fallen further behind because they have grown at a slower rate than the state and the nation. Unable to grow to a viable size, they are irrelevant to the needs of contemporary society and the contemporary economy. While they may linger on indefinitely, few can have much hope of significant growth.

These small places pose a major question of public policy. Certainly, we need to respect the wishes and tenacity of the people who wish to continue living in them, but we probably also need to ask whether we should continue further public or private investment in places that have little hope of significant future growth. Instead of continuing to subsidize them, perhaps we should subsidize people to move away.

Some romantics might fret that communities would be lost if small places were cashiered, but these places may never have been true communities. Residents of small places rarely use the term "community" unless academics from outside have taught them to do so. Of course, people in small places have ties with each other, but those ties often are less than affectionate. Negative attitudes may start in the schools, where the town kids poke fun at the country kids. Historically, Yankee businesspeople in the towns looked down on the immigrant farmers in the countryside, who may have spoken broken English, and the farmers suspected the townspeople of taking advantage of them.

Even larger towns in Minnesota must rethink their future. They will need to drop old competitive feelings and work together to create new "dispersed" cities in the countryside. The automobile, which killed Main Street, enables people to travel easily from place to place to find the things they need. They can live in one place, work in another, shop in a third, and go to a fourth when they are ill. No single place can expect to have everything, but each place can have a specialized function, just as the various parts of cities have their own specialized functions.

The future of small towns depends on their leaders. In some places people seem content to grumble about how terrible things are, but no one is prepared to accept the risk of change. All too often both the ranks and the minds of small-town leaders are closed. They must become more open to women, to young people, to newcomers. In the places that are growing today, civic leaders understand that change is inevitable, and they willingly accept new ideas. Small town leaders need to realize that their town is no longer an agricultural service center. Attempts to pump life into moribund Main Street probably are doomed, and the town must find a new and different function.

In the long run, a town is what its people make it. It has grown over the years because of the dedication and commitment of its individual citizens, their belief in their town, and their willingness to commit their time, money, and energy to helping it prosper. The decision to incorporate, which is closely related to the present size of the town, was a major act of civic commitment. The construction of a solid block of buildings on Main Street, the hallmark of towns that have continued to grow, was a major commitment by individual citizen investors. Attracting new industry and encouraging local industrial entrepreneurs requires another major commitment, and civic leaders must think clearly about how the function of their town has changed and what they want it to become.

Kittson
Roseau
Lake of the Woods
Marshall
Beltrami
Koochiching
Pennington
Red Lake
Polk
Clearwater
Itasca
Norman
Mahnomen
St. Louis
Lake
Cook
Hubbard
Cass
Clay
Becker
Wadena
Aitkin
Carlton
Wilkin
Otter Tail
Crow Wing
Pine
Todd
Grant
Douglas
Morrison
Mille Lacs
Kanabec
Traverse
Stevens
Pope
Benton
Big Stone
Stearns
Sherburne
Isanti
Chisago
Swift
Anoka
Kandiyohi
Meeker
Wright
Washington
Lac Qui Parle
Chippewa
Hennepin
Ramsey
Yellow Medicine
Renville
McLeod
Carver
Scott
Dakota
Sibley
Lyon
Redwood
Nicollet
Le Sueur
Rice
Goodhue
Wabasha
Brown
Murray
Cottonwood
Watonwan
Blue Earth
Waseca
Steele
Dodge
Olmsted
Winona
Rock
Nobles
Jackson
Martin
Faribault
Freeborn
Mower
Fillmore
Houston
1. Tower
2. Ely
3. Grand Rapids
4. Virginia
5. Crosby
6. Duluth
7. Lake Superior
8. Lake Vermilion
9. Winston
10. Two Harbors
11. Soudan
12. Boundary Waters Canoe Area Wilderness
13. Mountain Iron
14. Biwabik
15. Eveleth
16. Cloquet
17. Hibbing
18. Calumet
19. Silver Bay
20. Hoyt Lake
21. Chisholm
22. Trommald
23. Riverton
24. Manganese

12

The Iron Ranges

Minnesota's iron ranges are a world unto themselves. Long, empty miles of coniferous forest isolate them from the rest of the state. They were the last part of the state to be occupied, and immigrants from different areas settled them. The ranges have more Slavs and Finns than the rest of Minnesota and fewer people of British, German, or Scandinavian ancestry. The ranges are transitory, even though they have been occupied for more than a century, because all mining areas are transitory. Some survive longer than others, but all will eventually be abandoned or completely made over when the last shovelful of mineral has been taken from the ground.[1]

Minnesota has three iron ranges. The Vermilion Range extends twenty-five miles from Tower to Ely and had ore that was 63 to 70 percent iron. The Mesabi Range runs more than one hundred miles from Grand Rapids to Babbitt, with a great elbow south of Virginia, and had ore that was 55 to 60 percent iron. The Cuyuna Range is entirely in eastern Crow Wing County, near Crosby. Its ore ranged from 45 to 63 percent iron, but it was rich in manganese, which is an essential mineral for making steel. These three ranges have produced more than 3 billion tons of iron ore, an almost incomprehensible amount.

Prospectors and engineers had long suspected that the tough old rocks that lay beneath the trackless wilderness north of Duluth might contain valuable deposits of metallic ores. They saw similar rocks south of Lake Superior in northern Michigan, where Indians had found almost pure copper on the Keeweenaw Peninsula, and at Marquette, Michigan, where a mine had begun producing iron ore in 1854.

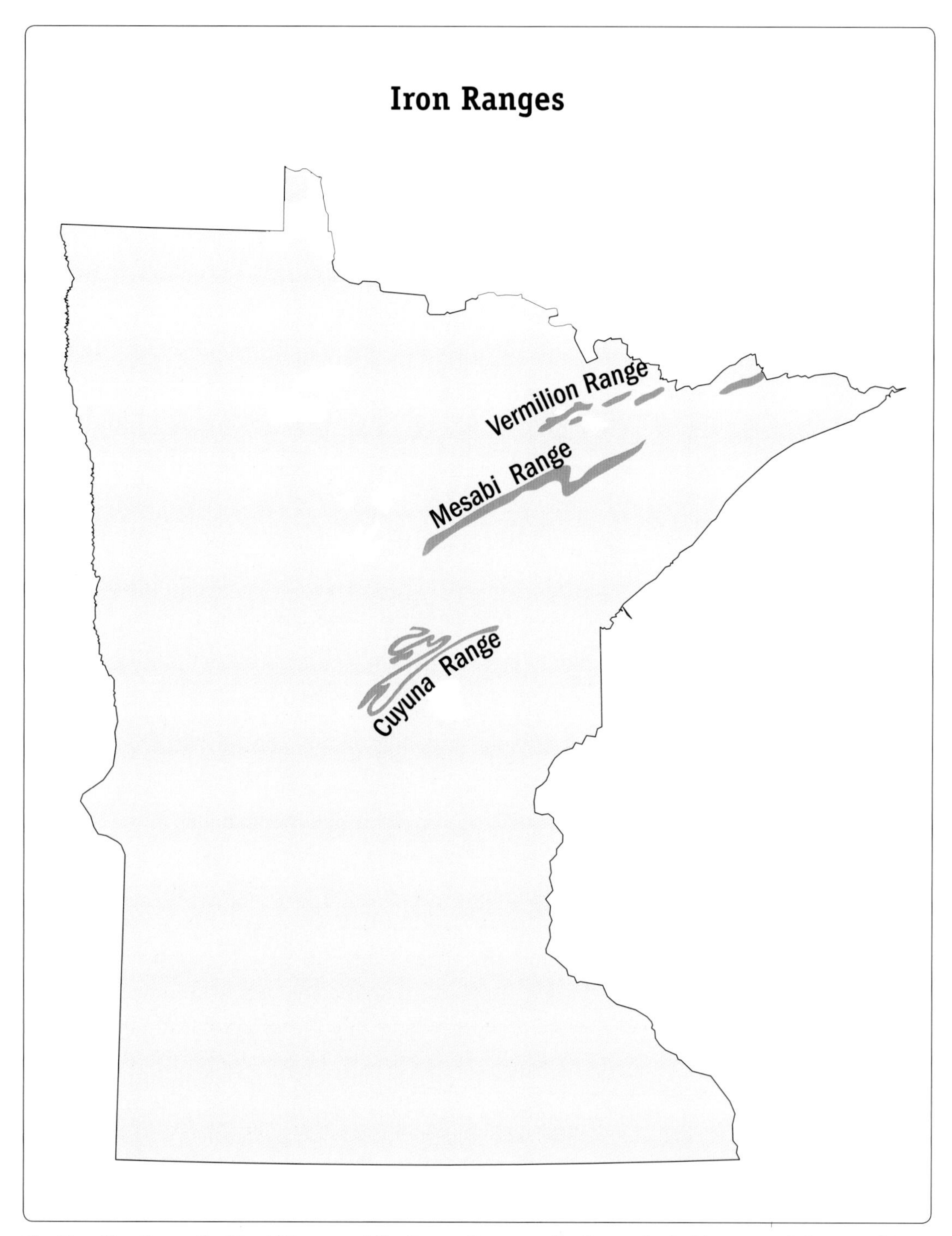

The Vermilion Range, the Mesabi Range, and the Cuyuna Range are the three principal iron-ore mining areas in Minnesota.

Initially, Minnesota prospectors and engineers focused on copper, gold, and silver. Although they suspected that large areas were underlain by highly magnetic iron because their magnetic compasses behaved so strangely in these areas, they were uninterested. In 1865 state geologist H. H. Eames described the iron ore deposit near Lake Vermilion but is reported to have snorted, "To hell with iron. It's gold we're after."

During the winter of 1865, Minnesota enjoyed its first gold rush. Samples of quartz said to have been taken from rocks near Lake Vermilion were sent to an assay office and tested out rich in gold. Hundreds of people swarmed to the area in freezing cold and sank shafts into quartz veins, but they found nothing. Eventually they realized that no one actually knew the source of the samples that had been assayed, and they suspected that some dastard had "salted" them by blasting them with a shotgun loaded with pellets of gold instead of buckshot. The town site of Winston on Lake Vermilion became Minnesota's first mining ghost town.

The first iron mines in Minnesota opened on the Vermilion Range. George C. Stone, who settled in Duluth in 1869, became convinced that the ore at Lake Vermilion was worth mining. He sold this idea to Charlemagne Tower, a wealthy lawyer in Pottsville, Pennsylvania, who agreed to finance a mining operation after a survey in 1880 proved the existence of a valuable ore body.

Tower began buying up land in the area. By 1883 he owned 17,000 acres. He was reluctant to start mining until he knew how he would be taxed, so Stone got himself elected to the Minnesota legislature and sponsored a law exempting iron ore from all taxes except one cent for each ton mined and shipped. He also secured a liberal land grant for construction of the railroad necessary to haul in equipment and to haul out the ore.

At the time Lake Vermilion was almost inaccessible. Long stretches of swamp were nearly impassible in summer, and everything—machinery, supplies, food, even miners—had to be hauled in by sled across the frozen ground in winter. In summer all supplies had to be carried on the backs of men or by Indians in canoes over a circuitous route. In the winter of 1884 the railroad was completed from Two Harbors on Lake Superior to the new town of Tower. It actually crossed the Mesabi Range at Mesaba, but no one thought the ore there was worth mining.

The first mine at Tower was an open pit at Soudan. The miners dug the ore with picks, shoveled it into wheelbarrows, and pushed them to the railroad cars. As the pits grew deeper, they shoveled the ore into buckets that were hoisted to the surface by windlasses, but eventually, because the ore body is nearly vertical, the ore had to be mined from deep underground shafts.

As soon as word of the Soudan mine got out, people began rushing in to the Vermilion Range and battling viciously for rights to land that might contain iron ore. Hundreds of people filed claims, taking advantage of every possible law by which land could be alienated from the public domain, and Tower himself had not set an admirable example. Inevitably some of these claims overlapped.

Section 30, three miles east of Ely, became notorious. The rock looked just like the rock that had produced ore at Soudan, and in 1886, thirty-one claims totaling 2,180 acres had been filed in this section, although it only contained 640 acres. Lawsuits over these overlapping claims slowly worked their way up through the courts until 1902, when the U.S. Supreme Court finally ruled, after more than a million dollars of litigation, that the local land office adjudication of claims in 1886 had been correct all along. No iron ore has ever been found in section 30.

At the other extreme was the "sliver mine" on the Mesabi Range. In 1908 a lawyer realized that the boundaries of two adjacent mines did not quite touch. He filed a claim on the sliver of land between these two mines, and the royalties from it made him a millionaire.

The Soudan mine was followed by several new mines near Ely in 1888. Within a decade the Vermilion Range, which had shipped 62,124 tons of ore in 1884, was shipping more than 1 million tons a year. The mines continued to go deeper and deeper, until they were no longer economical, and the last Vermilion mine closed in 1967. Ely has evolved from a mining camp into a major tourist outfitting center because it is the gateway to the Boundary Waters Canoe Area Wilderness, the largest wilderness area in the United States east of the Rocky Mountains, where mining, logging, and most uses of motorized vehicles are not permitted.

In 1962 the United States Steel Corporation gave the original Soudan mine to the state of Minnesota on the condition that it be made the centerpiece of a state park. Visitors can take the three-minute elevator ride down 2,400 feet to see what the mine was like when it was operational. A deep part of the mine has been converted into a huge underground laboratory to capture subatomic particles called neutrinos that are shot from the Fermilab near Chicago, 450 miles away. The dense rock above the underground lab shields it from the high-energy interference that is common at the earth's surface.

MINNESOTA'S MESABI RANGE was not mined until eight years after the Vermilion, but it quickly became the state's leading iron range. For many people today, it is simply "The Range."

The seven Merritt brothers of Duluth deserve credit for having opened up the first Mesabi mine. Other prospectors, including Frank Hibbing and the Longyears, had been sniffing around the area, but the Merritts had been prospecting it persistently for fifteen years before they finally found ore worth mining.

Leonidas Merritt would strap a hundred-pound pack on his back and hike seventy-five miles north from Duluth through an area that had no trails, staying for thirty days at a time. He claimed that he had mapped the entire range by crisscrossing it on foot and measuring the magnetic lines with a dip needle. Like other prospectors, he was looking for an ore body of hard rock that would be mined from deep underground shafts, and the idea that he was actually standing on iron ore did not occur to him.

The iron ore of the Mesabi Range was at the surface or was covered only by a thin veneer of shallow glacial deposits.[2] The ore formation was nearly horizontal, but it dipped slightly to the south toward Lake Superior. The ore was not a long, continuous body. It consisted of a string of many separate pockets, of irregular size and shape, that were embedded in tough and less valuable taconite. The individual ore pockets were 200 to 500 feet deep.

When it was originally formed, the ore was taconite, containing only 25 to 30 percent iron. It was rich in silica, a highly undesirable impurity, but at some time in the geologic past this silica was leached out by oxygen-rich alkaline groundwater, leaving rock that was 55 to 60 percent iron. This rock was so "soft" and crumbly that miners could dig it right out of the ground or easily blast it loose.

Each of the separate ore pockets of the Mesabi Range had to be discovered independently, and in the early days many were discovered only when an accident exposed the rich red dirt. It clung to the roots of trees blown down by strong winds. It stained the water that flooded a prospector's firepit after a heavy rain. The Merritts made the first discovery on the entire range at Mountain Iron in 1890 in the ruts cut by the wheels of a heavy wagon, but soon other pockets were discovered at Biwabik, Eveleth, Hibbing, and many other places.

Iron ore, when it comes from the ground, is simply rock. It must be smelted in a blast furnace to remove its impurities and convert it into raw pig iron. A blast furnace needs four tons of coal for each ton of iron ore, so it makes sense to transport iron ore to coal rather than vice versa. Because the principal coalfields of the United States are in western Pennsylvania and northeastern Ohio, Mesabi ore had to be hauled to ports on Lake Superior and loaded onto the ore boats that would carry it to ports on Lake Erie. (In 1915 the United States Steel Corporation finally yielded to the importunities of Duluth politicians and built two blast furnaces and a steel mill in Morgan Park, a planned suburb. This plant was a high-cost producer, it was too remote from the principal markets for its products, and it made so little sense geographically that it had to be closed in 1971.)

To haul their ore the Merritts built a railroad line south from Mountain Iron along the present route of State Road 7 to Brookston, thirteen miles north of Cloquet, where it tied in to a railroad with ore docks in Superior, Wisconsin. In 1892 they shipped ore over the docks at Superior, and in 1893 they built their own railroad to ore docks they had constructed in Duluth.

Ore docks are gigantic railroad trestles that extend out for hundreds of yards into the deep waters of the lake. Ore boats tie up alongside the trestle, and trains of loaded ore cars are backed out onto it. When the cars' hopper bottoms are opened, gravity carries the ore through chutes into the holds of the waiting ore boat.

The Merritts were heroic prospectors, but they were not good businessmen, and they did not have the enormous amounts of money they needed to develop their mines and to build railroads, ore docks, and ore boats. In 1893 they borrowed money from John D. Rockefeller to build their railroad, but they made the mistake of tangling with him instead of working with him. Within a year Rockefeller had taken control of their mining properties, their railroad, and everything else they owned.

The iron ore of the Mesabi attracted other magnates. Henry W. Oliver of Pittsburgh, who had grown wealthy making farm machinery, visited the Range in 1892 and decided to form the Oliver Mining Company. He sold a half interest in his company to his fellow steelmaker Andrew Carnegie. They were so concerned that Rockefeller might try to compete with them that in 1896 they leased all his mines and agreed to ship all their ore on his railroads and fleet of ore boats.

In 1901 J. Pierpont Morgan, the famous New York banker, combined the Rockefeller and Carnegie interests to form the United States Steel Corporation. It assigned the Oliver Mining Company responsibility for managing its operations on the Range and continued to buy many small mining properties. It was worried about possible competition from James J. Hill of St. Paul. In 1899 Hill had added to his empire a small railroad on the Range that owned extensive ore land. In 1907 the corporation leased Hill's ore land, in which he was not particularly interested, but he drove a hard bargain and got a very good price for it.

Some of the early mines on the Range were underground, but most were vast open pits that kept getting bigger and deeper as larger and more powerful machinery was developed. Oliver laid railroad lines right into the open pits, where enormous power shovels tore tons of ore from the ground in a single bite and loaded it directly into the waiting railroad cars. In 1892 the Merritts had proudly shipped 4,245 tons of iron ore from Superior. In 1906 the United States Steel Corporation shipped 27,492,949 tons.

In the early days most of the workers in the mines were young, single men. They had to be strong, but they did not need to know how to read or to speak English, and men of more than forty nationalities, mostly from eastern and southern Europe, were employed on the Range. They lived in boardinghouses, and the bars that proliferated in all Range towns were their social clubs and their anodyne.[3] Many bars served pasties, the traditional dish of the Range, which consist of coarse ground beef and pork, potatoes, carrots, onions, and the essential rutabagas wrapped in pastry dough. They have the lumpy look of half-deflated footballs.

Work in the mines was dirty and dangerous. The mining companies paid the workers low wages and drove them hard, with working days up to fourteen hours. The Finns, who had brought with them advanced socialist ideas, organized labor unions as well as cooperatives. The first strike was in 1907, but the unions did not secure their first company contract until 1943. The Range remains a stronghold of labor unions and the Democratic Party. Politics on the Range is a blood sport, and Range politicians have become adept at ensuring that their constituents get their fair share, plus a bit more, of the public trough, both state and national.

MOST TOWNS ON THE RANGE started as "locations," where the mining companies provided housing for their workers. None has grown large enough to become dominant, although Hibbing has become the largest place in the west, and Virginia in the east. The smallest places still are mere locations, with few urban services except for the ubiquitous bars. They have an air of impermanence, and much of the housing is unpretentious, although towns such as Hibbing and Virginia have historic houses that architects admire greatly. All housing on the Range has to be solid against the bitter winters, and much of it is well maintained.

Hibbing nicely illustrates three distinctive features of Iron Range towns. First, the original town had to be moved because it sat on a valuable deposit of iron ore. In 1918 the Oliver Mining Company laid out a completely new town a mile to the south and moved all the houses and buildings to the new town. The bit of the old town that still remains has city blocks, streets with curbs, and street signs, but not a single building. During the move, an enterprising young Hupmobile auto owner made money by charging people to taxi them back and forth between the old and new towns, and from this slender start he developed the Greyhound Bus Company, which has a museum in Hibbing.

Second, like most Range towns, Hibbing used tax revenues from the iron mines to spend lavishly on schools, libraries, municipal services, and other public improvements. Hibbing High School, for example, is one of the most monumental structures in Minnesota. In 1921 it cost $4 million, the equivalent of more than $75 million in today's dollars. Its auditorium has imported cut-glass chandeliers, a stage large enough to hold a symphony orchestra, and 1,800 velvet-covered seats. Many people on the Range, unfortunately, have become accustomed to this now declining source of revenue, and they are reluctant to tax themselves to provide and maintain necessary public services.

Third, Hibbing has overextended its municipal boundaries to include valuable mining properties that increase its tax base. The city limits of Hibbing enclose an area nearly double the area of Minneapolis and St. Paul combined. Much of the incorporated area of Hibbing is rural, and like much of the rural landscape of the entire Iron Range it is pockmarked with deep abandoned mine pits and studded with giant flat-topped mesas of carefully contoured mine waste. The bottoms of the pits are below the groundwater table. They had to be pumped out regularly when the mine was working, and they have been flooded and become lakes since it was abandoned.

The Iron Range has overlooks with fine views of former mine pits. Viewpoint in the Sky, south of Virginia, was originally a vantage point from which pit foremen could survey mine operations that nearly undercut some of the city's streets. The municipal park north of Mountain Iron has a distant view of the massive Minntac plant and a well-preserved 1910 railroad switching locomotive that doubles in brass as a jungle gym. The Hill Annex Mine State Park in Calumet has bus tours of the pit, pontoon boat trips on the lake, and tours for fossil hunters.

The granddaddy of them all is the Hull Rust Mahoning (HRM) mine view north of Hibbing, which overlooks the world's largest open-pit iron ore mine, one of the most awesome human creations on the face of the earth. This site is managed by volunteers who once worked in the mine and who are delighted to regale visitors with their tales. It overlooks a pit in which more than thirty individual mining properties have been combined into a single pit 3.5 miles long, up to 2 miles wide, and 600 feet deep, covering 3,075 acres, or nearly five square miles.[4]

Hibbtac is still mining taconite in the pit, which is about as close as you can get to a modern mine operation. The mine works five fifty-foot-high benches. Rotary drills set back from the edge of each bench bore a line of parallel holes to the level of the bench below. These holes are filled with explosives, and most Wednesdays around noon the whole side of the bench is blown away in a spectacular blast. Power shovels lumber in, scoop up the ore, and in three or four sweeps fill the 240-ton production trucks that haul it to the Hibbtac plant on the far side of the pit. From the mine overlook, the trucks look like little toys scuttling around in the pit, but their tires are twelve feet high.

Iron ore mined from the pits of the Mesabi was used to make the guns, tanks, planes, and ships that won World War II, but by the end of the war nearly all the high-quality ore had been taken out of the ground, and American steelmakers thought they were going to have to depend on imported ore. The U.S. Steel Corporation built its new state-of-the-art postwar plant on the East Coast near Philadelphia, and the iron and steel industry in the Midwest, which had bitterly opposed the St. Lawrence Seaway because it could be used to import iron ore that would compete with Mesabi ore, suddenly decided that building it was a dandy idea.

The future for the Range looked bleak, but fortunately an engineer at the University of Minnesota named E. W. Davis had been working since 1913 to figure out a way to use taconite as iron ore. At first the hard-bitten miners dismissed him as an academic crackpot, but they began to pay more attention to him when they started to run out of high-quality natural ore, and eventually they adopted his process.

Groundwater has flooded part of the Hull Rust Mahoning iron ore mine north of Hibbing.

The Hull Rust Mahoning mine north of Hibbing is one of the most awe-inspiring human creations on the face of the earth.

The mine trucks are so huge that they can run over and crush a passenger automobile without even knowing that they had hit it.

Davis knew that a thick layer of taconite lies beneath the entire Mesabi Range and slopes gently southward beneath Lake Superior. It is the rock from which the high-quality Mesabi ore was formed when geologic processes leached out its silica, and it would be a virtually inexhaustible source of iron ore if only there were some economical way of mining it and removing its silica.

Taconite is one of the hardest rocks mined anywhere in the world. Some forms of taconite are so hard, in fact, that Native Americans used the rock to make arrowheads and projectile points. Extracting this hard material required the development of a drilling technique called "jet piercing." The jets use an incandescent flame of kerosene and oxygen to pierce a precise pattern of drill holes fifty feet into the ground. Then the holes are charged with a mixture of ammonium nitrate and fuel oil, and the whole side of the mine is blasted loose.

Enormous trucks haul the rock to the processing plant, which is a labyrinthine lattice of monster machines on steel decks connected by steel stairs. These machines crush and grind the rock until it is as fine as face powder.[5] This powder goes to separators, where powerful magnets extract the tiny particles of magnetite, and the silica is discarded as tailings. The powdery magnetite is mixed with a clay binder and rolled into marble-sized pellets, which then are baked in rotary kilns at temperatures up to 2,400° F. These pellets, which are about 65 percent iron, are expensive because processing taconite is brutally tough on equipment, but they work better than natural ore in blast furnaces because of their uniform shape and iron content.

Giant revolving drums crush taconite rock to fine powder, which is then pelletized in rotary kilns.

Three tons of taconite produce one ton of pellets and two tons of powdery tailings. A slurry of tailings is pumped into a vast tailings basin, which covers a square mile or more. The fine tailings eventually settle out, and the water is recycled through the processing plant. The tailings basin must be protected by vegetation to keep the dust from blowing. Tailings basins turn out to be good wildlife areas, because they are not accessible to the public, including hunters, although poaching can be a problem.

The first taconite plant, which was built at Silver Bay in 1956 by the Reserve Mining Company, was unusual because it hauled ore forty-seven miles by train from the mine at Babbitt. All other plants are so close to their mines that they haul ore by truck. The Silver Bay plant attracted much unfavorable publicity after it dumped 67,000 tons of tailings a day straight into Lake Superior, assuming that they would settle to the deep bottom of the lake.

Most did, but the very finest particles, which might be carcinogenic, were carried by currents along the lakeshore toward Duluth, and Reserve had to build a tailings basin seven miles inland, to which it pumped tailings slurry through a pipeline. The company went bankrupt in 1986, but the operation was reopened in 1990 as Northshore Mining.

A second plant was built at Hoyt Lake in 1957, but taconite plants were so expensive that companies were reluctant to make the enormous investment necessary until they could be sure that they would be taxed equitably. In 1964 the citizens of Minnesota voted to amend the state constitution to guarantee tax stability for taconite plants by giving them the same tax status as other forms of manufacturing, instead of taxing them as mining operations.

Five new taconite plants were built almost immediately after the constitution was amended, but only one new plant has been built since then. The taconite industry is extremely volatile, and its fortunes rise and fall like a yo-yo. The Butler plant closed in 1986, and the Hoyt Lake plant in 2001. Production peaked at 54 million tons in 1979, plummeted to 23 million tons in 1982, and ran around 42 million tons from 1990 to 2000 before dropping to 31 million tons in 2001.

The industry seems to be in a constant state of restructuring. The names of companies and owners change regularly, and the multiple ownership of some plants complicates decision making. In 2003, for example, Evtac declared bankruptcy, was bought jointly by a Chinese steel company and an American steel company, and was renamed United Taconite.

The future of the taconite industry does not look auspicious.[6] Expensive pellets must compete with cheaper foreign ore, and the entire steel industry is shifting

Taconite Plants, 2004

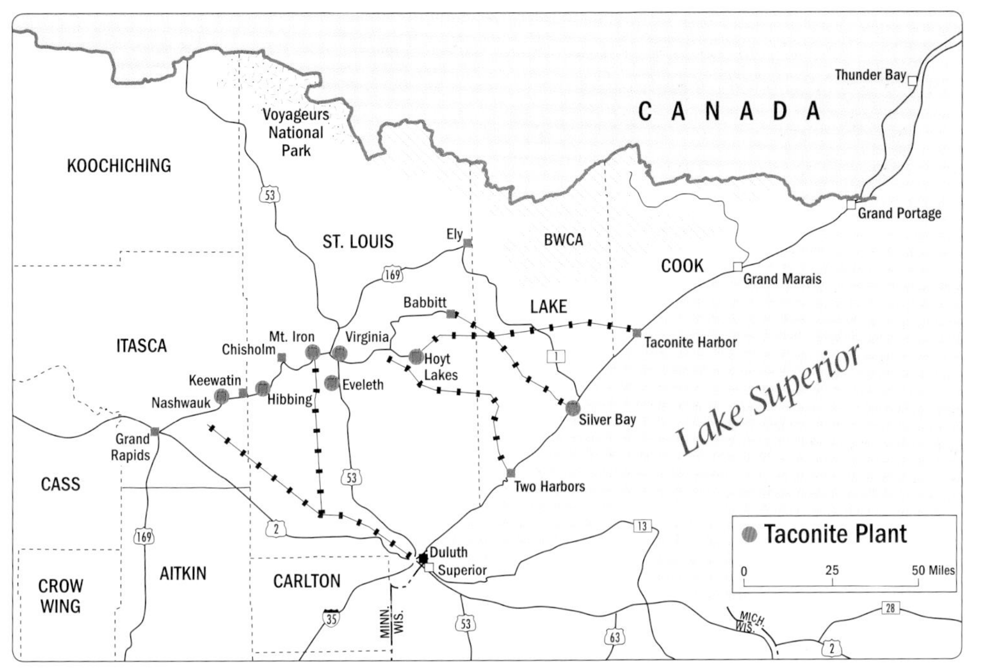

The number of taconite plants in Minnesota has fluctuated with the fortunes of the industry.

Taconite processing plants are enormous.

from massive blast furnaces to flexible mini-mills that use scrap rather than iron ore. Workers on the Range seem to recognize the gravity of the situation. They have never trusted the out-of-town companies that own the mines and plants, but their unions are beginning to cooperate with management to cut costs.

Despite these problems, the people on the Range have a dogged optimism that often verges on wishful thinking. In 1986, for example, Governor Rudy Perpich, the first Iron Ranger elected governor of Minnesota, announced a major step in efforts to revitalize the Range: a Canadian company in Hibbing would make chopsticks for export to Japan using Danish machines designed to make Popsicle sticks. The company went broke in 1989.

The state is trying to help. In 1941 the legislature created a state agency to strengthen and diversify the economy of the Range. This agency has enjoyed various names over the years. Its current incarnation—known as Iron Range Resources (IRR)—is funded by taxes on taconite production. It operates economic development and tourism programs; makes loans and grants for community improvement; pushes information technology and remote telephone customer-service centers; and has outstanding mineland reclamation, restoration, and reforestation programs.[7]

The IRR encourages the towns of the Range to work together. They have a long tradition of intense rivalry, but they are beginning to realize that they must cooperate for the common good. For example, Virginia, Eveleth, Mountain Iron, and Gilbert have formed the Quad City Alliance. Long-distance cross-commuting has become a fact of life, and the Range is evolving into a "conurbation," or network of communities. Hibbing and Virginia are only twenty-five miles apart, and are connected by an excellent four-lane highway. Anything that helps one town helps them all.

Some of the IRR tourism development programs have been more successful than others. IRR owns and operates the successful Giants Ridge Golf and Ski Resort in Biwabik. It also is developing old railroad lines for 132 miles from Grand Rapids to Ely into the paved Mesabi Trail, on which one can hike, bike, ski, snowshoe, and in-line skate. IRR owns and operates the Ironworld Discovery Center, where visitors must buy an expensive ticket to an amusement park before they can enter the museum, which deserves much better patronage than it currently receives.

MINNESOTA'S THIRD IRON ORE DISTRICT, the Cuyuna Range, is now a derelict mining area. Its iron ore was in vertical lenses of highly varied quality. These deposits lay deep beneath glacial drift and thus were hard to locate. The first mine was opened in 1911 by a prospector named Cuyler Adams, who named the range for himself and his faithful dog, Una. Cuyuna's ore is rich in manganese, an essential alloy metal, and the Cuyuna produced most of the nation's manganese during both world wars. In the 1990s the last mine closed.

The early mines were underground, but later mining used open pits. The flooded pits have precipitous rocky walls, and hills of unreclaimed mine waste surround them. Most of the mines were west and north of Crosby-Ironton in eastern Crow Wing County, where places such as Trommald and Riverton are nearly ghost towns today. The village of Manganese is a true ghost town. Its name appears on maps, and its streets and paved sidewalks are still in place, but the only remaining structure is the battered foundation of the old commissary store.

For more than a century prospectors fueled with true Iron Range optimism have never stopped trying to find gold, silver, copper, nickel, zinc, uranium, and other non-ferrous metallic ores in the tough old rocks of northeastern Minnesota. No one can predict when or even whether they might finally luck out, but so far they have not.

An abandoned headframe that was used to work an underground mine on the Cuyuna Range.

Duluth has played a critical role in Minnesota's iron mining story. The city originated as a sawmilling and lumber town near the mouth of the St. Louis River at the western end of Lake Superior, but it boomed as the principal port, banking, and administrative center for the iron ranges. The population of the city grew from 838 people in 1880 to 98,917 in 1920, but it has grown little since then. Duluth is the county seat of St. Louis County, the second-largest county in the United States east of the Rocky Mountains. The county, which is larger than the states of Connecticut and Rhode Island combined, contains most of the Mesabi Range and all of the Vermilion.

Duluth boasts that it is the world's largest inland port. It is 2,342 miles from the salt water of the Atlantic Ocean, at the westernmost tip of the Great Lakes. Each year it ships 30 to 40 million tons of iron ore from the Range. It also ships wheat hauled in by train and truck from the Red River Valley and the plains to the west. It is one of the nation's leading bulk cargo ports. Bulk cargo is loaded and unloaded mechanically, however, and it generates precious little port income. The shipping season is only eight months long, and the opening of the St. Lawrence Seaway failed to produce the eagerly anticipated increase in general cargo, which is far more profitable for a port.

Duluth has an excellent harbor. It is protected from storms on Lake Superior by a long, narrow sandbar known as Minnesota Point, whose natural entrance is six miles south of the city near Superior, Wisconsin. In 1871 Duluth wanted to dig a canal through the north end of this sandbar to allow ships to sail straight into the harbor instead of having to make the long detour through the natural opening. The city of Superior secured a court order to block the canal, but the good citizens of Duluth got out their picks and shovels to dig it before the order became effective. A thick layer of frozen gravel on the final stretch almost stopped them on the last day before the deadline, but a couple kegs of gunpowder solved that problem just in time.

Grain from the prairies to the west is stored in grain elevators in Duluth until it is loaded onto lake boats to be carried eastward.

The "drowned" or submerged mouth of the St. Louis River gives Duluth a superb natural harbor.

Initially a ferryboat crossed the shipping channel at Canal Point to connect the homes on Minnesota Point with downtown Duluth, but in 1905 the city built massive truss towers on either side to carry an aerial gondola back and forth across it.[8] The gondola carried everything from pedestrians to streetcars. In 1929 it was replaced by a horizontal truss span that can be raised to allow ships to pass beneath it. The Aerial Lift Bridge, which has become the symbol of Duluth, was entered in the National Register of Historic Places in 1973. Bridge engineers raise and lower it twenty-five to thirty times a day during the shipping season to allow lake boats to reach the level waterfront land of the inner harbor southwest of the city, which has ore docks, grain elevators, coal and limestone dumps, and industrial areas.

Trains from the Iron Range bring ore to the docks that finger out into Duluth's harbor.

Trains on the ore docks unload ore by gravity into the ore boats that tie up beside the docks.

The waters of Lake Superior endow Duluth with cooler summers, milder winters, and heavier snowfalls than areas inland. Few American cities have such spectacular sites. The built-up area scales steep, rocky bluffs that soar more than 800 feet above Lake Superior. The city is sixteen miles long, but few parts are more than a mile from the lake. Magnificent mansions of wealthy families are along the shore at the northern end.

Fearfully precipitous streets, which are even more fearful when it snows, climb straight up the bluffs. The houses on the streets that parallel the lake look down on the roofs of houses on the next street below. The crest of the bluffs is crowned by dramatic Skyline Drive, which commands magnificent views of the city, the harbor, the lake, and Wisconsin in the distance.

Downtown Duluth still looks like a 1920 city, even though it is laced together with 3.5 miles of second-story skywalks. A new downtown has developed near the airport atop the bluffs at Hermantown. Duluth is home to the University of Minnesota–Duluth, has become a major regional medical center, and has developed a tourist destination at Canal Point near downtown, with fine motels and restaurants, a convention center, and the Great Lakes freshwater aquarium. Visitors can explore an old ore boat that is permanently moored at the waterfront, enjoy fascinating boat tours of the harbor, and be reminded of the glory days of Duluth in a fine museum in the historic 1892 railroad station, which also serves as a regional center for the arts, history, and culture.[9]

The fortunes of Duluth and northeastern Minnesota have risen and fallen with iron ore, and the future remains uncertain.[10] Taconite mining may be no more than a stopgap, because it is too expensive to compete with natural ores. Periodically, optimists extol prospects for new mineral deposits, but the excitement expires quickly, and even if deposits are discovered, they might not warrant the enormous capital investment necessary to develop them.[11] Tourism and recreation seem to offer brighter prospects, but they will require a new mind-set, and they will support only a greatly reduced population.

Kittson
Roseau
Lake of the Woods
Marshall
Beltrami
Koochiching
Pennington
Red Lake
Polk
Clearwater
Cook
Lake
Itasca
Norman
Mahnomen
St. Louis
Hubbard
Cass
Clay
Becker
Aitkin
Carlton
Wadena
Crow Wing
Wilkin
Otter Tail
Todd
Mille Lacs
Pine
Grant
Douglas
Morrison
Kanabec
Traverse
Stevens
Pope
Stearns
Benton
Big Stone
Sherburne
Isanti
Chisago
Swift
Anoka
Lac Qui Parle
Chippewa
Kandiyohi
Meeker
Wright
Hennepin
Ramsey
Washington
Yellow Medicine
Renville
McLeod
Carver
Scott
Dakota
Sibley
Lincoln
Lyon
Redwood
Nicollet
Le Sueur
Rice
Goodhue
Wabasha
Brown
Pipestone
Murray
Cottonwood
Watonwan
Blue Earth
Waseca
Steele
Dodge
Olmsted
Winona
Rock
Nobles
Jackson
Martin
Faribault
Freeborn
Mower
Fillmore
Houston

13

A Demographic Mini-Atlas

Most social and economic maps of Minnesota have remarkably similar patterns. Subtle variations in the rest of the state are overshadowed by the sprawling metropolitan area of the Twin Cities and, to a lesser degree, by the great metropolitan arc that extends from St. Cloud to Rochester. The Twin Cities of Minneapolis and St. Paul are the metropolis of Minnesota and the entire Upper Midwest. This metropolis can be defined in many ways, and the two incorporated cities of Minneapolis and St. Paul, which have spread far beyond their boundaries, are but a small part of it.[1] Perhaps the best statistical definition of the Twin Cities is the "Urbanized Area" delineated by the U.S. Bureau of the Census because it includes the entire built-up area of the two central cities and all their suburbs. This Urbanized Area contains only 1 percent of the state's land area but nearly half of its people (2000 population census).

The Census Bureau also defines metropolitan areas, which consist of entire counties and thus include large areas that are still rural. It is incorrect to assume that "metropolitan" is the same as "urban," or that "nonmetropolitan" is the same as "rural" because the rural parts of metropolitan areas may contain urban places.

The definition of the Twin Cities metropolitan area confuses many people because this area has been defined officially both by the Minnesota legislature and by the U.S. Census Bureau, but these definitions have never agreed. In 1967 the legislature defined a seven-county (Anoka, Carver, Dakota, Hennepin, Ramsey, Scott, and Washington counties) Metropolitan Council area, which remains unchanged. The federal government, however, has never officially recognized this seven-county area, although many Minnesotans think of it as the real "metro area."

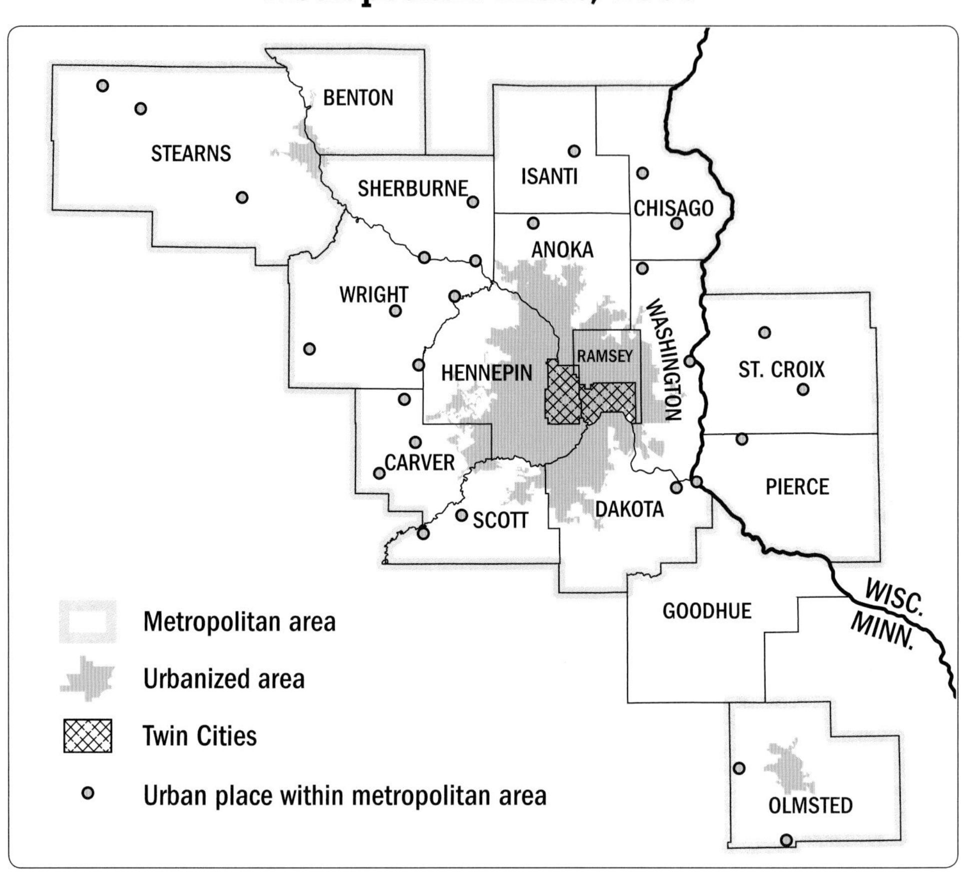

The metropolitan areas of Minneapolis–St. Paul, St. Cloud, and Rochester include a number of smaller urban places and extensive rural areas.

As the metropolis has continued to expand, the Census Bureau has added new counties to the official Twin Cities federal metropolitan area. In 1950 the first census metropolitan area included only five counties (Anoka, Dakota, Hennepin, Ramsey, and Washington). In 1972 the census added five more counties (Carver, Chisago, Scott, and Wright in Minnesota and St. Croix in Wisconsin) to create a new ten-county metropolitan area, and in 1983 it added Isanti County in Minnesota and Pierce County in Wisconsin to create a twelve-county metro area.

In 2000 the Census Bureau wanted to add Sherburne County, but St. Cloud objected strenuously because the county includes a small part of the incorporated city of St. Cloud. Furthermore, because many federal funding programs are tied to population, the St. Cloud metropolitan area would lose money if it "lost" Sherburne

County to the Twin Cities. The Census Bureau then decided to create a combined metropolitan statistical area that includes both the Twin Cities and St. Cloud metropolitan areas by adding Stearns, Benton, and Sherburne counties. St. Cloud will probably not be able to maintain its statistical independence much longer.

St. Cloud is the northern end of a metropolitan arc that extends south through the Twin Cities to Rochester. St. Cloud is a diversified commercial and industrial center, and the home of St. Cloud State University. Rochester has a large IBM plant but is best known for the world-renowned clinic, founded by the Mayo Brothers after a tornado devastated the town in 1883. This metropolitan arc is the economic core of Minnesota. It has slightly less than one-tenth of the area of the state, but it has almost two-thirds of its people, a share that will continue to increase.

The state's metropolitan counties and its lake-and-resort counties have the highest rates of natural increase and the highest in-migration rates. The metropolitan population is younger and has a smaller percentage of people aged sixty-five and older. The metropolitan arc has lower percentages of people who have not completed high school, and higher percentages of people who are college graduates.

The metropolitan population has higher income, a lower percentage of people receiving Social Security income, and a lower percentage of families in poverty status. The median value of housing units in the metro area is more than double that in the southwestern corner of the state, which had the greatest percentage of houses built before 1940 and the lowest percentage of new houses. The metropolitan areas, conversely, had the greatest percentage of new houses and the lowest percentage of old houses.

Pore over the maps for yourself. On some, the lake-and-resort counties stand out; on others, counties with small towns, especially small towns that boast four-year colleges, are prominent. The northeastern and southwestern corners are the least privileged parts of Minnesota, and their lack of privilege is manifest in many different ways. Studying maps is an adventure in geographic exploration and explanation.[2]

Minnesota has eighty-seven counties.

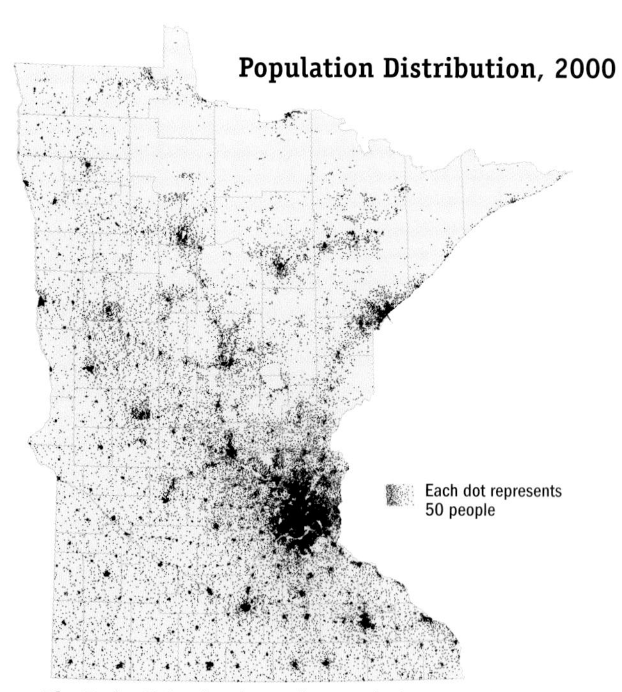

The Twin Cities dominate the population map of Minnesota.

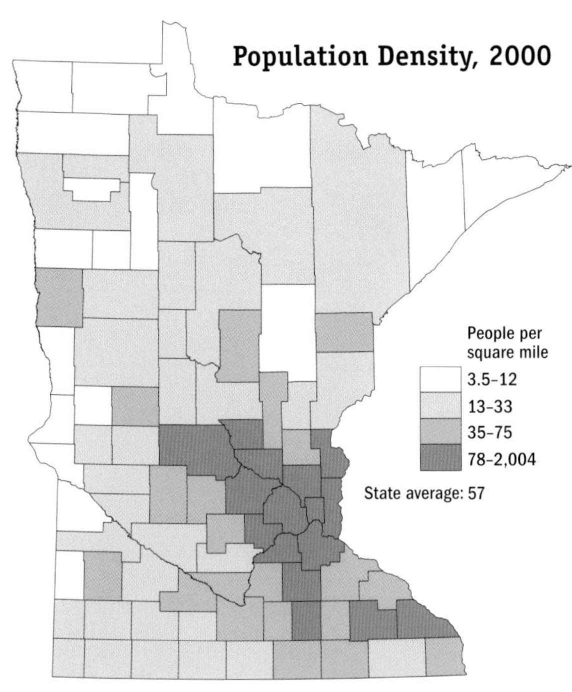

The metropolitan arc extends from St. Cloud to Rochester.

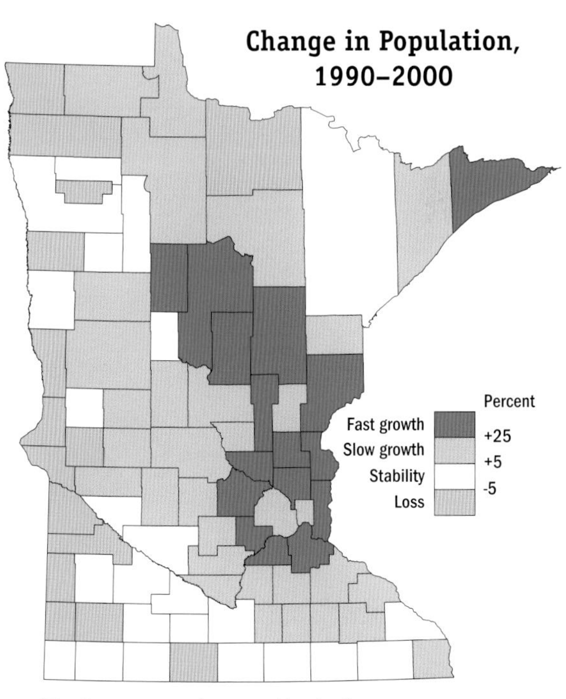

The fastest-growing counties in the 1990s were around the metropolitan core and in the northern lakeshore resort and retirement area.

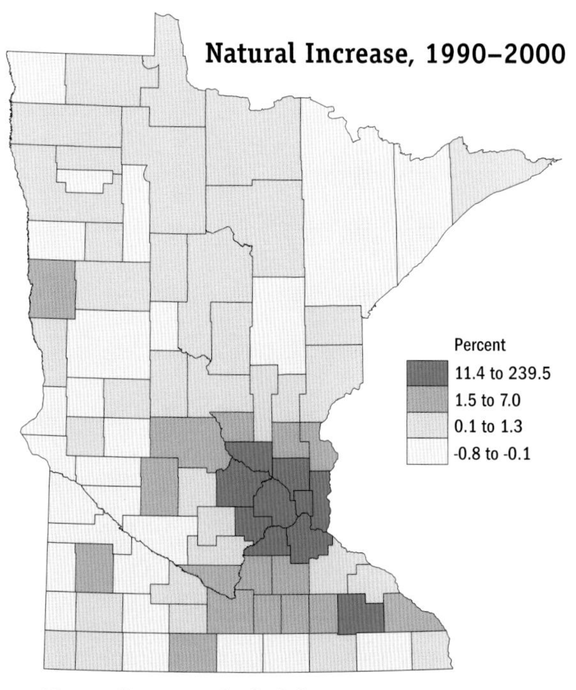

Metropolitan counties had the greatest population increase in the 1990s.

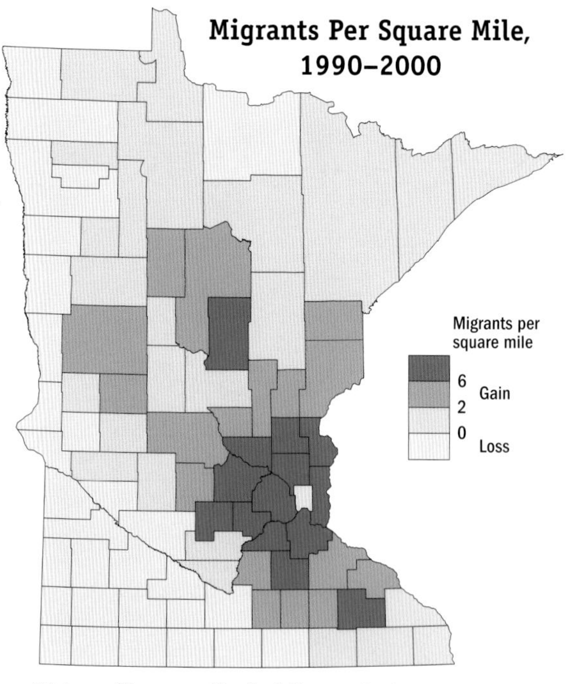

Metropolitan counties had the greatest in-migration in the 1990s.

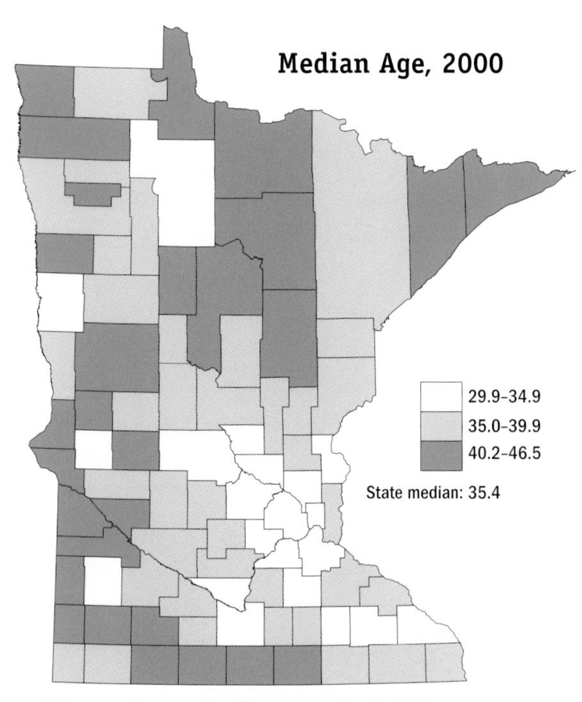

Metropolitan and college counties had the youngest population in 2000.

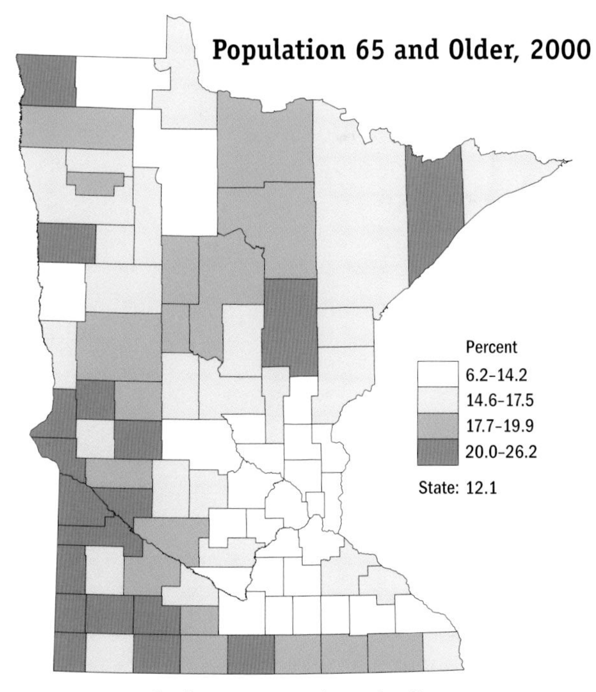

East central Minnesota counties and college counties had the lowest percentages of people aged 65 and over in 2000.

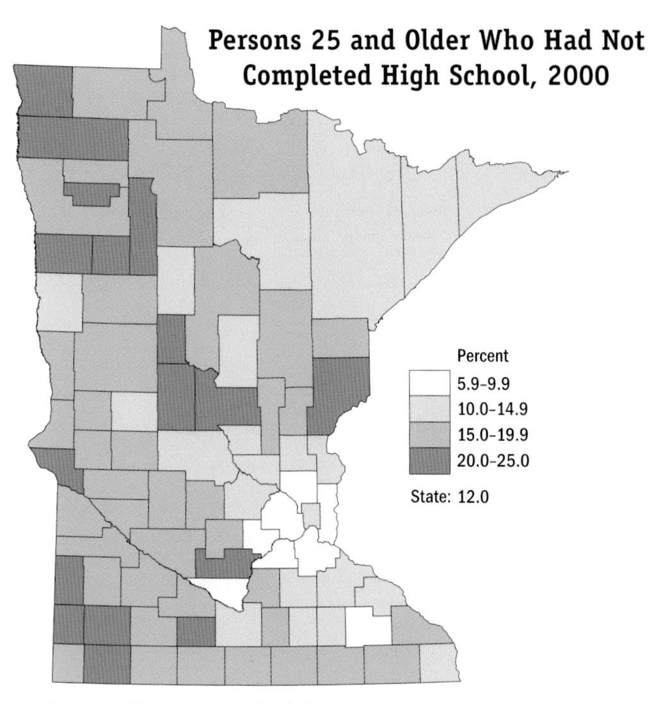

Metropolitan counties had the lowest percentages of people who had not completed high school.

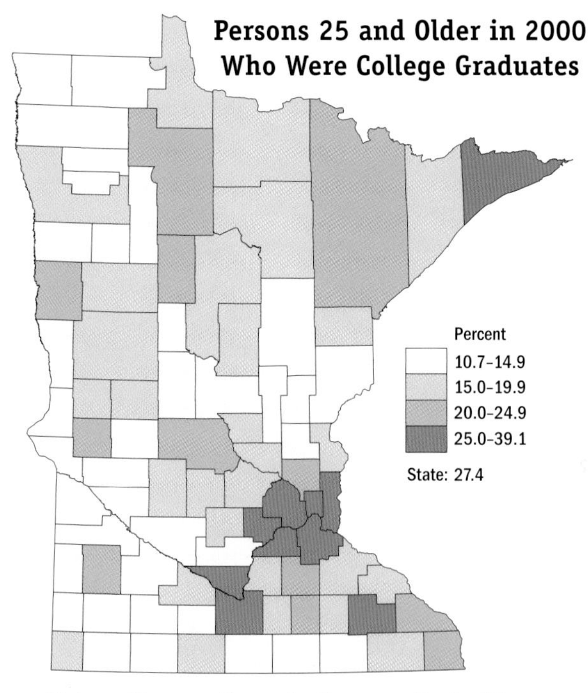

Metropolitan counties and college counties had the highest percentages of college graduates.

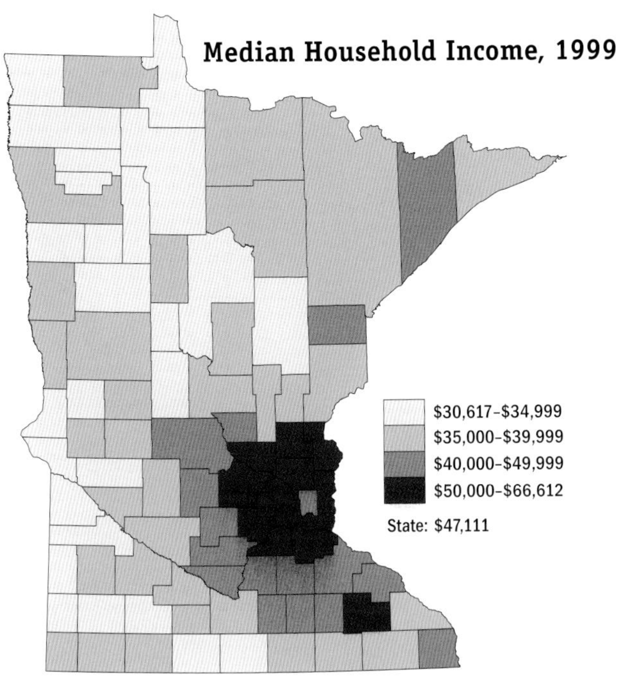

Metropolitan counties had the highest median household income.

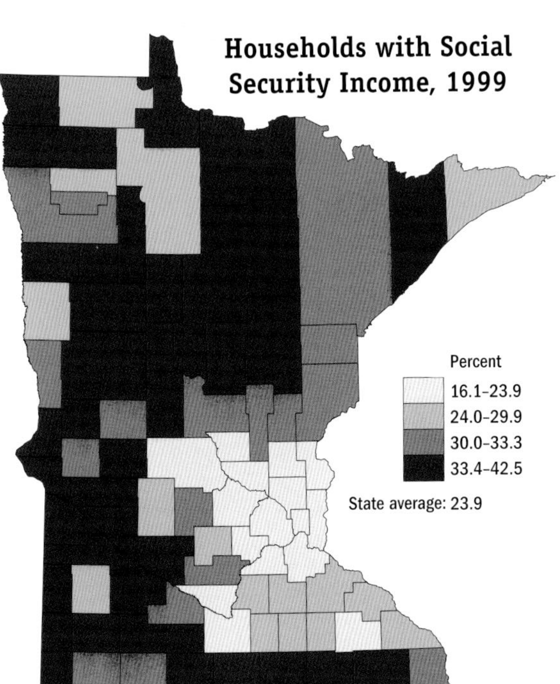

Metropolitan counties had the lowest percentages of households with Social Security income.

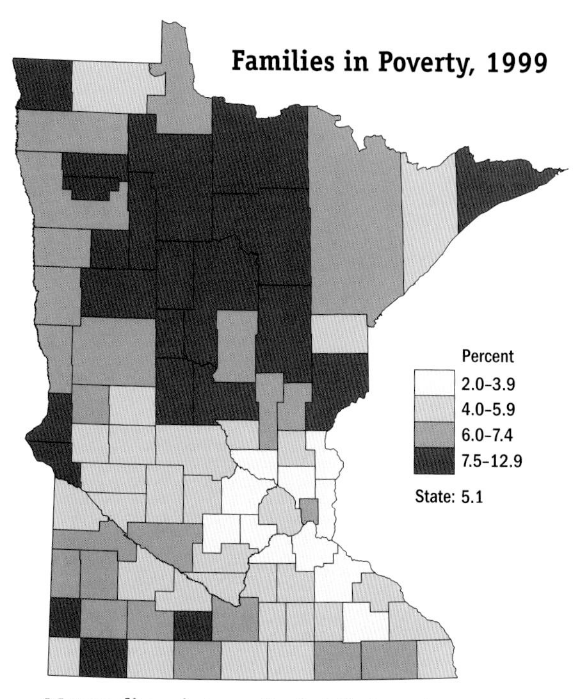

Metropolitan ring counties had the lowest percentages of families in poverty status.

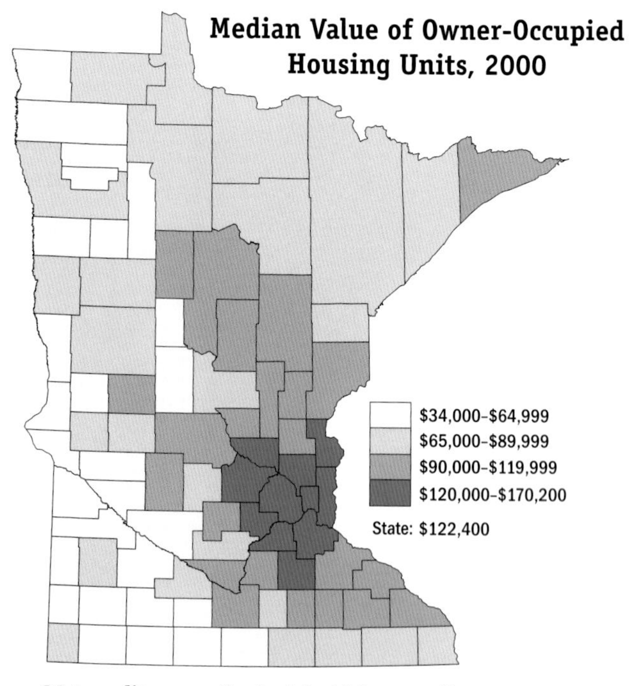

Metropolitan counties had the highest median value of housing units.

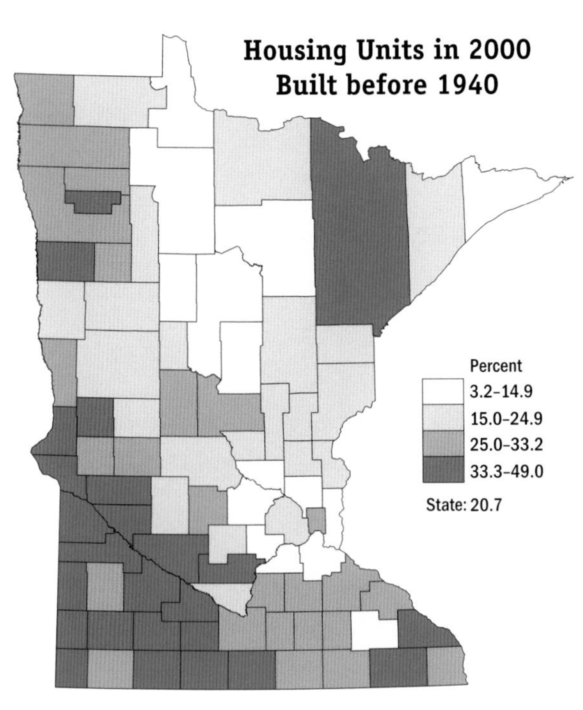

Metropolitan counties and lakeshore resort counties had the lowest percentages of older housing.

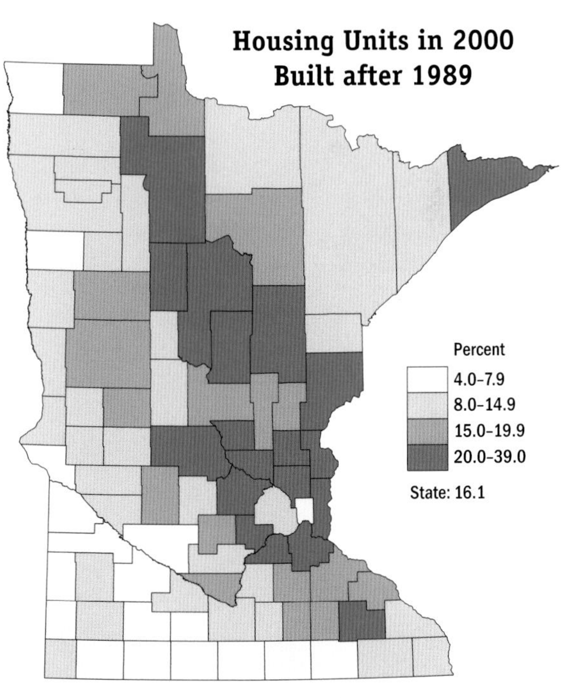

Metropolitan counties and lakeshore resort counties had the highest percentages of newer housing.

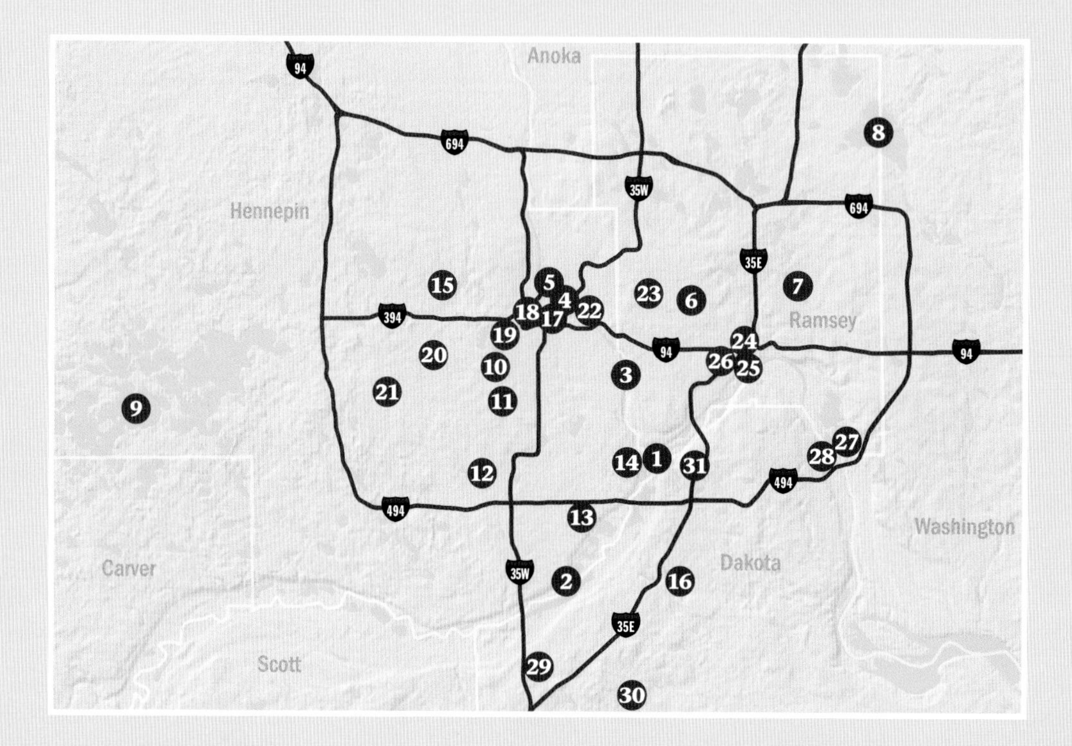

1. Fort Snelling
2. Minnesota River
3. Mississippi River
4. Falls of St. Anthony
5. Nicollet Island
6. Lake Como
7. Lake Phalen
8. White Bear Lake
9. Lake Minnetonka
10. Lake Calhoun
11. Lake Harriet
12. Southdale
13. Mall of America
14. MSP Airport
15. Golden Valley
16. Eagan
17. Metrodome
18. Convention Center
19. Lowry Park
20. St. Louis Park
21. Hopkins
22. University of Minnesota–Minneapolis Campus
23. University of Minnesota–St. Paul Campus
24. Capitol
25. Rice Park
26. Cathedral of St. Paul
27. North Star steel mill
28. Stockyards
29. Burnsville
30. Apple Valley
31. Mendota Heights

14

A Twin Cities Primer

The Twin Cities began in 1819, when the U.S. Army began to build Fort Snelling. The strategic fort on the bluffs overlooked the confluence of the Minnesota River, which drains the prairies to the southwest, and the Mississippi, which flows south from the northern coniferous forest. The Mississippi was the lifeline that first brought Europeans and manufactured goods into Minnesota, but its deep, trenchlike valley was and still is a major barrier to movement overland.[1]

The valley of the Minnesota River, and the valley of the Mississippi River below its junction with the Minnesota, is a glacial spillway. It was carved by the overflowing waters of ice-dammed glacial Lake Agassiz in the Red River Valley to the northwest. Sheer bluffs one hundred feet high loom over the half-mile-wide marshy bottomland of the valley, which floods regularly when heavy spring rains coincide with rising temperatures to melt the winter's accumulation of snow.

At the end of the glacial epoch, the Mississippi River thundered over the brink of the bluff in a magnificent waterfall. This falls eroded its way seven miles upstream to the present site of downtown Minneapolis by 1680, the year when Father Hennepin first saw it and named it after his patron saint. The forty-foot-high Falls of St. Anthony are the largest natural waterpower site west of Niagara Falls. Downstream the river flows through a deep gorge, but just upstream from the falls it divides into two shallow channels near Nicollet Island, which boasted the first bridge ever built across the Mississippi River in 1855.

Fur traders built their cabins near the walls of Fort Snelling, and some of them sold grog to Indians and soldiers. In 1840, when the exasperated commandant ordered them to clear out, they moved six miles downstream to Pig's Eye Landing, where a fur trader named Pierre "Pig's Eye" Parrant had set up shop in a cave. The next year they changed the name of their settlement to St. Paul, in honor of the log chapel that Father Lucien Galtier had named for his patron saint.

St. Paul is perched high on the bluffs overlooking the floodplain. For all practical purposes, it is the head of river navigation because farther upstream the Mississippi is blocked by the Falls of St. Anthony, and the shallow Minnesota River has treacherous shoals and sandbars. Just east of modern St. Paul's downtown a small stream called Trout Creek, long since filled in and built over with railroad yards and factories, had a broad, open valley that led to the uplands back of the bluffs.

Site of the Twin Cities

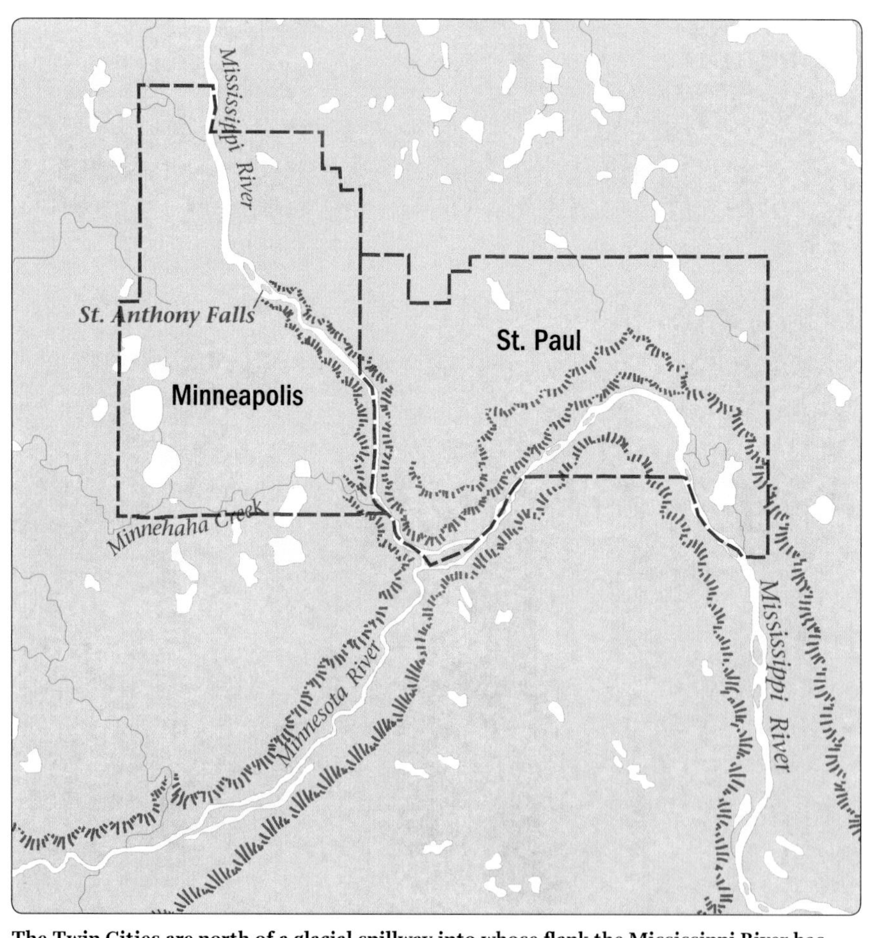

The Twin Cities are north of a glacial spillway into whose flank the Mississippi River has carved a gorge seven miles upstream to the Falls of St. Anthony.

Minneapolis grew at the Falls of St. Anthony power site and the Nicollet Island bridging point.

Downtown St. Paul is on the high bluffs overlooking the Mississippi River.

St. Paul is essentially the head of navigation on the Mississippi River, although some barges do forge upstream.

In the early days steamboats crowded the waterfront in St. Paul. It was outside the great bend of the Mississippi, with access to the countryside in three directions. It was both the headquarters of the fur trade and the river terminus of the oxcart trail that ran northwestward along the edge of the deciduous forest to the settlements in the Red River Valley.

St. Paul, the largest occupied settlement in the area, was the obvious choice for the capital when Minnesota became a state in 1858. Stillwater, the second-largest place, was awarded the state penitentiary, and Hastings received the state insane asylum. St. Anthony (which was incorporated into Minneapolis in 1872) received the state university as a consolation prize.

The first sawmill was constructed at the Falls of St. Anthony in 1848. Others soon followed, but Minneapolis did not really begin to take off until after the Civil War, when a flurry of railroad construction connected it with Chicago, Duluth, and the productive plains to the west. The railroads hauled lumber from the sawmills to the treeless prairies and returned with wheat to feed the flour mills that were beginning to replace the sawmills at the falls.

Populated Places in 1860

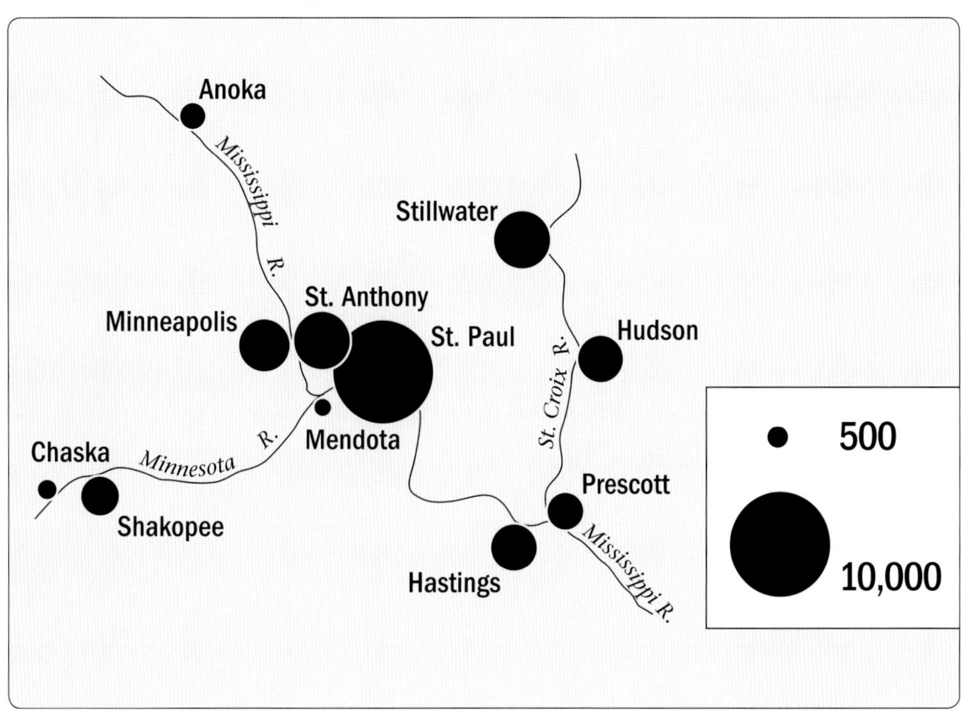

In 1860 St. Paul was larger than Minneapolis and St. Anthony combined, which they later did.

The sawmills could leave the falls and move upriver to new sites because they were shifting from waterpower to steam power, which they generated by burning their sawdust and other scrap. The new flour mills were massive, boxlike structures that covered entire city blocks and were flanked by batteries of towering cylindrical elevators where grain was stored until the mill was ready for it. The mills were powered by water diverted into canals that paralleled the river.

In the 1880s the flour business was consolidated into a few large companies. John S. Pillsbury, on the east bank of the river, had the largest mill. Second largest was the mill built on the west bank by entrepreneurs Cadwallader Washburn and John Crosby, whose Washburn-Crosby Company later became General Mills, but not before it had given its acronym to radio station WCCO. Another well-known figure in the early grain business was William W. Cargill, a son of a British sea captain who came to Minneapolis in 1865 and started a grain trading company that has become a major international corporation.

The railroads created twin cities, because the power site and bridging point at the falls were too far from the head of navigation in St. Paul to be served by a single central station in the horse-and-buggy days. The railroads were forced to develop separate facilities at each place, and thus was born a rivalry that still simmers.

This rivalry flamed into white heat when the census was taken in 1890, because St. Paul feared that Minneapolis had become the larger. Both cities inflated their numbers so flagrantly that a recount was mandatory. The recount found, among many other curiosities, that 75 people had been reported as residents of a newspaper office in St. Paul, and 275 as residents of the Union Depot. Minneapolis had 18,299 fictitious folk to a mere 9,425 in St. Paul, but the recount still showed that Minneapolis was larger, and so it has remained ever since.

Minneapolis grew faster because its sawmills and flour mills were booming, while the small factories in St. Paul were still tied to the languishing fur trade. Minneapolis also grew faster because it was on the less competitive side of the metropolis. To the east St. Paul had to compete with Chicago, Milwaukee, and other major cities, but to the west Minneapolis was able to extend its banking and wholesale dominance westward over a vast trade area extending to the Rocky Mountains. Although it is vast, this trade area is sparsely populated and has been losing population steadily for half a century or more.

Massive immigration in the 1870s and 1880s stoked the growth of both cities. Most of the Germans and Irish were Catholics who settled in St. Paul. Most of the Scandinavians were Lutherans who settled in Minneapolis. Each nationality group brought its own particular variety of Lutheranism, and the process of trying to consolidate them into a single church has been long and painful. The Twin Cities attracted fewer immigrants from southern and eastern Europe, whose main concentration was in Northeast Minneapolis, near blue-collar jobs in the sawmills and flour mills along the river.

Until about 1880, most people lived in tightly packed housing near the central cities because they had to live within walking distance of their jobs, and most factories, stores, and offices were in or near downtown. Residential areas were further constrained by the railroad lines that sliced through the cities. These lines were major barriers, with up to ten trains an hour hurtling along the tracks. Crossing them was an adventure, and only the poorest people lived on the other side of the tracks.

Railroad lines east and north of downtown St. Paul channeled the growth of residential areas toward the west, where high bluffs provided fine sites for the mansions of the wealthy. The railroad lines north of Minneapolis channeled residential growth to the south, and the chain of lakes attracted it toward the southwest.

Farsighted planners in both cities ensured that the land around their lakes would be preserved as public parks by acquiring it before it was developed for residential use. St. Paul created parks around Lakes Como and Phalen, and Theodore Wirth, park commissioner in Minneapolis from 1905 to 1935, developed a superb urban park system around the chain of lakes in the western part of the city. These lakeshore parks have greatly enhanced the attractiveness and value of the residential areas around them.

The residential areas of both cities began to expand rapidly after 1885, when Thomas Lowry began developing an electric streetcar system that was one of the finest in the country. His lines laced both cities and allowed people to live much farther from downtown, although only a single line connected the two central cities. The streetcar lines, which ran out to White Bear Lake and Lake Minnetonka, enabled day-trippers from the cities to enjoy amusement parks and excursion boats on the lakes. The streetcar lines were the spokes along which the residential areas of both cities first developed.

Merchants opened stores along the streetcar lines, which became shopping strips. Larger commercial districts developed where lines intersected and riders transferred from one line to another. These shopping strips and commercial districts have suffered in the age of the automobile because they lack adequate parking space, and in many cases their stores have been put to such use as discount and secondhand stores. Lake Street, which runs across southern Minneapolis, for example, once boasted the city's leading automobile dealerships. For many years, the street languished with mostly low-rent stores, but it has evolved into a lively ethnic shopping street. In 2004 the former Sears and Roebuck store and distribution center on Lake Street was renovated at a cost of $190 million into the Midtown Exchange office tower, with headquarters for 1,600 hospital employees, 360 housing units, a parking ramp, a transit hub, and a "global market" on the ground floor with many shops selling ethnic food, arts, and crafts.[2]

Although the population of the Twin Cities has been and remains predominantly white, immigration from non-European areas has slowly changed its composition. African Americans first arrived in the nineteenth century to work as cooks, waiters, and sleeping-car porters on the railroads or as domestic servants in the mansions of the wealthy, and many more arrived after World War II. Mexican immigrants came to Minnesota in the early twentieth century to toil in the sugar-beet fields of the Red River Valley; some of them wintered in St. Paul, where they became the nucleus of a rapidly growing community. Vietnamese and Hmong, who aligned with the United States during the Vietnam War, began arriving in the mid-1970s, and Minnesota is now home to the nation's second largest Hmong community. Somali refugees, who began arriving in the 1990s, make up the country's largest Somali concentration.

Many minorities live in the two central cities, often in deteriorating single-family houses. The principal concentrations of new immigrants are in North Minneapolis, the Phillips neighborhood south of downtown Minneapolis, the Summit-University area northwest of downtown St. Paul, the Dayton's Bluff and East and West Side neighborhoods of St. Paul, and West St. Paul.

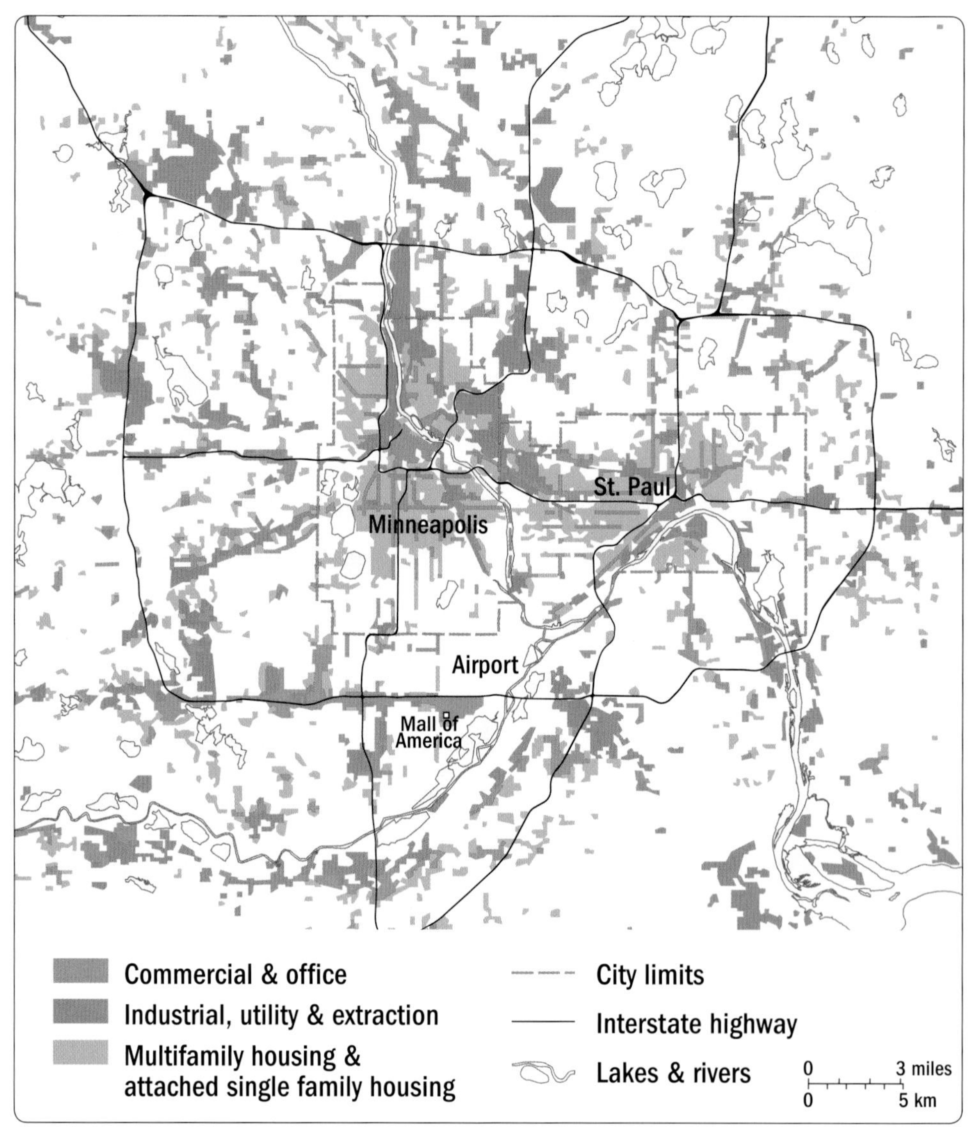

Major land-use areas in the Twin Cities.

Ethnic Minority Areas, 2000

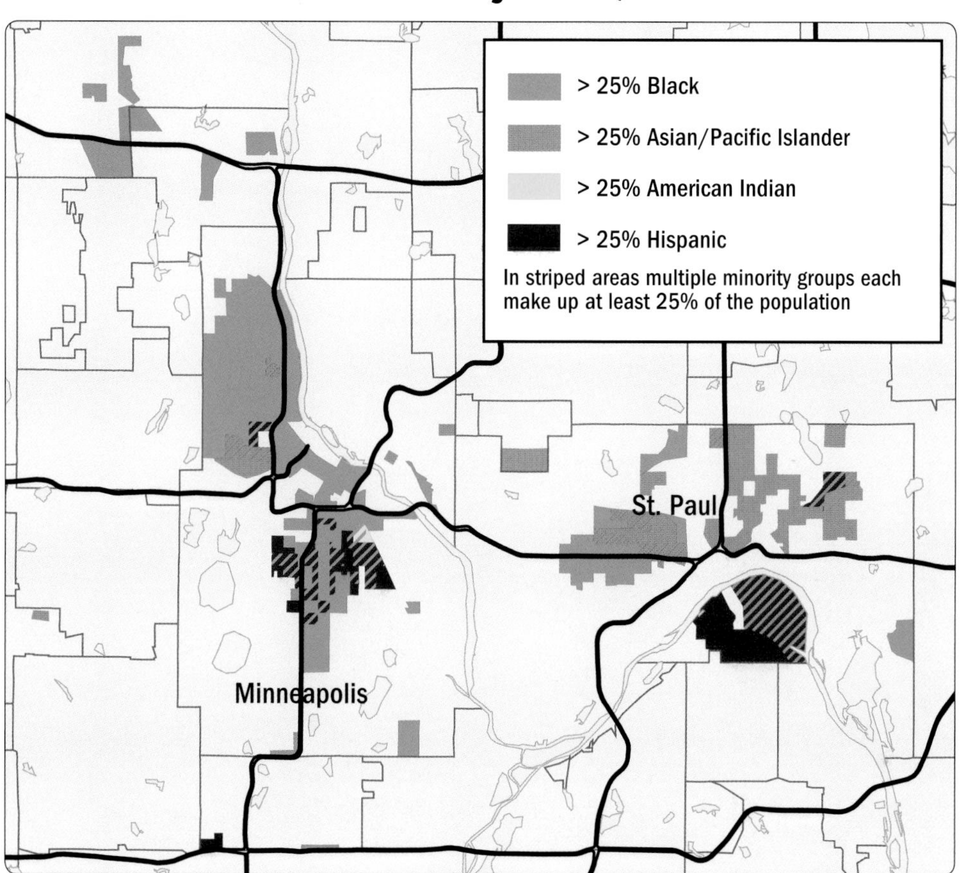

Ethnic minority residents are concentrated in the older residential areas of Minneapolis and St. Paul.

Number of Persons in Selected Minority Groups, 2000

	Minnesota	Hennepin and Ramsey Counties	Percent of state total
Black or African American	202,972	159,710	**78.7%**
Asian	162,414	109,673	**67.5%**
Hispanic or Latino	143,382	72,418	**50.5%**
American Indian or Alaska Native	81,074	26,021	**32.1%**

The economy of the Twin Cities began to change around the time of World War I. Sawmilling peaked in 1899 but then plummeted, and the last sawmill closed in 1919.[3] Flour milling topped out in 1915, but then it too began to decline because wheat could be shipped to mills in Buffalo, New York, more cheaply than flour and because competing flour mills opened in Kansas City. Major milling companies still have offices and research laboratories in the Twin Cities, but their plants are scattered around the globe.

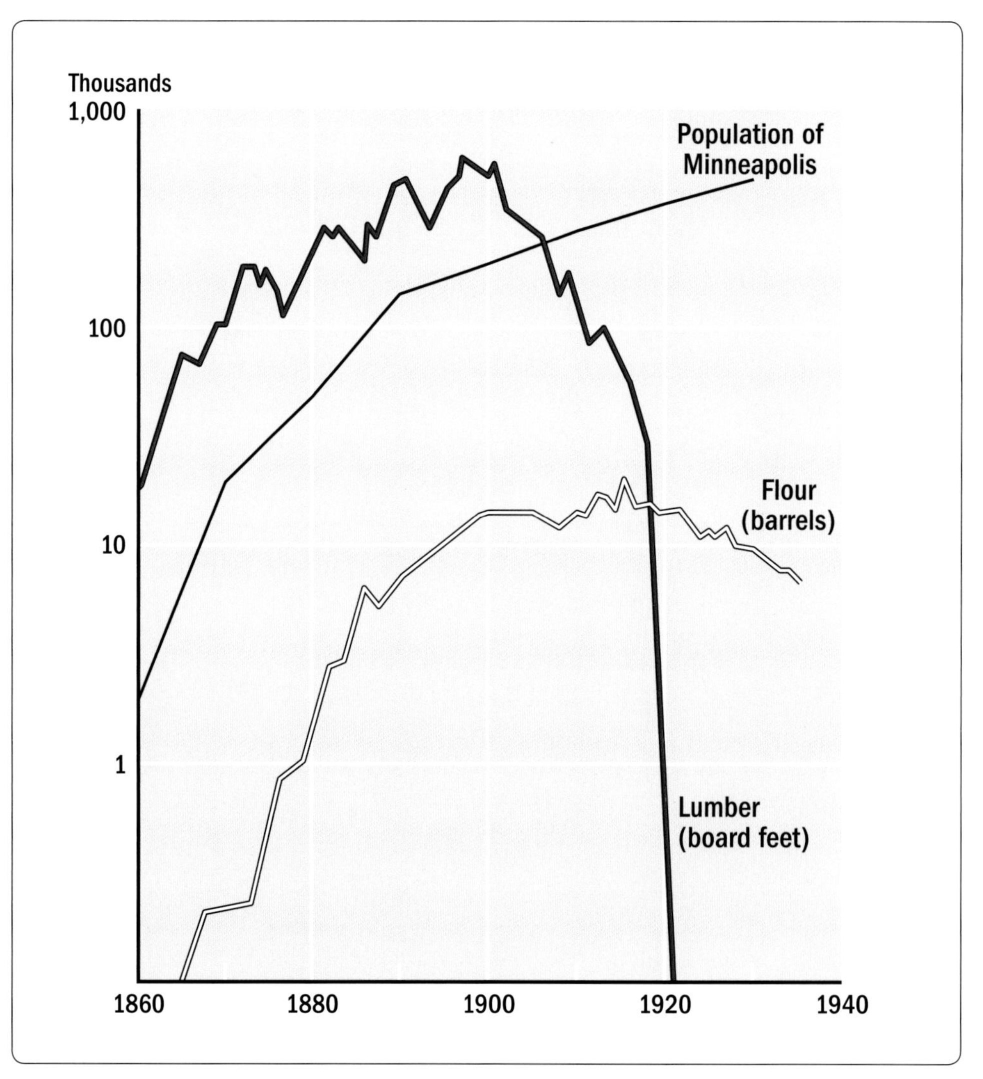

Sawmilling peaked in 1899, and flour milling had peaked by 1920.

The necessity of diversifying the economy became apparent just at the time when the nation's electrical equipment industry was beginning to grow, and the Twin Cities' economy has grown with it. Local fortunes provided capital, local entrepreneurs introduced significant innovations, and the mechanization of farms released hardworking mechanics for new plants. The availability of highly skilled labor was, and still is, one of the principal factors that has attracted new industries to the Twin Cities and maintained them as a major electronics and computer-manufacturing center. Today the Twin Cities are a national leader in the manufacture of medical devices and equipment, but the economy is not dominated by any single firm or industry. It is increasingly similar to the diversified service, information processing, and management economies of other second-order regional capitals, such as Denver, Kansas City, Dallas, and Atlanta.

The leadership of the Twin Cities is growing more cosmopolitan and further from its local roots, and the strong local tradition of philanthropy, with some exceptions, seems threatened. In the past, most leaders of business, government, education, labor, and nonprofit organizations were homegrown, and they knew each other well, often from childhood. Massive corporate reorganizations in the 1980s, however, brought in new executives from other places who are committed to their organizations and careers rather than to the community in which they happen to be employed.

For its first one hundred years, practically the entire built-up area of the Twin Cities was contained within the boundaries of the two central cities. Since 1950, however, the suburbs have expanded dramatically, aided and abetted by the availability of high-speed automobiles and the construction of highways and expressways.

The Twin Cities are at the intersection of Interstate 35 and Interstate 94. I-35 from Kansas City and Des Moines bifurcates south of Minneapolis into I-35W, which passes through Minneapolis, and I-35E, which passes through St. Paul. They converge north of St. Paul and head for Duluth. I-94 from Chicago and Madison runs through both central cities. It also bypasses them with I-694 on the north side and I-494 on the south side before it heads northwest for St. Cloud and Moorhead. I-394 has been built straight west from downtown Minneapolis to Lake Minnetonka to serve the affluent western suburbs. The catastrophic collapse of the I-35W bridge across the Mississippi River near the University of Minnesota in August 2007 created colossal congestion that clearly showed the importance of the Interstate highway system to the Twin Cities.

Interstate highways were built after the suburbs were already well developed. The initial spokes of suburban growth were along major arterial highways that already existed, and the interstices were soon filled in with vast expanses of single-family homes on spacious lots. Most people who move out to the suburbs remain in the same general sector of the metropolitan area, so the suburbs reflect the residential patterns of the central cities. The northern suburbs generally are blue-collar areas with less expensive housing, whereas the south and southwest generally have white-collar suburbs with more expensive housing and considerably more economic activity. The state's glacial topography also influences the character of suburban residential areas. Level, well-drained, sandy outwash plains have large-scale tract housing, whereas rolling, wooded, lake-studded morainic areas have larger and more expensive homes.

The first suburban business areas were strips along major arterial highways that extended well-known city streets; otherwise their customers would have had trouble finding them in the mystifying maze of new streets in the rapidly growing suburbs. In 1956 Southdale, the first enclosed shopping mall in the United States, was built southwest of Minneapolis in Edina at the center of a superblock with unlimited free

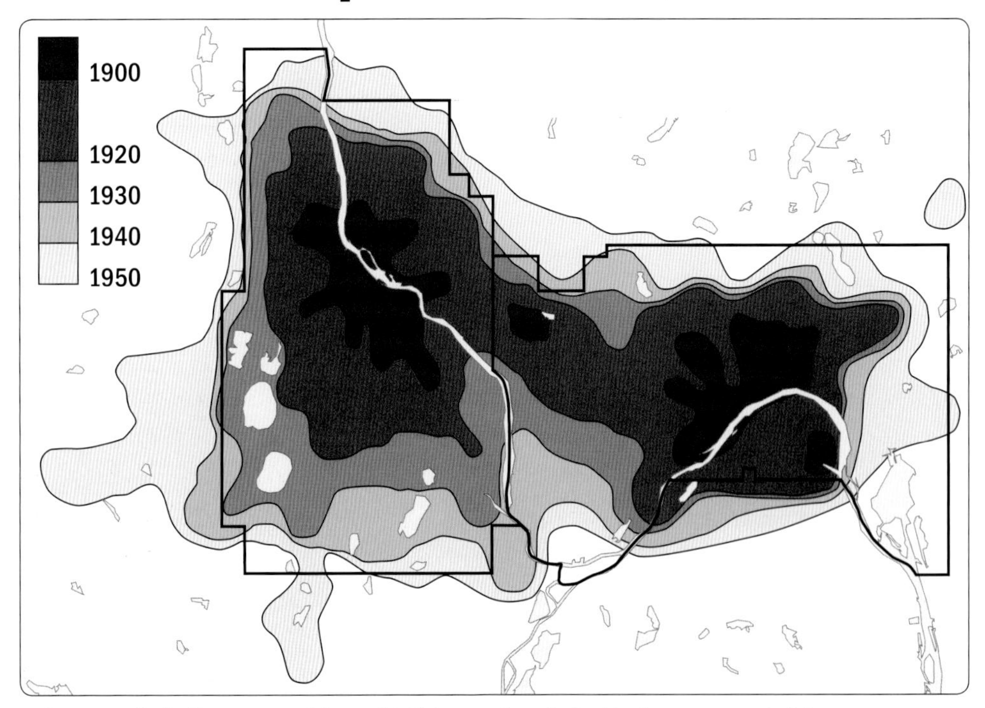

Before 1950 the built-up area of the Twin Cities was largely inside the two central cities.

Major Highways near the Twin Cities

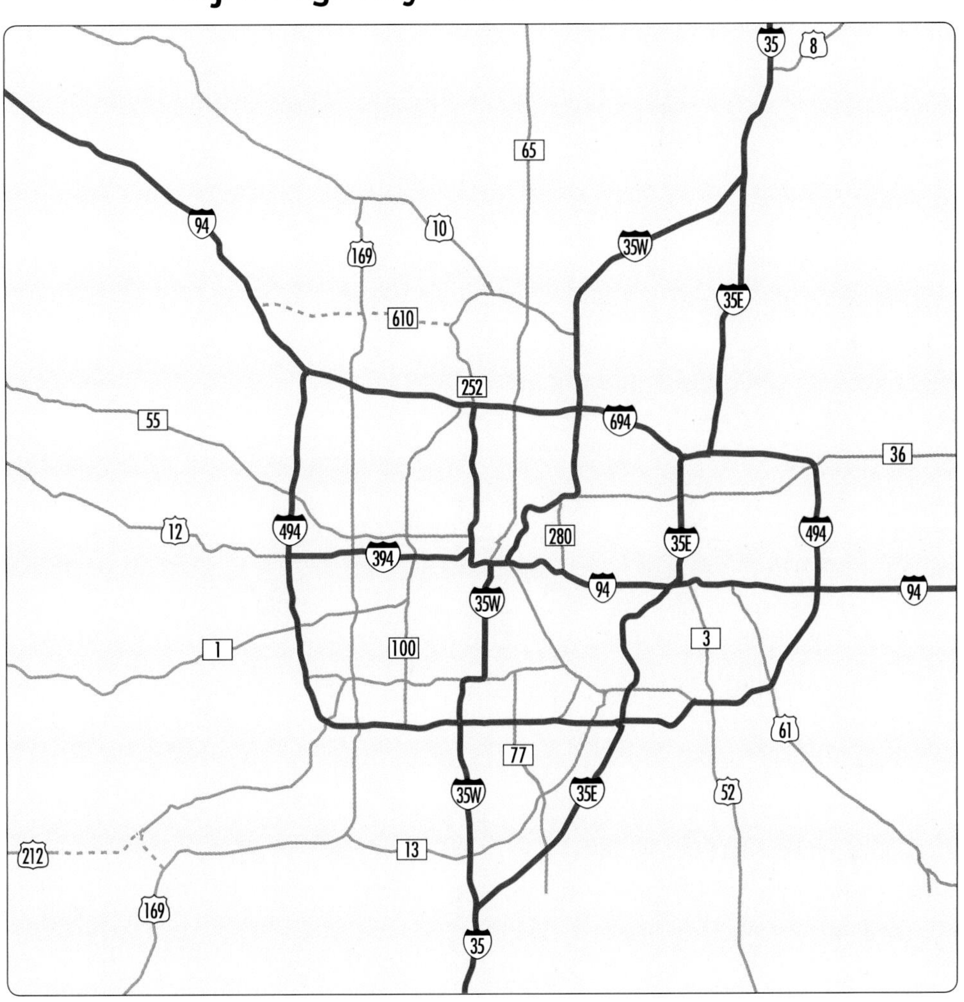

Interstate highways 35 and 94 intersect in the Twin Cities.

parking. This concept was so successful that it was replicated at a number of other suburban sites, known collectively as "the Dales."

In 1992 the Mall of America opened on the site of the former Metropolitan Stadium near the Twin Cities international airport. The center of this megamall is an eight-acre indoor amusement park complete with rides. Around the park are four tiers of shops, eating places, and other commercial activities. At each corner is an anchor department store, and at each end is an enormous free parking ramp. The mall attracts families with children, and it is a popular teenage hangout. It has been a tourist destination that attracts chartered international flights, as well as

By 1980 Southdale had spawned a ring of new apartment, office, and retail buildings.

The Mall of America has massive parking ramps at either end.

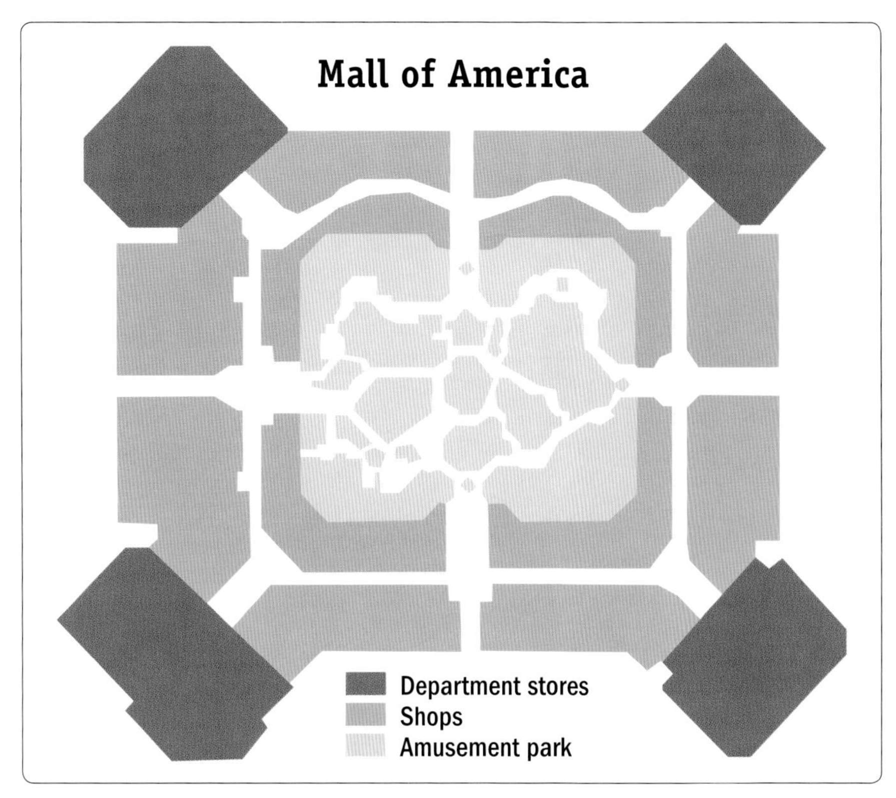

The Mall of America is anchored by department stores at each corner, with an amusement park in the center.

busloads of people from neighboring states. The Mall has more than 500 shops, employs more than 11,000 people, attracts more than 40 million visitors a year, and contributes more than $1.8 billion annually to the economy of Minnesota.

The Mall of America and the airport are the eastern end of a "new downtown" along Interstate 494 that does more retail business than downtown St. Paul and nearly as much as downtown Minneapolis.[4] This bypass strip is a jumble of shopping malls, office buildings, automobile dealerships, motels, restaurants, light industrial plants, and warehouses. With abundant parking, they are easily accessible both from within the metropolitan area and from outside it. The activities serving the central cities still remain in the traditional downtown areas, but many activities that serve the entire metropolitan region and beyond have migrated to the I-494 strip.

Since 1970, most new job growth in the metropolitan area has been in the suburbs, which have the most desirable new jobs and new office buildings. Major companies have established impressive headquarters at suburban sites: General Mills in Golden Valley, Cargill and Carlson in Minnetonka, and Thomson West and

The Interstate 494 strip is the new "downtown" of the Twin Cities.

Northwest Airlines in Eagan. The southwestern suburbs have extensive areas of attractive, new one-story buildings that combine office space in front with light manufacturing or warehouse space in back.

The postwar suburbs had expanded so haphazardly that in 1967 the state legislature created the Metropolitan Council to control and manage their growth. The governor appoints the members of the council, which has cognizance over sewers, airports, transit, and regional parks in a severely fragmented area that includes seven counties, nine metropolitan agencies, twenty-two special districts, forty-nine school districts, fifty townships, and 138 cities, each of which fiercely cherishes its independence.

Much of the Metropolitan Council's power derived from its control over the requests of local units of government for federal funding. The council attempted to control growth by creating a metropolitan urban service area outside which no sewers would be provided. Planners, unfortunately, are prone to tell people what they ought to want instead of giving them what they do want, or educating them about what they ought to want. Developers have been forced to leapfrog to areas outside the Metropolitan Council area, where the population has grown most rapidly in recent years. Federal funding cutbacks have reduced the council's power.

The Metropolitan Council has also had difficulty in providing acceptable transit services. In 1954 the cities' efficient streetcar system was dismantled and replaced by buses. The new bus service was so unsatisfactory that in 1970 the Metropolitan Transit Commission took over the bus company, but it has had difficulty increasing ridership despite improved equipment, better route patterns, and park-and-ride lots.

The Twin Cities' suburbs, which have developed in the era of the automobile, consist mostly of single-family homes on large lots. They lack the concentrations of people that could be served effectively by public transit systems. Most suburban people commute to work by car, often from suburb to suburb rather than to downtown, and at rush hour many highways are seriously congested. The Twin Cities need an innovative system for getting individuals to and from their jobs efficiently.

In the early 1970s, some citizens began lobbying for a light-rail transit (LRT) system. They ignored the geography of jobs and people, and based their emotional

Metro Area Cities and Townships, 2000

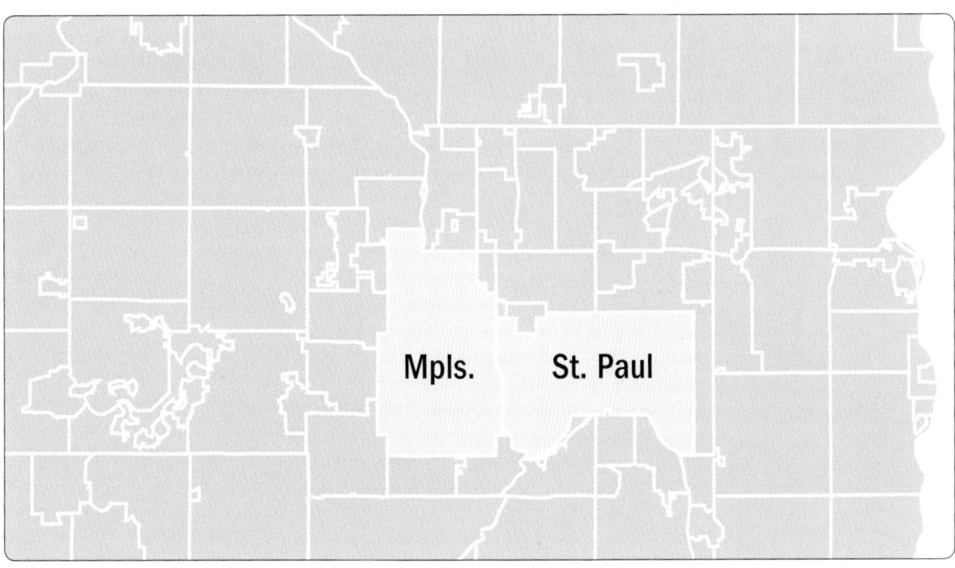

The urbanized area of the Twin Cities is severely fragmented into many small political units.

The suburbs of the Twin Cities have developed in the era of the automobile.

appeal primarily on light rail's glamorous cosmopolitan image rather than on rational cost-benefit analysis. Skeptics argued that low-density residential areas cannot support high-capacity transit systems; a few inflexible spokes radiating from central cities will not alleviate highway congestion; and a good bus service would be far more effective and less expensive.

Enthusiasts nevertheless managed to push through the construction of a single LRT line, which opened in 2004, from downtown Minneapolis to the airport and the Mall of America, and a central-corridor line connecting downtown Minneapolis and St. Paul has been proposed. Skeptics say light rail is an inordinately expensive, heavily subsidized white elephant, but only time will tell.

Despite the growth of suburbs, most people still think first of the two traditional downtown areas when they think of the Twin Cities. The streets downtown sometimes seem deserted, especially in the winter, because most pedestrian traffic is contained in the glass-enclosed second-story skyway system that laces them together. Skyways lined with small shops connect nearly all downtown buildings. Minneapolis skyways are owned jointly by the owners of the buildings they connect, but the city owns the skyways in St. Paul. Visitors often remark that "downtown" is smaller and less impressive than they had anticipated because they forget that there are two downtowns.

The retail core of Minneapolis centers on six blocks on Nicollet Avenue, which were converted into a mall for buses and pedestrians in 1965 to maintain vitality. For many people, downtowns are less attractive than suburban malls because of crime, urban grittiness, and difficult parking, but some people have a vested interest in maintaining the vitality of the central cities.[5] Their efforts are encouraged by the news media, most of whose offices are downtown. Downtown areas are becoming convenience shopping areas for the people who work in offices there, although downtown Minneapolis is also a popular entertainment and nightlife area.

The office core of downtown Minneapolis is four blocks wide on either side of the Nicollet Mall. Marquette Avenue, which parallels the mall a block to the east, is synonymous with banks and financial institutions. Seven blocks to the east is the Metrodome, opened in 1989 as a venue for baseball and football, and equally unsuitable for both. South of the retail core, Nicollet Avenue enters a convention and hospitality area, with major hotels, Orchestra Hall, and the Convention Center, opened in 1989. Although Minneapolis attracts many conventions, its hotel and other facilities are inadequate for meetings of more than around 2,000 people.

The south end of downtown Minneapolis has more than 2,000 high-rise luxury apartments and condominiums for empty nesters from the suburbs and affluent

Land Use in Downtown Minneapolis

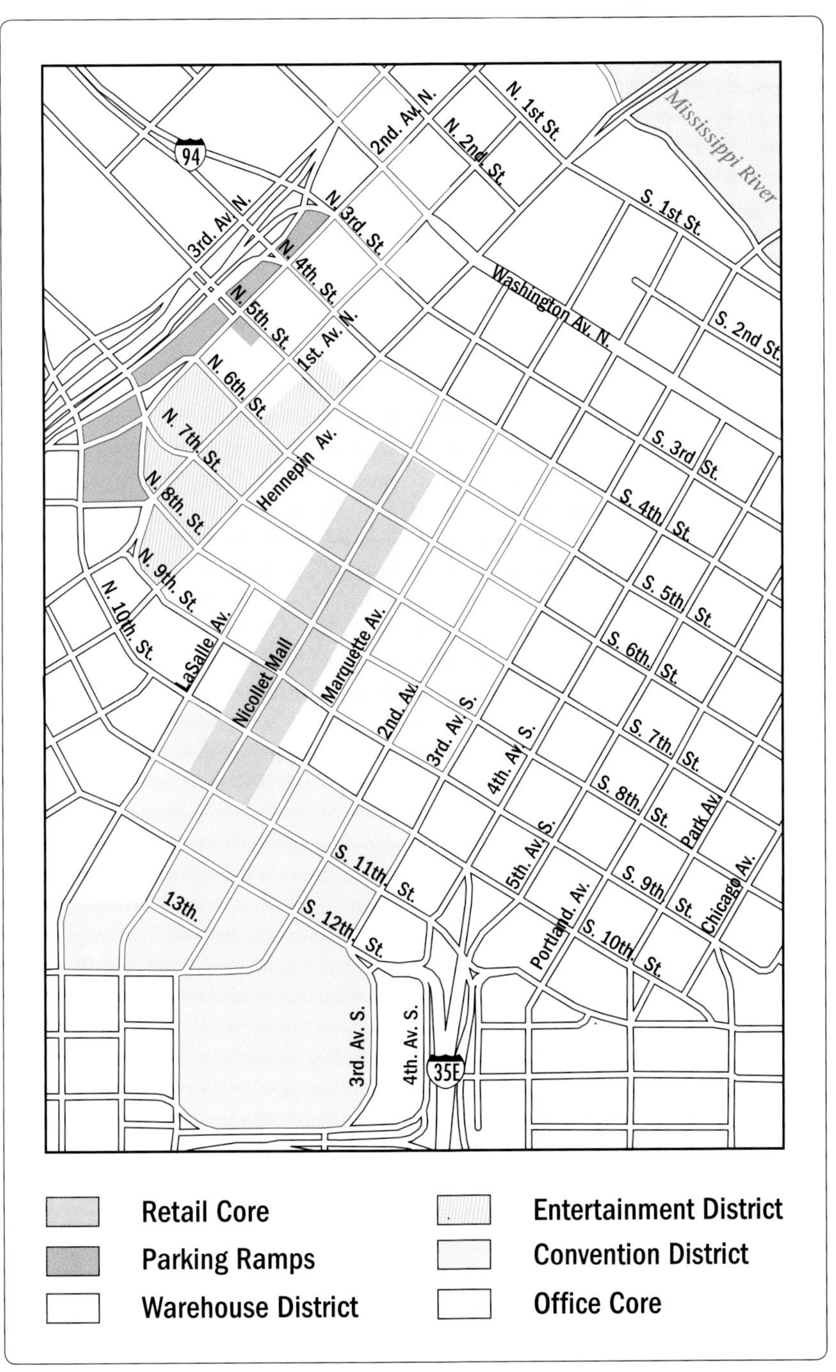

The retail core of Minneapolis is flanked by office blocks, with a convention district to the south and an entertainment and warehouse district to the north.

singles who like to live near downtown. Just to the west is Loring Park, with its statue of Ole Bull, and beyond, on the lower slopes of Lowry Hill, are the Walker Art Center, founded by lumber baron Thomas B. Walker, and the former site of the Guthrie Theater. Lowry Hill, the first affluent residential area in Minneapolis, is crowned with some magnificent mansions.

Hennepin Avenue, which parallels Nicollet Avenue on the west, marks the edge of an old warehouse and light manufacturing area that has become the principal entertainment district of Minneapolis. Beginning in the 1980s the old loft buildings were remodeled into galleries, restaurants, and avant-garde offices that have become the center of the city's nightlife. The southern end of this district houses the Target Center arena, which is used for basketball, rock concerts, and political rallies, and to the west is the new baseball park.

For many decades downtown Minneapolis turned its back on the river that made it, but the city has rediscovered the Mississippi and is developing three miles of riverfront parkland and housing. The Pillsbury A Mill on the north side of the river, which closed its operations in 2003, is being redeveloped for residential, retail, and office uses. The Washburn-Crosby A Mill on the south side was gutted by fire in 1991, but the Minnesota Historical Society has developed a fascinating Mill City Museum within its limestone ruins. The museum, which opened in 2003, has a unique auditorium on an elevator that carries seated visitors up through the mill tower to learn the history of flour milling.

The white Metrodome is east of downtown Minneapolis.

The warehouse district northwest of downtown Minneapolis serves the Upper Midwest.

High-rise apartment complexes have been built on both sides of the river, and upscale specialty shopping centers have been developed in some of the old industrial buildings north of the river. These centers have not prospered, however, perhaps because the river is a psychological barrier. Upscale shopping centers have also been developed in other parts of Minneapolis, but few have flourished. Perhaps there are too many centers, perhaps they are in poorly chosen locations.

An industrial area extends up the Mississippi River from downtown Minneapolis. Barges carry sand and gravel, grain, scrap metal, and road salt to and from lumberyards, scrap dealers, a power plant, and a defunct brewery. Another industrial area with food processors, warehouses, and machine industries extends along the railroad line southwest from Minneapolis through St. Louis Park to the once outlying village of Hopkins.

Fifteen blocks east of downtown Minneapolis is the main campus of the University of Minnesota, separated from the university's St. Paul campus by the Midway industrial district. This district, which straddles the boundary between St. Paul and Minneapolis, developed when the area at the falls had become too congested. The western edge of St. Paul still was sparsely populated and had abundant level land that could be used for railroad yards, factories, and warehouses.

The Midway district was the first major industrial area in the Twin Cities that depended on truck transportation and automobile commuting, but many of the buildings are showing their age. Most of the companies that once occupied them have moved to newer facilities in the suburbs, and the buildings have gone to lower-rent uses. New loft-style housing, however, has begun to line the street in anticipation of a light-rail transit line scheduled to connect the two old downtown areas at some future date.

The Midway district is traversed by University Avenue, which was the principal street connecting downtown Minneapolis with downtown St. Paul before Interstate 94 was completed. The western end of University Avenue is in an industrial area, but the street becomes more of a tired commercial strip as it approaches the state capitol in St. Paul.

Downtown St. Paul looks like an older eastern city. Wabasha Street, a principal business street, has empty storefronts. Redevelopment of Lowertown, the historic warehouse and manufacturing area east of Wabasha Street, is slow.

West of Wabasha Street is Rice Park, a Twin Cities jewel surrounded by handsome buildings: the grand old St. Paul Hotel, completely refurbished; the Ordway Theater, opened in 1985; and Landmark Center, a splendid Victorian courthouse built in 1902 and redeveloped in 1978 as a civic center. Nearby, Xcel Center, opened in 2000, hosts sports competitions, including Minnesota Wild hockey games, and cultural events. Many smaller food and drink establishments have developed in its wake along West Seventh Avenue.

St. Paul is dominated by the state capitol, which crowns a hill across the Interstate 94 gulch from downtown. On both sides of the vast lawn that sweeps down from the capitol are massive state office buildings. Across the freeway, on West Kellogg Boulevard, is the large, new Minnesota History Center on the lower slope of Summit Hill.

Up the hill from the History Center is the Cathedral of St. Paul, completed in 1906. The cathedral, which sits across the street from the enormous mansion built by railroad magnate James J. Hill, is at the beginning of Summit Avenue, the best-preserved Victorian boulevard in the United States. The mansions on Summit Avenue are opulent. Some have been converted to institutional uses or subdivided into apartments.

East of St. Paul, in isolated grandeur just across the Ramsey County line in Washington County, is the luxurious campus of 3M. In 1902 four businessmen in Two Harbors formed a company to make abrasives from a local mineral they thought to be corundum. When they discovered it wasn't, they moved to Duluth and started making sandpaper. In 1905 they sold the company to Lucius Ordway, who had made a fortune selling plumbing supplies. Ordway moved the company to St. Paul in 1910 and hired William L. McKnight and Archibald Bush to run it. These two Scots gave their name to Scotch tape, a well-known 3M product, and built 3M into a major international corporation with an enviable reputation for creative innovation.

The state capitol of Minnesota faces the Cathedral of St. Paul across the freeway gulch.

The Mississippi River downstream from St. Paul holds the principal "nuisance industries" of the metropolis, including the North Star steel mill, a mini-mill that recycles scrap metal, such as crushed automobiles, into usable steel products; a chemical plant; and two oil refineries that receive crude oil by pipeline from western Canada. Packing plants were built at the nearby stockyards of South St. Paul in the 1890s, but the last one closed in 1984, and the stockyards area is still waiting for redevelopment.

The area south of the Minnesota River from Minneapolis lagged until the construction of freeways made it accessible. Burnsville and Apple Valley were the first suburbs to develop, and Mendota Heights and Eagan boomed after the completion of Interstate 35E south from St. Paul. The area west of Burnsville has a number of commercial entertainment enterprises, including Valley Fair Amusement Park, the racetrack at Canterbury Downs, and the Mdewakanton Sioux gambling casino at Mystic Lake. Shakopee has massive riverside elevators that store grain trucked in from farmland to the west until it can be loaded onto barges and taken down the river.

The Twin Cities developed as the head of navigation outside the great bend of the Mississippi River. They grew with sawmilling and flour milling at the Falls of St. Anthony. Over a century and a half they have become an economically diversified regional capital at the eastern edge of a sparsely populated trade area that extends west to the Rocky Mountains. They have grown large in the age of the automobile, and their housing consists primarily of single-family homes on spacious lots. They enjoy the perquisites of being a metropolis and are fortunate to have only some of the challenges that face older, larger, high-density cities.

Kittson
Roseau
Lake of the Woods
Marshall
Beltrami
Koochiching
Pennington
Red Lake
Polk
Clearwater
Lake
Cook
Itasca
Mahnomen
Norman
St. Louis
Hubbard
Cass
Clay
Becker
Crow Wing
Aitkin
Carlton
Wadena
Wilkin
Otter Tail
Pine
Todd
Mille Lacs
Grant
Douglas
Kanabec
Morrison
Traverse
Benton
Stevens
Pope
Stearns
Chisago
Big Stone
Sherburne
Isanti
Swift
Anoka
Kandiyohi
Meeker
Wright
Washington
Lac Qui Parle
Chippewa
Hennepin
Ramsey
Yellow Medicine
Renville
McLeod
Carver
Scott
Dakota
Sibley
Lyon
Redwood
Nicollet
Goodhue
Rice
Le Sueur
Wabasha
Brown
Pipe-stone
Murray
Cottonwood
Watonwan
Blue Earth
Waseca
Steele
Dodge
Olmsted
Winona
Rock
Nobles
Jackson
Martin
Faribault
Freeborn
Mower
Fillmore
Houston
1. Brainerd
2. Alexandria
3. Detroit Lakes
4. Park Rapids
5. Walker
6. Grand Rapids
7. Aitkin
8. Cuyuna Range
9. Gull Lake
10. Little Falls
11. Pequot Lakes
12. Breezy Point
13. Lake Mille Lacs

15 The Lakeshore Resort and Retirement Belt

Since World War II, increasing numbers of Minnesota city folk have flocked to the state's rural areas in search of recreation and retirement. Cities large and small have sprawled into the adjacent countryside as residents have sought larger homes on larger lots. In addition, greater affluence, longer vacations, early retirement, faster highways, and better cars have enabled city people to move to or spend weekends in more distant rural areas blessed with amenities such as lakes, woods, scenic views, and wildlife. An extraordinary new form of "dispersed" city has evolved in the lakes area north of Brainerd in north central Minnesota, a city that has a population of 300,000 to 500,000 people on summer weekends.

Every summer Friday afternoon, highways leading north from the Twin Cities are jammed with vehicles, many of them towing boats, as people flock to the lake. It is always "the lake," as though Minnesota had only one, rather than 12,000 or more. For many Minnesotans, in fact, the state does have only a single special lake, the one graced by the family's second home. A summer cottage on a lake symbolizes the good life for many Minnesotans. It means escape from the heat of summer pavements and from the daily routine of office or shop. It also symbolizes having "arrived" economically because lakeshore property is expensive almost anywhere in Minnesota.

The ability to afford a second home is a powerful status symbol. Flaunting wealth is socially unacceptable, but people manifest their money more subtly—by living in the right suburb, driving the right car, wearing the latest fashions, sporting a midwinter tan, and having a cottage on the lake, preferably on the right lake, because some lakes are "better" than others.

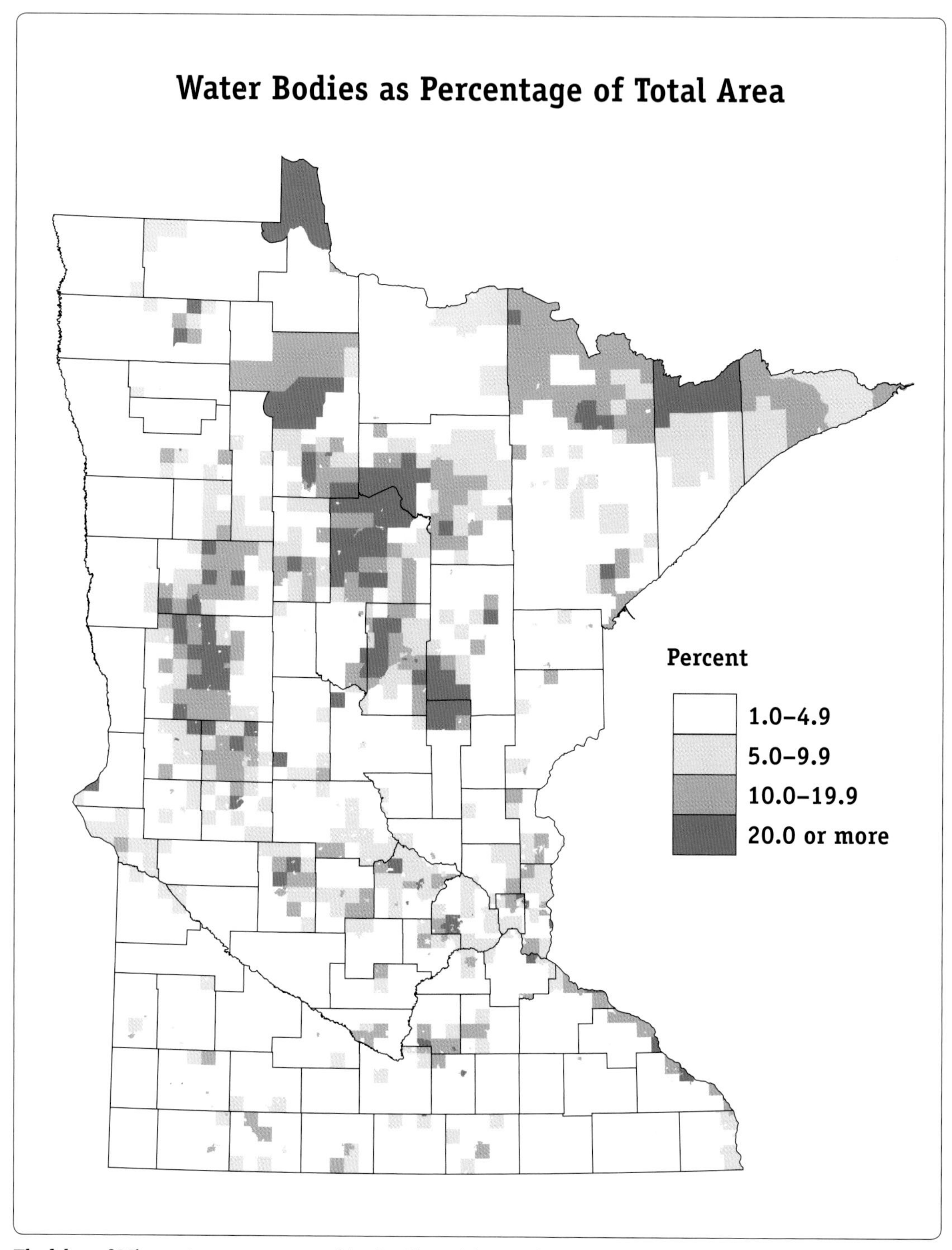

The lakes of Minnesota are concentrated in the Alexandria Moraine, the area north of Brainerd, and the Boundary Waters Canoe Area Wilderness.

Most parts of Minnesota have lakes, and rows of summer cottages festoon most Minnesota lakes of any size. The greatest lakeshore summer-cottage belt, however, arches across the north center of the state. It is anchored by Alexandria, Detroit Lakes, Park Rapids, Walker, Grand Rapids, Aitkin, and Brainerd and includes parts or all of Douglas, Otter Tail, Becker, Hubbard, Cass, Crow Wing, Aitkin, Mille Lacs, Kanabec, and Pine counties. The keystone and the epitome of this arch is the area north of Brainerd in Crow Wing and southern Cass counties.

The Brainerd lakes area has 466 "ice-block" lakes. These lakes originated when enormous chunks of stagnant glacial ice were left in a moraine and then buried by sandy outwash. When the ice blocks finally melted, they left lake basins with sandy shores. Wave and current action has given the lakes their rounded shorelines. In addition, some of the lakes of eastern Crow Wing County are the flooded pits of abandoned iron ore mines in the Cuyuna Range. These lakes are deep, with precipitous shorelines that crumble easily.

The Brainerd lakes area did not come into its own as a popular lakeshore resort area until after World War II, despite its attractiveness, because it was too far and inaccessible for most Minnesotans even though Minnesotans have sought summer shores almost from the very first. Before 1900, affluent residents of the Twin Cities took interurban railways and trolleys to resorts and summer cottages on Lake Minnetonka and White Bear Lake, areas that have long since become incorporated into the built-up fabric of the metropolitan area. Affluent residents of smaller places also sought summer solace on the shores of nearby lakes, but few people had the necessary free time or the money to vacation at more distant lakes.

Logging railroads first opened up the north woods lake country to visitors. Local boat owners met trains at lakeside stations and ferried passengers to lakeshore camps and cottages. People who liked to hunt and fish organized rod and gun clubs, bought cheap cutover lakeshore land, and built rustic clubhouses for their members. Sufferers from asthma and hay fever came north to stay in old logging camps or in the rough hotels that the lumber companies had built for visiting executives.

Some summer visitors paid farmers to let them sleep in the haymows of their barns, and farm wives made money by serving them home-cooked meals in the farmhouse kitchen. Some shrewd farmers saw the chance to make more money by building rough cabins they could rent to summer visitors, and thus originated the traditional north woods lodge-and-cabins resort, where city families came to spend their entire vacation.

Eventually resort owners upgraded their cabins into light housekeeping units, but the early cabins were little more than wooden tents where people could sleep.

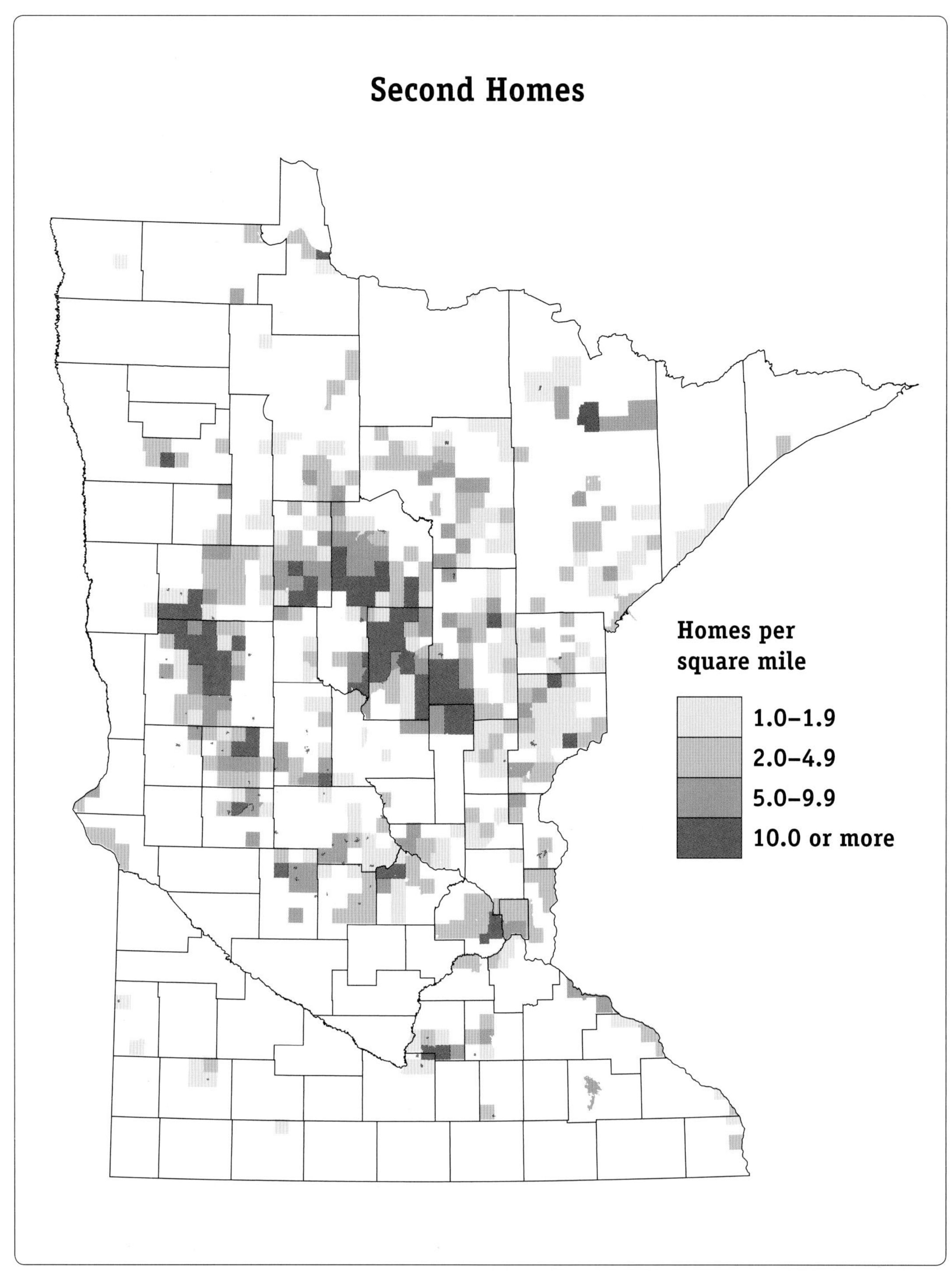

Second homes are concentrated near the lakes of the Alexandria Moraine and the area north of Brainerd.

They were rude shacks with four walls and a roof, a door and perhaps a window or two, but no plumbing, running water, or electricity. The heart of the resort was the large central lodge, where meals were served and where people could gather in the evening or in inclement weather to play games, socialize, and entertain each other.

Soon some regular summer visitors decided that they would prefer to have their own private cottages instead of paying to stay in a small, cramped resort cabin. After World War I, developers in the lakes area north of Brainerd began to subdivide lakeshore land into lots they could sell to people who wanted to build their own homes. In 1903 Thomas W. Harrison of Topeka, Kansas, bought a large tract of land near the southern end of Gull Lake, ten miles northwest of Brainerd, from the Northern Pacific Railroad; he hoped to develop it into a residential area, but he was ahead of his time.[1]

Twenty years later his son, John, joined C. H. Start of Kansas City and formed a real estate company to subdivide the area and sell individual lots with at least 100 feet of shorefront to "a very desirable class of people." In 1926 they developed a golf course to make the area even more attractive, and they advertised that it offered fishing, boating, swimming, and golf, plus dancing and bridge in the clubhouse. Their brochure boasted that "by the end of 1928 there will be continuously surfaced roads from Brainerd to Texas, through Kansas City, all-weather roads!"

The trek from the Twin Cities to the Brainerd lakes area took five hours or longer and was always an adventure. Automobiles were slow and undependable, and most roads were unpaved, unmarked, and unfenced. They were choked with great clouds of dust in dry weather and turned into muddy quagmires after every rain. Highways were neither marked nor numbered, so taking a wrong turn and getting lost were a predictable part of the trip. Drivers had to be alert for farm animals, which routinely wandered onto the unfenced highway, and at one time or another virtually everyone hit and killed a chicken, a dog, or an even larger animal.

Running a resort was equally adventurous. The summer resort season lasted only from Memorial Day or the Fourth of July to Labor Day, so the operator had only a dozen frantic weeks in which to make his total income for the year. Over time many resorts have gone broke or been sold to others, and the names of some resorts have changed more than once when new owners have taken over. In 1935, Harrison and Start sold their resort to Jack Madden, and it still bears his name.

Lakeshore lots sold slowly during the Great Depression that began in 1929, and even more slowly when gas was rationed during World War II, but people with money still could get to the Brainerd lakes area, and in the 1930s it became notorious as a den of iniquity. Local legend holds that movie stars such as Clark Gable and gangsters such as John Dillinger came to the area to disport themselves. Nightclubs brought in big bands for dancing and sold bootleg liquor during the Prohibition era. Gambling has always been popular in the lakes area, and during the 1930s many resorts, taverns, and stores had illegal slot machines. Periodic crackdowns by law enforcement officers confiscated more than 8,000 of these one-armed bandits.

The summer resort season is all too short.

When gas rationing ended after World War II, the Brainerd lakes area was ripe for rapid growth. The population of Crow Wing County, in fact, nearly doubled between 1950 and 2000. Faster cars and better highways cut the travel time from the Twin Cities in half. For many families, Mom and the kids could enjoy the cottage all week long, and Dad could drive up on weekends. Better pay and longer vacations enabled many more middle-class people to afford lakeshore cottages, although even today most cottage owners still have higher than average income and education.[2]

Over the last half century, highways have steadily been upgraded, and in 2005 the last remaining two-lane stretch, a nerve-rattling white-knuckle thirty miles north from Little Falls to Brainerd, became a four-lane road. The city of Brainerd, where summer weekend traffic often backed up bumper to bumper for five miles or more, was one of the few remaining bottlenecks. It was so bad that even the merchants in

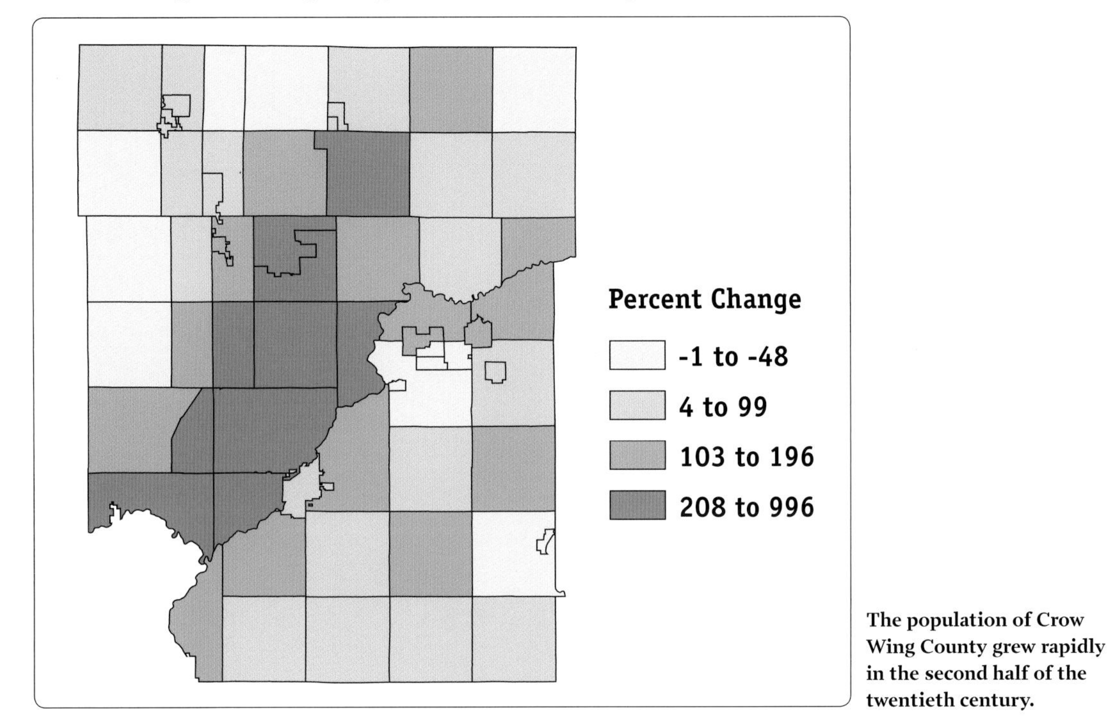

The population of Crow Wing County grew rapidly in the second half of the twentieth century.

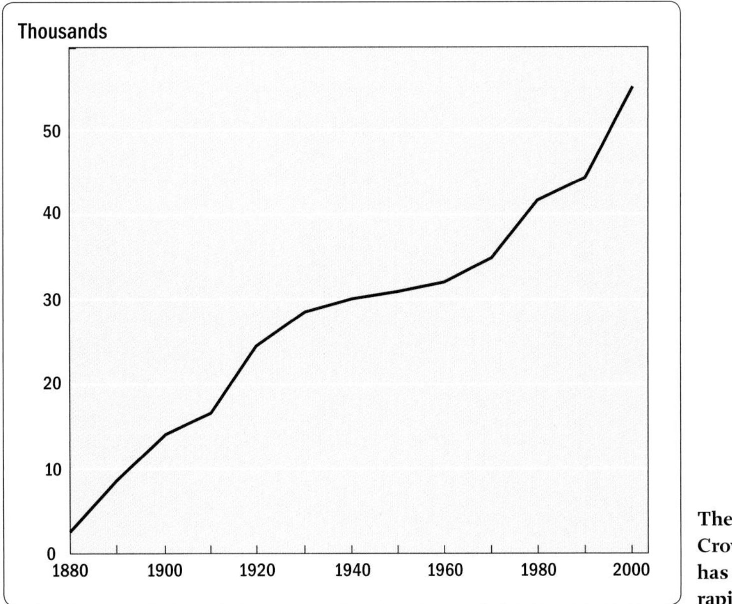

The population of Crow Wing County has grown even more rapidly since 1970.

Brainerd lobbied for a four-lane bypass to run west of Brainerd to Baxter. It opened in 2000, and big-box retailers immediately started to move in. Today the four-lane strip north of Baxter is rapidly becoming the principal commercial center of the dispersed city that is evolving in the Brainerd lakes area.

In 2000 the Brainerd lakes area claimed more than 15,000 lakeshore houses. On many lakes the houses are packed so close together that they almost touch each other on lots no more than 100 feet wide. Driving around a lake the traveler sees endless name signs and unpaved roads disappearing back into the underbrush, where owners try to protect their privacy with dense vegetation screens.

Cabin owners treasure their vacation time. They come to the lake to relax, and many happily pay someone else to do the heavy lifting, such as hooking up the plumbing, cleaning the yards and shorelines, putting in the docks and boat lifts in the spring, and reversing the process in the fall. They also need urban services such as trash and garbage collection, and some hire services to watch their homes in winter. Occasionally cabin owners need carpenters, electricians, or pest exterminators, and of course they need groceries, hardware, medicines, and the whole range of public and private services.

Today the lakes area has enough cottage owners to provide jobs for many local people for a significant part of the year. Cottage owners departing in the fall simply tell contractors to have particular jobs done by the time they return in the spring. Contractors are able to schedule jobs to keep crews busy through the winter, and many construction jobs are easier and faster to complete when the owner is absent.

New jobs have kept young people in the lakes area. In most rural Minnesota areas, young people must pack their bags and move to a city to find work after they graduate from high school or complete a stint in the military, but the availability of jobs in the lakes area not only has stemmed the out-migration of local young people but has actually attracted young in-migrants from other areas.

THE NUMBER OF GOOD JOBS IN LAKE COUNTRY has also increased because it has become a popular place to retire. Many people who bought lakeshore cottages in the 1950s and 1960s have winterized them for use as permanent homes when the owners reach retirement age. These people bring more money into the area. In 2000, more than one-third of the families in Crow Wing County were drawing Social Security checks, and about half of these families had additional retirement income.

Official census statistics today underestimate the number of retired people in Crow Wing County because many of them are "snowbirds" who can own or rent winter homes in Sun Belt states and miss the April census count. Snowbirds often stay in states with no state income tax for at least six months and one day each year, thereby establishing their official legal residence elsewhere. They pay no state income tax in Minnesota.

The population of Crow Wing County grew rapidly during the second half of the twentieth century. The age-sex pyramid showing ages and sexes in Crow Wing in 1950 is tall and lean, typical of most rural areas in Minnesota, where a common high-school graduation present was a suitcase. The county had lots of children, but few people remained after the age of twenty.

The county's population pyramid in 2000 is plumped out and prosperous, showing the effects of massive in-migration, especially among seniors. The number of baby boomers bulged predictably, and the declining numbers of children in the youngest age group showed that the baby boomers had already passed beyond their prime childbearing years. Even the very youngest age group or cohort, however, was as large as the same cohort had been in 1950.

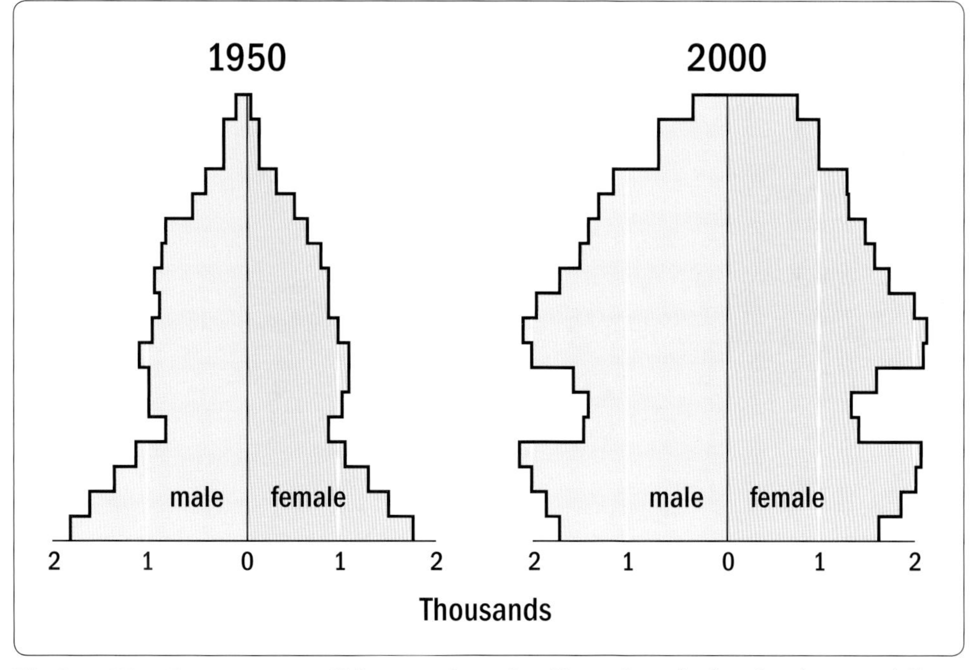

The Crow Wing County age pyramid for 2000 shows the effects of massive in-migration, especially among seniors.

Cohort survival ratios are the most effective method for showing migration into and out of an area by people of any given age group during each census decade.[3] A cohort survival ratio value of 100 shows no migration. Values higher than 100 show that people of that age group migrated into the area during the census decade, and values lower than 100 show that people of that age group migrated out.

During the 1950s, all Crow Wing County age groups declined through out-migration, as was true for most nonmetropolitan counties in Minnesota. The group aged twenty to twenty-four lost the largest numbers, because these young people had left for college or gone to the military or to seek their fortunes in the big city, and precious few of them ever returned. Even in the older cohorts, small numbers of people were leaving the county, and none were moving in.

In the 1960s, Crow Wing County continued to lose young people of college age, but some of them had started to come back home after they had graduated, and the working-age cohorts aged thirty to sixty-four enjoyed a slight increase. Crow Wing County was not reclaiming all its young people who had gone away, but it was doing a far better job than most nonmetropolitan counties in Minnesota.

Crow Wing County Cohort Survival Ratios

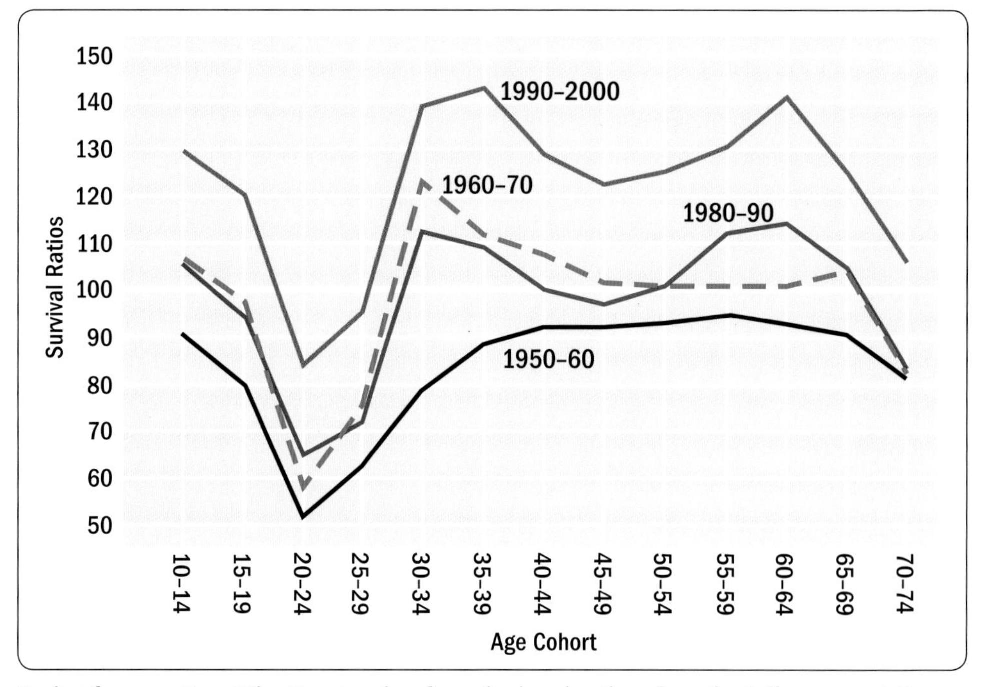

During the 1990s Crow Wing County enjoyed massive in-migration of people of all ages except the college-age cohorts.

The profiles for the 1970s and 1980s are nearly identical. The county continued to attract young working-age people, and it also began to attract significant numbers of in-migrants of retirement ages fifty-five to sixty-nine.

In their lifetimes most Americans make three major migrations: around age eighteen, when they graduate from high school; around age twenty-five to thirty, when they take their first long-term job; and sometime after age fifty-five, when they retire. In the 1990s, Crow Wing County benefited from the second two migrations, and the county even managed to attract significant numbers of people in the middle working-age cohort aged thirty-five to sixty, people who are usually less mobile.

During the 1990s, the county attracted large numbers of retired people, and this growth provided jobs that attracted large numbers of people of working age, especially people aged thirty to thirty-nine. These people brought their children, which explained the large influx of youngsters aged ten to nineteen. The rapid growth of the school-age population was manifest in one of the most impressive construction projects in the county in 2003, the huge new Pequot Lakes Elementary School a few miles west of the Breezy Point resort.

THE INFLUX OF COHORTS AGED FIFTY-FIVE AND OLDER indicates that more people are retiring at an earlier age, and many of them want a more active retirement lifestyle. The resort and retirement areas of the lakes country have adjusted to serve them. A generation ago the area offered mostly good fishing and cool breezes. Today, only a handful of visitors need to catch fish. Truly ardent fisherfolk venture farther north to Canada in search of less developed lakes.

Contemporary visitors want canoes, sailboats, pontoon boats, powerboats, water skis, and jet skis. They want indoor and outdoor swimming pools, plus sandy lakeshore swimming beaches. They want golf courses, tennis courts, and riding stables. They want nature trails, bike trails, and snowmobile trails. They want bingo, movies, and video game rooms to keep them entertained in the evening, and they want to sleep in rooms that their parents would have considered downright luxurious.

Resort owners strive to give their guests everything because they realize they are competing for tourist dollars in a global market that includes not just Wisconsin, but Hawaii and Acapulco and Europe. As a result, resort owners have encouraged the development of commercial recreational facilities such as amusement parks, water slides, go-kart tracks, miniature golf, and stock car and drag races at the Brainerd International Raceway. Unlike private cottages, these commercial establishments crave as much attention as they can get, and the gaudier the better. Many affect "north woods rustic" architecture, with as much log and stone and wooden shingle as possible.

Larger resorts in the Brainerd lakes area have capitalized on their location at the geographic center of Minnesota by developing conference centers and meeting rooms. They promote their golf courses and advertise the Brainerd Golf Trail, which boasts 300 holes of golf on eighteen courses within a forty-five minute drive. New golf carts have global positioning systems that tell the golfer the direction and distance to the next hole and tell the course manager the precise location of the cart at all times. The carts also have phones for ordering food and drinks and for requesting prompt medical assistance, a service much appreciated by seniors.

Brainerd Golf Trail

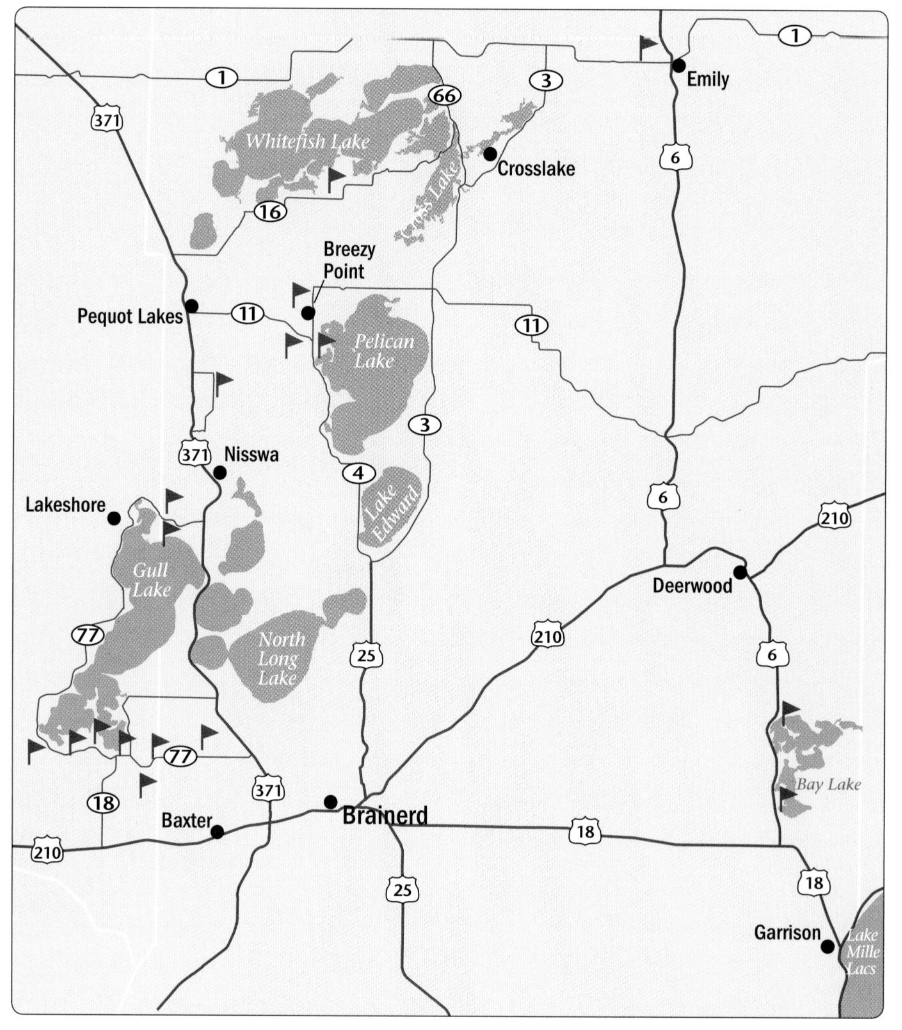

The Brainerd Golf Trail boasts 300 holes of golf on 18 courses within a 45-minute drive.

Many resorts now are open year-round. Older couples enjoy the spring and fall "shoulder" seasons when children are in school and the trees may be just leafing out or turning colors. Many resorts have facilities for cross-country and downhill skiing, snowmobiling, and ice fishing. The old railroad right-of-way that loops north from Brainerd has been paved and turned into the Paul Bunyan Trail for snowmobiles in winter and bicycles in summer. Minnesota has 7,000 miles of major snowmobile trails.[4]

People who neither ice-fish nor golf debate which is the sillier activity.[5] As one cynic observed, "In winter we fish through holes in the ice. Actually we just do a lot of drinking in huts on frozen lakes. Then, come spring, we drive cars out onto the ice and are surprised when they fall through. People will even drive around a tow truck pulling a car out of a hole in the ice."

The Quadna Mountain resort offers outdoor recreational activities for all four seasons.

Each winter the ice on Mille Lacs, some twenty miles east of Brainerd, is speckled with more than 5,000 ice-fishing houses. Set on timber frames that hold the floors six inches off the ice, they may have four to six fishing holes eight to sixteen inches in diameter and offer propane gas heat, chemical toilets, and battery-powered stoves, microwaves, and color television. Resorts along the shore rent them out in winter and cram them into storage lots in summer, but some of the houses are so nice that people rent them for summer cottages.

The Brainerd lakes area has hundreds of stores and eating places that cater to visitors. It also has thousands of real estate agents. Lakeshore property sells by the front foot, at prices of $500 to $1,000 or more per lakeshore foot. Many buyers like pine trees because they give the property a north woods look. Buyers come from Iowa, Nebraska, and Kansas, as well as Minnesota.

The resort economy of Crow Wing County has matured, and land prices are so high that development has spilled over into less expensive lakeshore areas in adjacent counties. There is still land available. In 1985 the Minnesota Department of Natural Resources estimated that the state had enough good lakeshore for 5 million homes spaced at 200-foot intervals, but development on many lakes is slow because

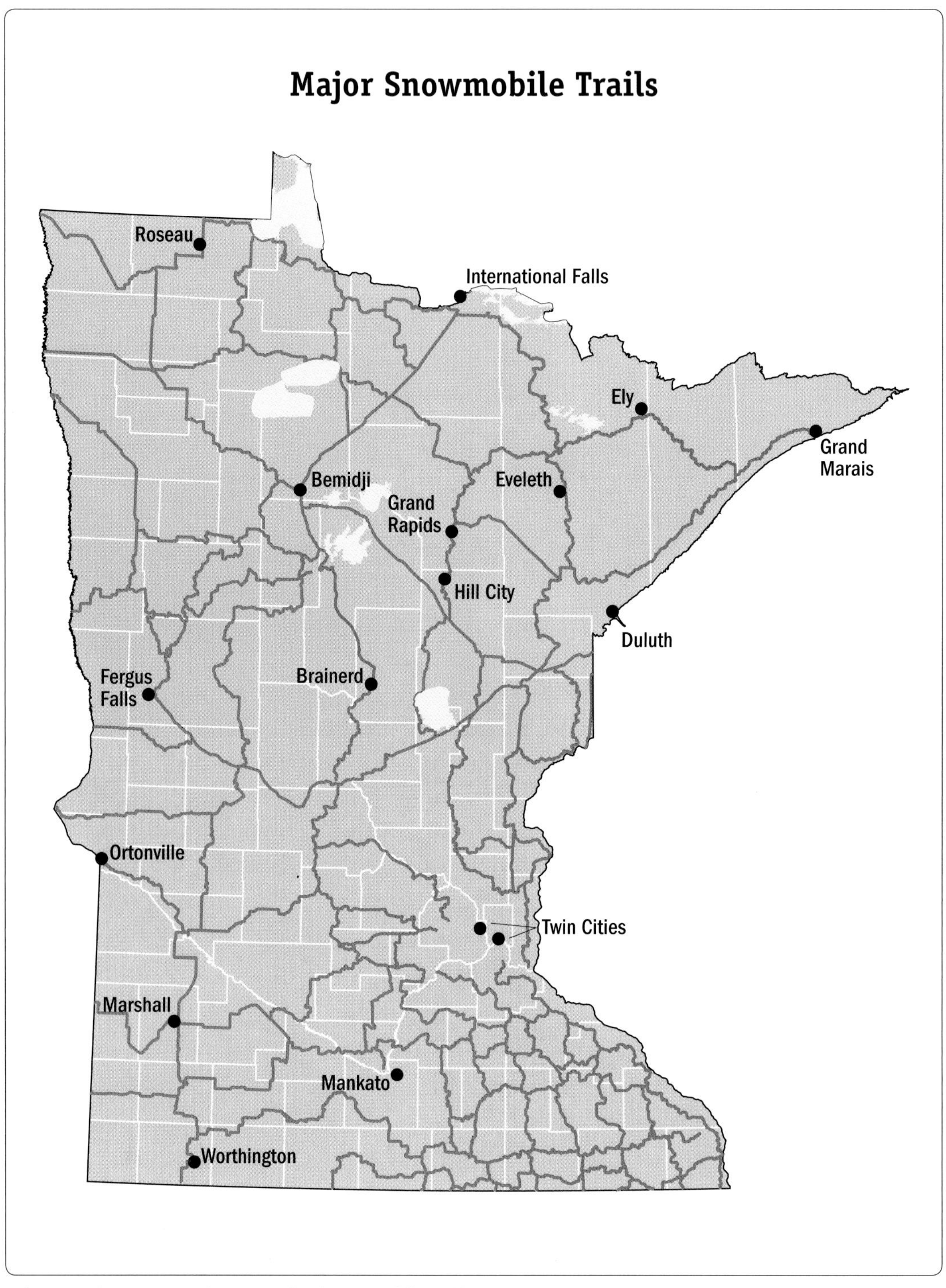

Minnesota has an extensive network of snowmobile trails.

Some people think ice fishing is fun.

most people, and especially the elderly, want to be within fairly easy driving distance of stores, hospitals, and other urban services and amenities.

People who move to the lakes are destroying the lakes they love. All lakes are doomed to die eventually as a result of natural processes, but lakeshore residents are accelerating their demise with leaky septic systems and runoff from fertilized lawns that encourage algal blooms. Runoff from roofs, roads, and parking lots adds metals, chemical solvents, road salt, and street dirt.

More boats, bigger boats, and louder boats increase noise pollution, and few lakes restrict their size, speed, hours, or wakes. Even good boating rules and regulations are poorly enforced because enforcement agencies are inadequately funded and management responsibility for many lakes is split between state, county, and local authorities. Lakeshore residents need to organize to protect their own lakes and to keep them as attractive and free of pollution as possible.

IN 2003 THE BRAINERD LAKES AREA had twenty motels that welcomed overnight visitors and seventy-five resorts, ranging from palatial establishments to small family operations with three or four simple cabins and outdoor toilets. Resorts reserve units by the week. Some families are regulars who have been staying for the same week in the same unit at the same resort for many years, and at the end of one year's stay they reserve the unit for the same week the following year.

The number of resorts is decreasing, but the size of the remaining resorts is increasing. Small resorts are closing because they are family-owned operations unable to generate capital to make the necessary improvements to keep them competitive. The owners' children are often unwilling to accept the headaches of the business when their parents die, so they subdivide the property and sell the lots for private lakeshore homes.[6]

Modern resorts have become so large and elaborate that they need maps to show visitors where to find their rooms and facilities. Two of the largest are Madden's on Gull Lake and Breezy Point on Pelican Lake.

In the early 1930s Jack Madden rented space for a soda fountain, a cigar counter, and a slot machine on the porch of a resort hotel near the failing Harrison and Start golf course on Gull Lake. He leased and soon purchased the course. In 1936 Madden expanded the hotel and began to build housekeeping cottages, and the resort has continued to grow ever since. In 1940 Madden talked his college friend Merrill Cragun into starting a resort of his own, and Cragun's has also continued to grow. In 2003 Madden's had three separate eighteen-hole golf courses plus a nine-hole course, and Cragun's had three eighteen-hole courses.

In the twenty-first century, Madden's is no longer just for golfers. It is a veritable city, with 294 hotel rooms and cabins, eight dining areas and a cookout pavilion, convention facilities, a shopping center, sports courts, beaches, activity programs, and a marina.

Breezy Point Resort on Pelican Lake has a much more checkered history. In 1921 Captain Billy Fawcett began to develop a flamboyant resort with accommodations for 1,000 guests. It flourished during the 1930s, but gas rationing closed it down during World War II. After the war, it changed owners and managers many times, and the original lodge was destroyed by fire. Ginny Sims, a well-known singer of the big-band era of the 1930s, bought the resort in 1963 and built some of the state's first condominiums there before declaring bankruptcy in 1966. In 1976 investors brought in Bob Spizzo, a recreational land development specialist, to rescue the resort. By 2002 Breezy Point employed 135 people year-round and three times that many in summer. Permanent workers had been at Breezy Point an average of seventeen years.

Among its many other firsts, Breezy Point was the first resort in Minnesota to develop time-share units, which by 2002 numbered 400. Time-share owners buy a specified week each year and pay an annual maintenance fee. Time-sharing spreads building and operating costs over many owners for the entire year and enables the resort to keep the same employees year-round. The resort also has a convention center, $9 million lodge, restaurant, and an all-year ice arena. Breezy Point features the lake and the links, but it also has a busy daily schedule of organized activities, live evening entertainment, and special children's programs. The philosophy is that happy children will want to come back when they grow up.

Minnesota's other renowned resort area is the North Shore of Lake Superior, one of the most scenic areas in the state. Bold, rocky cliffs rising from the lake command spectacular views across its waters, and picturesque waterfalls punctuate the trout-fishing streams that tumble down the steep slopes. Magnificent fall colors attract visitors each year, and in winter skiers enjoy some of the best downhill skiing in the state. While the North Shore has fewer homes than the lake country, its plethora of resorts, private campgrounds, and state parks serve visitors and enhance its importance as one of Minnesota's premier resort areas.

Clearly, the lakeshore economy of north woods Minnesota has matured. Mom-and-pop cabin and lodge operations have given way to palatial year-around establishments that compete in an international market. Summer cottages have been winterized or torn down and rebuilt as permanent retirement homes. Resorts and retirees generate enough jobs to attract young workers, big-box businesses are moving in, and a distinctive new form of dispersed city is making over the north country.

Madden's Resort occupies a peninsula on Gull Lake northwest of Brainerd.

Red River Valley
Northeast
Iron Range
Lakeshore Resort Area
Dairy Region
Metropolitan
Arc
Southwest
Kittson
Roseau
Lake of
the Woods
Marshall
Beltrami
Koochiching
Pennington
Red
Lake
Polk
Clearwater
Itasca
Lake
Cook
Mahnomen
Norman
Cass
St. Louis
Becker
Clay
Hubbard
Wadena
Crow
Wing
Carlton
Wilkin
Otter Tail
Aitkin
Pine
Todd
Morrison
Grant
Douglas
Kanabec
Benton
Mille
Lacs
Traverse
Pope
Stevens
Stearns
Big Stone
Isanti
Chisago
Sherburne
Swift
Wright
Anoka
Washington
Lac Qui
Parle
Chippewa
Kandiyohi
Meeker
Hennepin
Ramsey
McLeod
Carver
Yellow Medicine
Renville
Scott
Dakota
Sibley
Redwood
Lincoln
Lyon
Nicollet
Goodhue
Le Sueur
Rice
Wabasha
Brown
Pipestone
Murray
Cottonwood
Watonwan
Blue Earth
Waseca
Steele
Dodge
Olmsted
Winona
Rock
Nobles
Jackson
Martin
Faribault
Freeborn
Mower
Fillmore
Houston

16

What Next?

Geography is about becoming, and not simply about being. This geography of Minnesota—its land, weather, peoples, livelihoods, urban centers, and recreation and retirement regions—must conclude by asking what the state is becoming. Certain trends can be extrapolated, although trends are notoriously subject to unanticipated shocks. Some themes seem significant for the future of the state and the way in which these might play out in its diverse regions.

Minnesota's population is aging. The state will have to expand the medical and social services that an aging population will need, perhaps even at the expense of education. Where will elderly people choose to live? Retired people are no longer tied to jobs, so many may live wherever they choose. They can be a boon, bringing their pensions and transfer payments into an area, but they can be a burden as they become more elderly and need more services. Can and should Minnesota try to retain and recruit retired people in competition with areas where living is perceived to be easier?

Minnesota needs immigrants. The birthrate of the native-born white population of the state is below its replacement rate, and Minnesota must attract immigrants from other states and other countries to sustain its population growth. The state has always depended on immigration to power the growth of its population, but its new immigrants are Hmong, Somalis, and Mexicans, rather than Norwegians, Swedes, and Germans. Native Minnesotans will need to overcome any latent distrust for people of color who speak foreign languages in order to avoid tensions as the number of immigrants increases.

The scale of society and the economy is increasing. The scale of just about everything in modern life is growing rapidly, and we must add a zero or two to the way in which we think about business, farming, population, prices, speed, or just about anything else. Today one person can do what ten people once were needed to do, but a business needs ten customers where one once would suffice. Improved technology of transportation is shrinking distances, and improved technology of information management is encouraging the consolidation of control in larger organizations and larger places.

Economic and political forces will remain in constant tension. In a democratic capitalist society, the role of politics is to protect citizens from the worst excesses of the marketplace, and people will mobilize their political muscle to control and thwart economic developments they abhor, be it the spread of suburban residential areas ("sprawl") or the growth of large farms ("sustainability").

We are more interdependent. The modern economy and society are global, and no part of the world can be completely self-sufficient. Each place must provide some goods or services for other places in order to secure the goods and services it cannot generate from its own resources. Each place must ask of other places what they need instead of dictating what it will give them.

Resources remain essential. Each place has a unique suite of resources, human as well as natural. People must capitalize on these resources to supply the goods and services that they can produce most competitively. They must support entrepreneurs to develop these resources as effectively as possible, but they must also be sensitive to the limitations imposed by their resources. Human will has surprising power to triumph over obstacles, but it cannot work miracles.

HOW DO THESE TRENDS AND THEMES PLAY OUT in Minnesota's diverse regions? The people of northeastern Minnesota are its principal resource, because it has few natural resources. High-grade iron ore is exhausted, and taconite is an expensive substitute that competes with cheaper ore mined elsewhere. The region has no fossil fuels, and energy must be imported at considerable expense. Northeastern Minnesota lacks good local markets because it is remote from major concentrations of population that might be customers for whatever it might produce. Except for the North Shore of Lake Superior, it has few amenities that might attract retired people and few employment opportunities that might attract immigrants. Out-migration is far more probable, and the area faces the challenge of retaining its own young people.

The dogged optimism of Northeastern Minnesota's people is the region's principal resource, although this vision often transcends reality. Iron Range Resources

(IRR) has spearheaded an effort to develop information technology systems that will employ large numbers of people in outsourced back-office operations. The principal attraction for these operations is cheap labor, but other areas in the United States and abroad can offer even cheaper labor. Northeastern Minnesota may face a future of painful retrenchment.

Prospects seem brighter for the northern lakeshore resort and retirement areas, including the North Shore of Lake Superior. These areas cater to the aging population. They are enjoying a flush of growth and upgrading their infrastructure to cope with it. A new kind of dispersed city is evolving in the lakes area north of Brainerd. Many residents have two "second homes," a summer place in Minnesota and a winter place in the Sunbelt.

Lakeshore resort areas are pleasant places to live, at least in summer, or they will be until they become too crowded and too expensive. Development is already spreading to new lake areas that hitherto have been ignored, and some of the older areas have critical shortages of affordable housing for low-income workers who provide essential services. Minnesota resort areas also face increasing competition from distant areas around the globe that are becoming more easily accessible. For more than three decades doomsayers have been predicting that rising fuel prices will impact the lakeshore resort areas, but people have blissfully continued to ignore their predictions.

The Red River Valley is a frontier area where people are still trying to figure out how best to capitalize on agricultural resources. The level land is superbly suited to large, modern farm machines, and the soil is deep and free of stones, but the growing season is short and cool, and precipitation is meager. Competitive production of most commercial crops is difficult.

The farms of the valley will grow even larger, but what can they produce better than other areas? Farmers have experimented with a variety of crops. The current favorite, sugar beets, has been extraordinarily successful, but beets are heavily subsidized and protected, and they could be wiped out overnight by a Congress unsympathetic to agricultural subsidies. Wheat is the standby crop of expediency in the valley, but it does not pay particularly well, and no major wheat-producing area is farther from export markets for the crop.

Minnesota's central dairy region might become a major problem area because its small farms cannot compete with the large, new dairy operations that have evolved in other parts of the United States. A handful of Minnesota's entrepreneurs will develop large, modern dairy farms, but they may occupy only small tracts of land because feed may be shipped in from areas where it is produced more efficiently.

Marketing milk may become difficult because the number of processing plants is declining.

The picturesque rolling topography of the dairy belt is not well suited to the continuous cultivation of crops, and much of the land must be kept in pasture to protect the soil from erosion. These pastures will grow up in brush and trees if they are not grazed by cattle. They would be suitable for beef ranches, but it is expensive to acquire the large acreage needed for an efficient modern beef ranch. Former dairy farm land near cities might provide attractive residential sites for commuters, but more remote land might simply revert to nature.

Southwestern Minnesota will remain a prosperous corn and soybean farming region that will continue to lose population. Modern cash-grain farms need at least two square miles of cropland, and technological innovation will keep pushing that minimum acreage higher and higher. The population will continue to decline on farms and in the small towns that once served small farmers.

Some people worry that continuous cultivation of corn and soybeans will eventually impoverish the soil, but economic competition from new soybean-producing areas in Brazil and other parts of the world is of more immediate concern. Minnesota farmers must add value to the commodities they produce by processing them into products such as ethanol, vodka, and meat. Producers are feeding more Minnesota-grown grains to hogs and poultry, which is a good sign, but Minnesota sells a goodly share of its corn and soybean crop to cattle feeders who raise beef in other parts of the country. People who oppose large-scale livestock operations must recognize the consequences for the state's economy.

Southwestern Minnesota has too many tiny towns. Places that originated as agricultural service centers have lost their reason for being, and they are too small to grow. Wise public policy might buy out their residents and allow them to live elsewhere. Larger places must enhance their nonagricultural activities. None of them can expect to offer every service, and they must cooperate with neighboring places to create new dispersed metropolises in the countryside. Their future growth will depend on their ability to become attractive places for immigrants to live.

TO AN EVER-INCREASING DEGREE, the metropolitan arc that runs from St. Cloud to Rochester is the future of Minnesota. This metropolitan arc will continue to extend out into adjacent rural areas. Traffic congestion is becoming a serious problem, and the metropolis needs creative thinking about how best to connect people with their jobs. Since the oil crisis of the 1970s, people have been predicting that the scarcity and rising price of oil would slow or halt the expansion of residential areas into the adjacent countryside, but this slowdown does not yet seem imminent. Americans appear willing to make great sacrifices to keep their cars and single family homes. People in the metropolitan arc, many of whom no longer have rural roots, must reaffirm their commitment to and responsibility for people in other parts of the state. They need to be reminded that they are partners in the ongoing project known as "Minnesota."

Nicollet Mall and transit bus, Minneapolis.

NOTES

Chapter One

1. Bob von Sternberg, "-60°," *Minneapolis Star Tribune,* February 3, 1996, A1. Tower vies with Embarrass for the dubious distinction of being the coldest place in Minnesota. Both scorn International Falls, the self-proclaimed "Icebox of the Nation," where the temperature has never dropped below a balmy -55° below. Jim Ragsdale, "An Icy Relationship," *St. Paul Pioneer Press,* January 17, 2005, A1.

2. The standard reference work on the climate of Minnesota is Mark W. Seeley, *Minnesota Weather Almanac* (St. Paul: Minnesota Historical Society Press, 2006). The standard source of state climatic data is the *Minnesota Climatology Working Group Home Page,* http://climate.umn.edu (accessed June 2, 2006), the basis for maps in this chapter.

3. The theater of the seasons in Minnesota is described in Pete Boulay, "Hey, How's the Weather?" *Minnesota Conservation Volunteer,* July/August 2003, 32–43.

4. Ted Jones, "Ice Crashes Ashore at Lake Mille Lacs," *Star Tribune,* April 2, 1987, B1.

5. William Hoch, *Leaf Color Change in Autumn,* University of Wisconsin Garden Facts (Madison: University of Wisconsin Extension, 2004).

6. Carl Zimmer, "Those Brilliant Fall Colors May Be Saving Trees," *New York Times,* October 19, 2004, F1.

7. Chuck Haga, "Freezing Rainfall Leaves Ditches Full, Buses Empty," *Star Tribune,* November 29, 1987, A1.

8. Bill McAuliffe, "The Melting of Minnesota," *Star Tribune,* March 12, 2006, A1.

Chapter Two

1. The standard reference work is Richard W. Ojakangas and Charles L. Matsch, *Minnesota's Geology* (Minneapolis: University of Minnesota Press, 1982).

2. Larry Werner, "Gravel: Bedrock of Growth," *Star Tribune,* June 16, 2002, D1.

3. The classic work on the glaciation of Minnesota is H. E. Wright Jr., "Quaternary History of Minnesota," in *Geology of Minnesota: A Centennial Volume,* ed. P. K. Sims and G. B. Morey (St. Paul: Minnesota Geological Survey, 1972), 515–47.

4. Mary Lynn Smith, "Putting New Energy into Wind Power," *Star Tribune,* September 27, 2005, Bl.

5. Estimates of the amount of rent paid for wind towers seem to be about as reliable as fish stories. Robert Franklin, "Sowing the Wind," *Star Tribune,* February 19, 2006, D2.

6. B. A. Lusardi, *Minnesota at a Glance: Quaternary Glacial Geology* (St. Paul: Minnesota Geological Survey, 1997).

7. The most detailed source is R. K. Hogberg,

Environmental Geology of the Twin Cities Metropolitan Area, Educational Series 5 (St. Paul: Minnesota Geological Survey, 1971).

Chapter Three

1. John W. Williams, Bryan N. Shuman, Thompson Webb III, Patrick J. Bartlein, and Phillip L. Leduc, "Late Quaternary Vegetation Dynamics in North America: Scaling from Taxa to Biomes," *Ecological Monographs* 74 (2004): 309–34.

2. E. C. Pielou, *After the Ice Age: The Return of Life to Glaciated North America* (Chicago: University of Chicago Press, 1991); John R. Tester, *Minnesota's Natural Heritage: An Ecological Perspective* (Minneapolis: University of Minnesota Press, 1995), 16–17.

3. Margaret B. Davis, "Holocene Vegetational History of the Eastern United States," in *Late Quaternary Environments of the United States,* vol. 2, *The Holocene,* ed. H. E. Wright Jr. (Minneapolis: University of Minnesota Press, 1983), 166–81.

4. Francis J. Marschner, *The Original Vegetation of Minnesota, Compiled from U.S. General Land Office Notes,* 1:500,000 (St. Paul: North Central Forest Experiment Station, 1974); Marschner's wall map was generalized to produce a page-sized map published by the Minnesota Department of Natural Resources, Natural Heritage Program, Biological Report No. 1, *Natural Vegetation of Minnesota at the Time of the Public Land Survey, 1847–1907* (St. Paul: Minnesota Department of Natural Resources, 1988).

5. Classic ecological studies of the Big Woods include Rexford F. Daubenmire, "The 'Big Woods' of Minnesota: Its Structure and Relation to Climate, Fire, and Soils," *Ecological Monographs* 6 (1936): 233–68; Eric C. Grimm, "Fire and Other Factors Controlling the Big Woods Vegetation of Minnesota in the Mid-nineteenth Century," *Ecological Monographs* 54 (1984): 91–311; and Charles E. Umbanhowar, "Interaction of Fire, Climate and Vegetation Change at a Large Landscape Scale in the Big Woods of Minnesota, USA," *Holocene* 14 (2004): 661–76.

6. Interesting statistics on land-use change in the Big Woods are included in a booklet titled *Managing Landscapes in the Big Woods Ecosystem* (St. Paul: Minnesota Department of Natural Resources, n.d.).

7. Minnesota Natural Heritage Program Section of Wildlife, Minnesota Department of Natural Resources, Biological Report No. 18, *Minnesota Dwarf Trout Lily,* (St. Paul: n.d.)

8. There are ecological and aesthetic differences between the rare old-growth tall-pine forests and the more common uncut (or so-called virgin) pine forest that includes short-lived jack pine. See Clifford Ahlgren and Isabel Ahlgren, *Lob Trees in the Wilderness* (Minneapolis: University of Minnesota Press, 1984), 101–2.

9. See Kurt Rusterholz, "Minnesota's Old Growth Forests," *Minnesota Forests* 3 (1990): 12–16. John R. Tester also discusses remnant old growth in *Minnesota's Natural Heritage,* 105.

10. Dwight A. Brown, Philip J. Gersmehl, and Susy S. Ziegler, *Alternative Biogeographies of the Global Garden* (2nd ed., Dubuque: Kendall/Hunt, 2007), 262–65.

Chapter Four

1. An excellent and highly readable account of Dakota life is found in chapter 3 of Dave Kenney, *Northern Lights: The Stories of Minnesota's Past* (St. Paul: Minnesota Historical Society Press, 2003), 24–39.

2. A splendid description of the fur trade is found in Carolyn Gilman, *Where Two Worlds Meet: The Great Lakes Fur Trade* (St. Paul: Minnesota Historical Society, 1982).

3. *The Grand Portage Guide* (Grand Marais, MN: Grand Portage National Monument, 2002).

4. Grace Lee Nute, *The Voyageur* (New York: Appleton 1931; reprint, St. Paul: Minnesota Historical Society Press, 1987).

5. Grace Lee Nute, "Posts in the Minnesota Fur-trading Area, 1660–1855," *Minnesota History* 11 (1930): 353–85.

6. Rhoda R. Gilman, Carolyn Gilman, and Deborah M. Stultz, *The Red River Trails: Oxcart Routes between St. Paul and the Selkirk Settlements, 1820–1870* (St. Paul: Minnesota Historical Society, 1979).

7. Laura J. Smith, "Native American Trust Land Transfers in Minnesota," *CURA Reporter* 34, no. 2 (2004): 19–25.

8. Douglas Clement, "The Wealth (and Poverty) of Indian Nations," *fedgazette* 18, no. 2 (2006): 1.

9. Douglas Clement, "Indianpreneurs,"

fedgazette 18, no. 2 (2006): 4.

10. Patrick Sweeney, "Pawlenty Puts Price on Casino Monopoly," *St. Paul Pioneer Press,* October 23, 2004, A1.

11. Anthony Lonetree, "Tribal Land Issue Heating Up in Scott County," *Star Tribune,* February 11, 2006, B1.

Chapter Five

1. The classic reference work is Benjamin Horace Hibbard, *A History of the Public Land Policies* (Madison: University of Wisconsin Press, 1965).

2. Francis J. Marschner, *Land Use and Its Patterns in the United States,* Agriculture Handbook No. 153 (Washington, D.C.: U.S. Department of Agriculture, 1959).

3. Norman J. W. Thrower, *Original Survey and Land Subdivision: A Comparative Study of the Form and Effect of Contrasting Cadastral Surveys,* Association of American Geographers, Monograph No. 4 (Chicago: Rand McNally, 1966).

4. Hildegard Binder Johnson, *Order upon the Land: The U.S. Rectangular Survey and the Upper Mississippi Country* (New York: Oxford University Press, 1976).

5. Fremont P. Wirth, *The Discovery and Exploitation of the Minnesota Iron Lands* (Cedar Rapids, Iowa: Torch Press, 1937), 107.

6. Wirth, *Minnesota Iron Lands,* 122.

Chapter Six

1. The essential reference work on lumbering is Agnes M. Larson, *History of the White Pine Industry in Minnesota* (Minneapolis: University of Minnesota Press, 1949).

2. Theodore C. Blegen, *Minnesota: A History of the State* (Minneapolis: University of Minnesota Press, 1963), 315–37.

3. Rand E. Rohe, "Geographical Impact of Log Transportation: The River Driving Era in Wisconsin," *Wisconsin Academy Review* 27, no. 4 (1981): 17–24.

4. Marx Swanholm, *Lumbering in the Last of the White Pine States* (St. Paul: Minnesota Historical Society, 1978).

5. *Celebrating a Century of Papermaking in Grand Rapids* (Grand Rapids, MN: Blandin Paper Company, 2001); *Welcome to Boise Cascade in International Falls* (International Falls, MN: Boise Cascade Corporation, n.d.).

6. *Lockwood-Post's Directory of the Pulp, Paper and Allied Trades,* various years.

7. Structural Board Association, *OSB Guide,* http://www.osbguide.com/osbfacts.html (accessed June 4, 2007).

8. Bob Fick, "Boise Cascade to Sell Timber Assets," *Star Tribune,* July 27, 2004, D3; John Myers, "Breaking Up the Forest," *Minnesota Conservation Volunteer,* January/February 2006, 8–17.

Chapter Seven

1. Harold Austin Meeks, "The Growth of Minnesota Railroads, 1857–1957," master's thesis, University of Minnesota, 1957.

2. A massive cornucopia of information about migrants to Minnesota is in June Drenning Holmquist, ed., *They Chose Minnesota: A Survey of the State's Ethnic Groups* (St. Paul: Minnesota Historical Society Press, 1981).

3. Each census since 1850 has listed the number of Minnesota residents who were born in each state of the United States and in each foreign country. Subtracting the number listed in the preceding census produces an estimate of the net number of immigrants from each state for each decade. The actual number of immigrants was greater than this estimate, because some earlier immigrants had died during the decade.

4. Leonard S. Wilson, "Some Notes on the Growth of Population in Minnesota," *Geographical Review* 30 (1940): 660–64.

Chapter Eight

1. Edward Van Dyke Robinson, *Early Economic Conditions and the Development of Agriculture in Minnesota,* Studies in the Social Sciences, No. 3, *Bulletin of the University of Minnesota* (Minneapolis: University of Minnesota, 1915).

2. Hildegard Binder Johnson, "King Wheat in Southeastern Minnesota," *Annals of the Association of American Geographers* 47 (1957): 350–62.

Chapter Nine

1. John Fraser Hart, *The Changing Scale of American Agriculture* (Charlottesville: University of Virginia Press, 2003), 62–111.

2. Joy Powell, "Land O'Lakes 2002 Earnings Rise," *Star Tribune,* February 27, 2003, E3.

3. Robert Franklin, "Dairy Spurned by Township May Get Pulled into City," *Star Tribune,* April 25, 2006, B4.

Chapter Ten

1. John Fraser Hart, "Change in the Corn Belt," *Geographical Review* 76 (1986): 51–72.
2. Hart, *The Changing Scale of American Agriculture,* 14–39.
3. John Fraser Hart, "Part-Ownership and Farm Enlargement in the Midwest," *Annals of the Association of American Geographers* 81 (1991): 66–79.
4. John Fraser Hart and Lisa M. Rainey, "Redundant Farmsteads in Minnesota," *CURA Reporter* 28, no. 3 (1998): 1–6.
5. Tom Meersman, "Ethanol Plants OK New Rules," *Star Tribune,* February 4, 2004, A1; Alexei Barrionuevo, "For Good or Ill, Boom in Ethanol Reshapes Economy of Heartland," *New York Times,* June 26, 2006, A1; and "The Great Corn Rush," *Star Tribune,* September 24, 2006, A1, map on A18.
6. Matthew L. Wald, "Corn Power Put to the Test, as Experts Try to Find a New Recipe to Make It," *New York Times,* February 7, 2006, F3.
7. Greg Gordon, "Water Supply Can't Meet Thirst for New Industry," *Star Tribune,* December 26, 2005, A1.

Chapter Eleven

1. William Casey, *Trade Centers of the Upper Midwest: 1999 Update* (Minneapolis: University of Minnesota Center for Urban and Regional Affairs, 1999).
2. John C. Hudson, *Plains Country Towns* (Minneapolis: University of Minnesota Press, 1985).
3. John C. Hudson, "Towns of the Western Railroads," *Great Plains Quarterly* 2, no. 1 (1982): 41–54.
4. John Fraser Hart and Tanya Bendiksen Mayer, "Tough Times for Small Towns," *CURA Reporter* 21, no. 4 (1991): 8–11.
5. John Fraser Hart, "Small Towns in Minnesota Are Growing Again," *CURA Reporter* 32, no. 3 (2002): 8–11.
6. John Fraser Hart, "Small Towns and Manufacturing," *Geographical Review* 78 (1988): 272–87.
7. Dave Senf and Thomas Anding, "Retail Sales Trends in Minnesota," *CURA Reporter* 18, no. 5 (1988): 1–4.
8. John Fraser Hart, Neil E. Salisbury, and Everett G. Smith, Jr., "The Dying Village and Some Notions about Urban Growth," *Economic Geography* 44 (1968): 343–49.

Chapter Twelve

1. Our sources on the iron ranges include Williams Watts Folwell, *A History of Minnesota* (St. Paul: Minnesota Historical Society, 1930), 4:1–59; Fremont P. Wirth, *The Discovery and Exploitation of the Minnesota Iron Lands* (Cedar Rapids, Iowa: Torch Press, 1937); and Theodore C. Blegen, *Minnesota,* 358–83.
2. Ojakangas and Matsch, *Minnesota's Geology,* 125–38.
3. A local entrepreneur horrified the good merchants in Gilbert by proposing that the town should try to stimulate business and capitalize on its history by organizing a Whorehouse Days summer festival, because at one time nearly every building on Main Street was a speakeasy, a red-light house, or both; Bob von Sternberg, "Iron Range Town Wants to Red-Light Festival's Name," *Star Tribune,* January 28, 2005, A1.
4. *Hull Rust Mahoning: A National Historic Landmark* (Hibbing, MN: Tourist Center Seniors, n.d.).
5. Larry Oakes, "Taconite Is Back," *Star Tribune,* January 8, 1996, D1.
6. Larry Oakes, "North Country Blues," *Star Tribune,* May 20, 2001, D1.
7. IRR publishes a useful quarterly newsletter, *RangeView.*
8. Denis P. Gardner, *Minnesota Treasures: Stories behind the State's Historic Places* (St. Paul: Minnesota Historical Society Press, 2004), 107–10.
9. *Twenty-five Years of Sharing* (Duluth, MN: Depot Foundation, 2004).
10. For example, one wonders about the economic and environmental wisdom of hauling coal to northeastern Minnesota to build a power plant; Tom Meersman, "New Coal Plant Given Renewable Energy Grant," *Star Tribune,* February 18, 2005, A1.
11. A recent manifestation of Iron Range optimism is Dee DePass, "Revival on the Range," a

series of five articles about plans for development on the Range that was published in the business section of the *Star Tribune,* November 5–9, 2006; within ten days one ambitious project had already been scrubbed; "Range's Iron Nugget Project Called Off," *Star Tribune,* November 19, 2006, B4.

Chapter Thirteen

1. John Fraser Hart, "What Is the Population of St. Cloud?" *CURA Reporter* 15, no. 4 (1985): 6–9.

2. John R. Borchert and Donald P. Yaeger, *Atlas of Minnesota Resources and Settlement* (Minneapolis: University of Minnesota Center for Urban and Regional Affairs, 1968).

Chapter Fourteen

1. Sources for this chapter include Richard Hartshorne, "The Twin City District: A Unique Form of Urban Development," *Geographical Review* 22 (1932): 431–42; John R. Borchert, "The Twin Cities Urbanized Area: Past, Present, Future," *Geographical Review* 51 (1961): 47–70; John Fraser Hart and Russell B. Adams, "Twin Cities," *Focus* 20, no. 6 (1970): 1–7; Ronald F. Abler, John S. Adams, and John R. Borchert, *The Twin Cities of St. Paul and Minneapolis* (Cambridge, Mass.: Ballinger, 1976); Thomas J. Baerwald and Karen L. Harrington, eds., *AAG '86 Twin Cities Field Trip Guide* (Washington, D.C.: Association of American Geographers, 1986); and John S. Adams and Barbara J. VanDrasek, *Minneapolis–St. Paul: People, Place, and Public Life* (Minneapolis: University of Minnesota Press, 1993).

2. Chris Serres, "High Hopes, Higher Taxes," *Star Tribune,* January 30, 2006, D1; Steve Alexander, "Midtown Mania," *Star Tribune,* May 18, 2006, D1; Neal St. Anthony, "Midtown Renaissance," *Star Tribune,* June 26, 2006, D1; and Chris Serres, "A World of Hurt at Midtown Global Market," *Star Tribune,* January 4, 2007, A1.

3. Calvin F. Schmid, *Social Saga of Two Cities: An Ecological and Statistical Study of Social Trends in Minneapolis and St. Paul* (Minneapolis: Bureau of Social Research, Minneapolis Council of Social Agencies, 1937).

4. Thomas J. Baerwald, "The Emergence of a New 'Downtown,'" *Geographical Review* 68 (1978): 309–18.

5. Dorothy Kalins and Cathleen McGuigan, "Minneapolis: Design City," *Newsweek,* June 26, 2006, has glowing descriptions of the Guthrie Theater, the Central Library, the Walker Art Center, and the Minneapolis Institute of Arts.

Chapter Fifteen

1. Duane R. Lund, *Gull Lake Yesterday and Today* (Cambridge, MN: Adventure Publications, 1999).

2. Karren Mills, "Modern Recreation Complexes Replace 'Ma and Pa' Resorts," *Star Tribune,* August 22, 1982, 7E.

3. John Fraser Hart and Susy Svatek Ziegler, "Migration in Minnesota," *CURA Reporter* 33, no. 1 (2003): 1–6.

4. Doug Smith, "Trails of Change," *Star Tribune,* December 14, 2005, C14.

5. Larry Oakes, "It's a Pastime, Passion, Way of Life—and a Cause of Death," *Star Tribune,* February 16, 1997, A1; Neal Karlen, "Searching for Walleye: A Winter's Tale," *New York Times,* February 21, 2003, F1; and Roland Sigurdson and Bill Lindner, "The Wide, Wide World of Ice Fishing," *Minnesota Conservation Volunteer,* January/February 2005, 26–33.

6. John Fraser Hart, "Resort Areas in Wisconsin," *Geographical Review* 74 (1984): 192–217.

GLOSSARY

Age cohort All of the people born during the same period of time, usually five years.

Bearing tree A tree into which the surveyor chopped a blaze mark to identify the corner of a section.

Big Woods The maple-basswood forests of southeastern Minnesota.

Biome A large area with ecological communities of similar plants, animals, soils, and climate.

Cash-grain farming Production of crops such as corn, soybeans, and wheat for direct sale, usually at the local grain elevator.

Commutershed The area from which people commute to a central place.

Continental climate A climate with cold winters and hot summers.

Correction line An east-west line where the spacing of range lines was adjusted to correct for the northward convergence of meridians.

Croupier An employee of a gambling casino who collects and pays bets.

Disturbance An event that frees resources (e.g., light, space, water, nutrients) that were formerly used by the pre-disturbance vegetation.

Ecosystem An ecological community and its physical and chemical environments.

Ecotone A transition zone between two adjacent ecological communities.

Evapotranspiration The delivery of water to the atmosphere by evaporation from wet surfaces and transpiration from plants.

Family farm A farm that produces at least $500,000 worth of farm products each year.

Geography The study of organized knowledge about places, which emphasizes human occupancy of their natural and built environments.

Glacial drift The general name for all material deposited by glaciers.

Glacial spillway A valley eroded by a stream carrying the overflow water from an ice-dammed glacial lake.

Greenhouse effect Atmospheric warming caused by gases that absorb outgoing heat energy near the earth's surface.

Jet stream A great river of air six to eight miles above the ground that blows from west to east at average speeds of 110 miles per hour.

Lactation period The length of time a cow continues to produce milk after she has given birth to a calf.

Lake Agassiz A vast prehistoric inland sea in northwestern Minnesota and adjacent areas that was dammed by glacial ice to the north.

Lake effect Climatic amelioration caused by large water bodies, which modify the temperature of adjacent land bodies to produce cooler summers, milder winters, and increased snowfall.

Loess Fine particles of rock that were ground up by glaciers and then deposited by the wind.

Logging slash Logs, branches, and leaves left by lumbermen that is prime fuel for forest fires.

Mêti The mixed-blood child of a European father and an Indian mother.

Metropolitan area One or more entire counties that contain a central city with a population of at least 50,000 persons.

Mixed crop-and-livestock farming Production of crops such as corn, oats, and hay to feed livestock such as hogs and cattle to market weight.

Moraine Choppy topography formed by material deposited at the edge of a melting glacier.

Outwash Sandy water-sorted material deposited by meltwater flowing away from a glacier.

Part-owner farmer A farmer who owns part of the land he farms and enlarges his operation by renting land from neighbors who no longer farm it.

Pineries The pine forests of northern Minnesota.

Prior potior A basic geographic principle that says the earlier one is the more influential.

Scrip A document that entitled its owner to a specified acreage of land.

Section One of 36 one-square-mile subdivisions of a survey township, with an area of approximately 640 acres.

Snowbird A Minnesota resident who migrates to a warmer climate for the winter.

Solar radiation Energy received from the sun that drives the weather and climate of the earth.

Stumpage The right to harvest timber from the land without owning the land.

Till Heterogeneous mixture of rock fragments of all sizes deposited by glaciers.

Till plain Gently undulating land surface formed by the uneven deposition of till by a melting glacier.

Topography The shape of the surface of the land.

Township A six-mile square created by the Public Land Office (township and range) survey system. Most survey townships are also civil townships.

Urban place A place that has a population of 2,500 persons or more.

Urbanized area The built-up core of a metropolitan area, including both the central city and its densely populated suburbs.

Voyageur A man employed by a fur-trading company to transport goods and people to and from trading posts in the Northwest.

Weather front The boundary between two air masses, which is usually associated with changes in temperature and precipitation.

Wind turbine A tall tower with giant blades that rotate in the wind to generate electric power.

Witness tree A tree into which the surveyor chopped a blaze mark to identify the corner of a section.

Woodland Land covered by woody vegetation.

INDEX

Agriculture, 295–97
  See also Crops and cropland; specific crops
Aitkin, MN, 93, 277
Alexandria, MN, 208, 212, 277
American Indians. *See* Native Americans
Andersen Windows, 99
Angus, MN, ***197***
Anoka County, 239–40
Apple Valley, MN, 31
Austin, MN, 183, 193

Babbitt, MN, 219, 230
Barron, WI, 183
Baudette, MN, 98
Baxter, MN, 192, 282
Bayport, MN, 99
Beardsley, MN, 7
Becker County, 53, 277
Benton County, 241
Biwabik, MN, 223, 232
Big Sioux River, 79
Big Stone Lake, 40, 79
Big Woods, ***44***, 50–52, 110
Biological diversity
  "bearing trees" as historical record, 80
  bluff country, 54–55
  coniferous forest, 52–53
  deciduous forest, 50–51
  forest fragmentation and, 51–52, 103
  vegetation restoration for, 55

Biomes, 43, ***44***, 46, 50
Bluff country, 26, 29, 38, 43, 54–55
Borchert, John, 189
Boundary Waters Canoe Area Wilderness, 21, 53, 222
Brainerd, MN, retirement area, 283–91
  See also Crow Wing County
Browns Valley, MN, 39
Brule River, 60
Buffalo Ridge (Coteau des Prairies), 36–37
Bush, Archibald, 273

Calumet, MN, 226
Canada
  determining MN boundaries, 77–79
  fur trade, 58–67
  glacial periods, 29
  Lake Agassiz, 38
  tundra vegetation, 45
  weather and climate, 12
Canadian Shield, 23
Cannon River, 50–52, 105
Cargill, William W., 255
Cargill Corporation, 255, 265
Carver County, 239–40
Casey, William, 189
Cass County, 277
Cuyuna Range, 232–33. *See also Iron Range*
Chippewa River, 40, 92
Chisago County, 117, 240

Chisholm, MN, 98
Christensen, Bob, 185–87
Christensen Family Farms, 185–87
Cities and towns. *See* city and town by name; Population and diversity; Settlement
Clay County, 48
Clearwater County, 53
Climate. *See* Weather and climate
Cloquet, MN, 96, 98, 99, 208
Cold Spring, MN, ***25***
Coniferous forest. *See* Northern coniferous forest
Conservation efforts. *See* Environment
Corn production
  as % harvested (1880–2000), ***138***, ***170***
  as % harvested (2002), ***171***
  for ethanol, 177–79, ***180***
  future in Minnesota, 295–96
  in mixed crop-livestock farming, 137–40, 167–73
Coteau des Prairies (Buffalo Ridge), 36–37
Cottonwood County, 29
Cragun, Merrill, 290
Crookston, MN, 163
Crops and cropland
  agricultural regions, ***150***
  corn and wheat (1880–2000), ***135***
  cropland (1880–2000), ***132***
  cropland (2002), ***134***
  future of Minnesota, 295–96
  hay as % harvested, ***140***
  See also Corn production; Farms and farmers; Wheat production
Crosby, MN, 219, 233
Crosby, John, 255
Crow Wing County, 192, 204–05, 219, 233, 277–85
Cuyuna Range. *See* Iron Range

Dairy industry
  creameries (1880), ***146***
  early years, 141–47
  farm economics, 149–53
  future in Minnesota, 153–60, 295–96
  Little Pine Dairy, ***159***
  milk cows per farm, ***142–43***
  milk processing plants, ***154***
  sales of dairy products (1950–2000), ***151***
Dakota County, 239–40
Danvers, MN, 212
Davis, E. W., 227
Deciduous forest. *See* Eastern deciduous forest
Detroit Lakes, MN, 277
Dinosaur fossils, 28
Douglas County, 277
Douglas Machine, 212
Drayton, ND, 163
Dubuque, IA, 108
Duluth, MN, ***235***
  Fond du Luth Casino, 75
  population growth, 206
  pulp and paper mills, 99
  as railroad hub, 93, 96–97
  role in mining industry, 224, 234–37
Durst Brothers (Ron, Allen & Ken), 153–58
Durst Brothers Dairy, 153–58

Eagan, MN, 265
Eames, H. H., 221
East Grand Forks, MN, 163
Eastern deciduous forest, 43–45, 50–51, 81, ***87***
Economic development
  dairy industry, 141–47, 149–60
  early fur trade, 58–67
  egg production, 183–85
  employment and labor force, 208–17
  ethanol production, 177–79, ***180***
  flour milling industry, 255
  forest products, 99–103
  future needs of Minnesota, 293–97
  hog production, 185–87
  Indian gaming industry, 72–75, 273
  Iron Range Resources (IRR), 232
  lake and resort country, 287–91
  population and, 198–208
  Quad City Alliance, 232
  turkey production, 181–83
  Twin Cities, 260–61
Edina, MN, 262
Egg production, 183–85
Ely, MN, 219, 222, 232
Embarrass, MN, 299 (n1:1)
Employment and labor force, 208–17
Endangered species, 51, 55, 103
Energy, wind-generation, 36–37, ***38***, 299 (n2:5)
Entertainment industry, Indian gaming, 72–75
Environment
  air and water pollution, 155, 289
  biological diversity and, 54–55
  biome development and, 43

geography and, 1–2, 204, 293, 305
taconite industry and, 230
tree survival and dispersal, 45
Ethanol production, 177–79, ***180***
Ethnic identities and diversity. *See* Minority groups; Population and diversity
Eveleth, MN, 223, 232

Fairmont, MN, 31
Falls of St. Anthony, 40–41, ***41***, 92, 97, 126–27, 251
Faribault, MN, 50, 81, 183
Farms and farmers
average age (1940–2000), ***175***
average size by acres (1880–2000), ***175***
Cottonwood County (1996), ***178***
family farms, 153, 174, 304
family farms, example of, 179, 181
farm-cropland acres, 130–36
farmland (2002), ***131***
land and income for viable, 174–76
numbers sold (2002), ***177***
off-farm employment, 177, ***179***
woodland as % of farmland(2002), ***133***
Fillmore County, 110, ***112–13***
Fishing, forest management for, 102–03
Flour milling industry, 255, 260
Fond du Lac, MN, 60
Forest History Center, 95–96
Forest Product Mills, 2004, ***100***
Forest products, 99–103
Forests. *See* Big Woods; Eastern deciduous forest; Northern coniferous forest
Fort Snelling, 40–41, 66, 108, 251
Fort William, ***60***, 62
Fossils, 28, 43, 226
French-Canadian voyageurs, 58–59
Fur trade, 58–67, 256

Galtier, Lucien (Fr.), 252
Gaming industry, 72–75
General Mills, 255, 265
Geography, understanding, 1–2, 204, 293, 305
Gherty, Jack, 153
Gilbert, MN, 232, 302 (n12:3)
Glacial features, ***30***, ***32–34***
drift and till, 29
erosion and striations, 29
glacial lakes and inland sea, 28, 38–40
ice sheet formation, 28–29
ice-block lakes, 277
outwash deposits, 30
topographic landforms, 31, 35, ***37***
Glacial periods, ***44***
biome development and, 43
deposits and surface features, 26–29
ice advances and retreats, 35–41
lake formation, 26, 30–31
Global warming, 6, 21
Golden Valley, MN, 265
Goodhue County, 52
Goodview, MN, 192
Grand Portage trading post, 59–63
Grand Rapids, MN
Forest History Center, 95–96
Giants Ridge, 25
Iron Range and, 219
lakeshore resorts, 277
Mesabi Trail, 232
pulp and paper mills, 99
Granite Falls, MN, 70
Granite quarries, ***25***
Great Lakes, 58–67, 96–97, 223–24, 234
See also Lake Superior
"Greenhouse effect," 6
Greyhound Bus Company, 225
Gull Lake, 290–***91***

Harrison, Thomas W., 279
Hastings, MN, 192, 254
Hennepin, Louis (Fr.), 40, 251
Hennepin County, 208, 239–40
Hibbing, Frank, 223
Hibbing, MN, 84, 223, 225–26, 232
Hill, James J., 224, 272
Hillsboro, ND, 163
Hinckley, MN, 98
Hispanic population, 155, ***164***, 165, 185, 214, 257, 293
Historical sites
Forest History Center, 95–96
Grand Portage National Monument, 61–63
Hibbing, MN, 225–26
Ironworld Discovery Center, 232
Jeffers Petroglyphs, 29
Mill City Museum, 270
Minnesota History Center, 272
North West Company fur post, ***65–66***
History and heritage
boundaries of Minnesota, ***78***

establishing MN boundaries, 77–79
fur trade, 58–67
Indian land cessions and reservations, ***69***
Indian Wars, 68–70
Land Office locations, ***85***
land survey and settlement, 79–87
logging industry, 89–97
Native Americans, 57–58
Nerstrand Big Woods State Park, ***87***
nostalgia, 115, 206, 208, 216–17
Public Land Survey, ***80–82***
railroad construction, ***109***
Red River trails, ***67***
spread of settlement, ***111***
Hog production, 185–87
Hormel Foods Corporation, 183, 208
Holt, MN, ***215***
Hoyt Lake, 230
Hubbard County, 53, 277
Hudson Bay, 38, 60–63, 160
Hull Rust Mahoning mine, ***227, 228***
Hunting, forest management for, 102–03

Ice-fishing, 16, 287, ***289***
Immigration. *See* Minority groups; Population and diversity; Settlement
Indian gaming industry, 72–75, 273
Indian Gaming Regulatory Act of 1988, 72–75
Indians. *See* Native Americans
Industries
Andersen Windows, 99
Cargill Corporation, 255, 265
Christensen Family Farms, 185–87
Douglas Machine, 212
Durst Brothers Dairy, 153–58
forest product mills, 2004, ***100***
General Mills, 255, 265
Greyhound Bus Company, 225
Hormel Foods Corporation, 183, 208
Jennie-O, 181–83
Land O'Lakes dairy cooperative, 153
Little Pine Dairy, 159–60
Minnesota Mining and Manufacturing (3M), 212, 272–73
Marvin Windows, 99, 208
Northwest Airlines, 265
Pillsbury, 126–27, 255
Polaris, 208
Thomson West, 265
United States Steel Corporation, 222–25, 227
International Falls, MN, 97, 99, ***101***, 208, 299 (n1:1)
Iowa, 36, 78, 114
Ireland, John (Archbishop), ***116***
Iron ore industry. *See* Mining industry; Taconite industry
Iron Range, ***220, 231***
defined, 219
history and heritage, 221–25, 302 (n12:3)
planning for the future, 294–95
population declines, 206
resort economy, 277
role of Duluth, 234–37
towns, 225–26
See also Mining industry; Northeastern Minnesota
Iron Range Resources (IRR), 232, 294–95
Ironton, MN, 233
Isanti County, 240
Itasca State Park, 53, 89

Jennie-O, 181–83

Kanabec County, 277
Kellogg, MN, 54

La Crosse, WI, 105, 108
Lac Qui Parle, 40
Lake Agassiz, 38–40, 160, 251
See also Red River Valley
Lake and resort country
Brainerd Golf Trail, ***286***
Crow Wing County, ***281, 283–84***
demographics, 241
history and heritage, 275–82
in-migration and cohort survival ratio, 284–85
Madden's Resort, Gull Lake, ***291***
planning for future, 295
population and economic growth, 281–83
resort economy of, 289–91
retirement in, 285–87
second homes, ***278***
snowmobile trails, ***288***
water as % of surface area, ***276***
Lake Benton, MN, 36
Lake Calhoun, ***15***, 31
Lake Como, 257
"Lake effect," 6, 12
Lake formation, 26, 30–31, ***33***, ***35***, 277

Lake Harriet, 31
Lake Mille Lacs, 36, 287
Lake Minnetonka, 257, 277
Lake of the Woods, 39, 60, 77–79
Lake Pepin, 40, 58
Lake Phalen, 257
Lake St. Croix, 40
Lake Superior
  determining MN boundaries, 77–79
  Duluth, MN on, 234–37
  glacial meltwater drainage, 38
  "lake effect" temperatures, 6, 12
  North Shore, 53, 291, 295
  as syncline (geologic trough), 26
  taconite industry and, 230
  See also Great Lakes
Lake Traverse, 40
Lake Vermilion, 221
Land O'Lakes dairy cooperative, 153
Land Offices. *See* Public Land Survey
Lincoln County, 37
Litchfield, MN, 183–85
Little Pine Dairy, 159–60
Livestock and poultry production
  dairy farming, 141–47, 149–60
  evolution to mixed crop and, 130, 137–39, 181
  future of Minnesota, 296
  hay as % cropland harvested, ***140***
  hogs, 185–87
  poultry and eggs, 183–85
  turkeys, 181–83
Logging industry
  history and heritage, 89–98
  lakeshore resorts and, 277
  old-growth forest and, 53, 300 (n3:8)
  paper mills and forest products, 99–103
  protecting forest reserves from, 86
  on public lands, 82–84, 97–98
  sawmills, 1885, ***94***
  Twin Cities role in, 254–55, 260
Lowry, Thomas, 257

Mackinac, MI, 59
McKnight, William L., 273
Madden, Jack, 279, 290
Madelia, MN, 214
Magnus family (Clarence, Doug & Brenda), 179, 181
Manganese, MN, 233
Mankato, MN, 40, 50, 70, 78, 108, 192
Mantorville, MN, 153
Manufacturing, 208–15. *See also* Industries
Marine on St. Croix, MN, 90
Marquette, MI, 219
Marschner, Francis J., 46, 80
Marshall, MN, 214
Martin County, 31
Marvin Windows, 99, 208
Mendota, MN, 66–68
Merritt Brothers, 223–25
Mesabi Range. *See* Iron Range
Métis, 65–66
Michigan, 96–97
Mille Lacs County, 277
Mining industry
  federal land laws and, 84
  iron ore mining, 25, 226–27
  non-ferrous ores, 219, 221, 232–33
  sand and gravel, 30–31
  transition to taconite, 227–32
  See also Iron Range
Minneapolis, MN, 92–93, 97, 192–93, ***253***
  See also Twin Cities
Minnesota Historical Society, 62–63, 270
Minnesota History Center, 272
Minnesota Mining and Manufacturing (3M), 212, 272–73
Minnesota River
  Big Woods, 50
  determining MN boundaries, 79
  establishment of Fort Snelling, 66
  as settlement route, 108, 192–93, 206
  spread of settlement to, 110
  Twin Cities town sites on, 39–41, 251–54
Minnetonka, MN, 265
Minority groups
  African American, 257
  Asian, 119, 214
  Hispanic, 155, ***164***, 165, 185, 214, 257, 293
  Hmong, 257, 293
  racial tension toward, 214
  Somali, 257, 293
  Twin Cities population, 257–***59***
  See also Native Americans; Population and cultural diversity
Mississippi River
  Big Woods, 50
  determining MN boundaries, 77–78
  erosion and geologic activity, 28
  fur trade routes, 60

headwaters at Itasca State Park, 53
logging industry on, 90–92
as settlement route, 105–08, 192–93
Twin Cities town sites on, 251–54
Missouri River, 29, 77–78
Moorhead, MN, 7, 93, 163, 206
Morrill Act, 83
Morris, Larry, 162–63
Morton, MN, 40
Mountain Iron, MN, 223–24, 226, 232
Mountains, formation of, 25, 28
Myrick, Andrew, 70

National forests, 86
National Park Service, 63
Native Americans
Dakota (Sioux) heritage, 57–58
early fur trade, 58–67
Indian Gaming Regulatory Act of 1988, 72–75
Indian land cession and, 79
Jeffers Petroglyphs site, 29
land cessions and reservation era, 68, ***69***, 72, 77, 79, 105
Minnesota population (2000), ***71***
Ojibwe (Chippewa) heritage, 58
settlement and the Indian Wars, 70
Natural Areas. *See* Parks, state parks, and natural areas
Nerstrand Big Woods State Park, 51, 81, ***82***
New Ulm, MN, 70, 117
North Shore. *See* Lake Superior
North West Company trading post, 63–66
Northeastern Minnesota, 35–36, 84–86
See also Iron Range
Northern coniferous forest, 43–45, 52–53
Northfield, MN, 50
Northwest Airlines, 265
Northwest Angle, 77–78
Norwood, MN, 192

Oliver, Henry W., 224
Olmsted County, 212
Olson, Earl B., 181–83
Ordway, Lucius, 273
Otter Tail County, 277
Owatonna, MN, 110, 206

Park Rapids, MN, 277
Parks, state parks, and natural areas
Bluestem Prairie, 48
Hill Annex Mine, 226
Kellogg-Weaver Dunes, 54
Nerstrand Big Woods State Park, 51, 81, ***82***
Prairie Coteau, 48
Parrant, Pierre "Pig's Eye," 252
Pelican Lake, 290–91
Pembina, ND, 66, 79
Perham, MN, 159
Pierce County, WI, 240
Pigeon River, 60, 77
Pillsbury, John S., 255
Pillsbury Corporation, 126–27, 255
Pine City, MN, 63, 65–66
Pine County, 277
Pipestone County, 48
Polaris Industries, 208
Politics
early statehood, 114
mining industry, 223–25
planning for the future, 294
of small-farm economics, 149–53
of subsidies to agriculture, 295
of sugar quotas, 161–62
Population and cultural diversity
creation of trade centers, 189–92
declining farm numbers, 174–77
demographics, 241
employment and industry, 208–17
future needs of Minnesota, 293–94
Hispanic, 155, ***164***, 165, 185, 257, 293
immigration, 114–19, 301 (n7:3)
migrations by ethnic group, ***116***, ***118***
Native American, ***71***
place of birth, ***115***
population of Fillmore County, ***112–13***
role of rivers and railroads, 193–97
small town growth and decline, 198–208
"snowbird" phenomenon, 16, 282–83, 306
Twin Cities competition, 256
See also Minority groups
Population demographics
age (2000), ***245***
cohort survival ratios, ***284***
county, % change (1990–2000), ***243***
density (2000), ***243***
distribution (2000), ***242***
education (2000), ***246***
families in poverty (1999), ***248***
Hispanic (2000), ***164***
income (1999), ***247***

in-migration increase (1990–2000), ***244***
Native American (2000), ***71***
Poultry production. *See* Livestock and poultry production
Prairie du Chien, WI, 60, 66, 108
Prairie Island, MN, 70
Precipitation, ***17***
Public Land Survey
determining MN boundaries, 77–81
Homestead Act of 1862, 84–86
homesteading plat map, ***82***
Indian land allotments, 72
Indian land cessions, 105
Ordinance of 1787 grid system, 79–81
Preemption Act of 1841, 83–84
vegetation at time of survey, 46, ***47***

Quad City Alliance, 232

Radisson, Pierre Esprit de, 57
Railroads
construction, 108–10
conversion to recreational use, 232, 287
logging industry and, 93
in-migration and settlement, 117, 119
mining industry, 221, 224–25
modern-day unit trains, ***168***
population growth and, 193–97
public land grants, 83
role in Twin Cities, 255–56
Rainy River, 60, 63, 77, 97, ***101***
Ramsey, Alexander, ,105
Ramsey County, 208, 239–40
Recreation industry. *See* Lake and resort country; Tourism
Red Lake (Upper and Lower), 39
Red River Valley
agricultural diversity, 160–61
Duluth harbor and, 234–35
fur trade routes, 60, 65–66, ***67***
glacial advance and retreat, 36
planning for the future, 295
sugarbeet industry, 161–65
wheat in, 122, 125, 135
See also Lake Agassiz
Red Wing, MN, 70, ***107***, 192
Redwood Falls, MN, 70
Religion, migration and settlement, 114, 117, 256
Renville, MN, 163
Retirement. *See* Lake and resort country
Rice County, 52
Riverton, MN, 233
Rochester, MN, 193, 204–05, 208, 241
Rolette, Joe, 79
Root River, 105
Roseau, MN, 208
Roseau County, 204
Rum River, 92

Saginaw, MI, 97
Sayer, John, 63–65
Scott County, 239–40
Settlement
Crow Wing County, 281
cultural and ethnic migrations, ***116***, ***118***
employment and labor force, ***209–11***, 213
Fillmore County, ***112–13***
Homestead Act of 1862, 84–86
impact on forest biome, 50–51
Indian land cession and, 68, 77, 79, 105
land survey, 79–83
of logged and cutover land, 98
migrations, 110–19
Morrill Act of 1862, 83
population gains and declines, ***198–200***, ***205–07***
Preemption Act of 1841, 83–84
railroad construction and, 108–10
rivers as routes for, 105–08
town design, ***196***
town size, ***194–95***, ***201–02***, ***204***
trade center (town) hierarchy, ***191***
Shakopee, MN, 70, 75
Sherburne County, 240–41
Silver Bay, MN, 230
Sims, Ginny, 290
Skiing, 16, 102–03, 232
Skyline, MN, 192
Slayton, MN, 179
Sleepy Eye, MN, 185–87
Snake River, 90
"Snowbirds," 16, 282–83, 306
Snowmobiling, 16, 102–03, 208, 287–***88***
Soudan Underground Laboratory and Mine, 221–22
Soybean production, 135, 163, 167–74, 177, 181, 187, 296
Sparboe, Bob, 183–85
Spizzo, Bob, 290
St. Anthony, MN, 254

St. Anthony Falls. *See* Falls of St. Anthony
St. Cloud, MN, 50, 78, 117, 240–41
St. Croix County (Wisconsin), 240
St. Croix Dalles, 39
St. Croix River
  determining MN boundaries, 78
  fur trade routes, 60
  as "glacial spillway," 39–40
  logging industry on, 90, 97
  as settlement route, 192–93
St. Lawrence Seaway, 227, 234
St. Louis County, 234
St. Louis, MN, 66
St. Louis River, 77, 234–***35***
St. Paul, MN, ***108, 253***
  determining MN boundaries, 78–79
  early fur trade, 66, 256
  railroads, 93, 108
  as river navigation hub, 108
  as state capital, 254, 272, ***273***
  as urban center, 192–93
  See also Twin Cities
St. Peter, MN, 79
Start, C. H., 279
State Parks. *See* Parks, state parks, and natural areas
Stearns County, 117, 241
Steele County, 52
Stillwater, MN, 90–91, 192, 254
Stone, George C., 221
Sugar beet production, 160–65, ***162, 163,*** 295
Superior, WI, 224, 234
Swift County, 212

Taconite industry, 226–32
  See also Iron Range; Mining industry
Tallgrass prairie, 43–50
Taylors Falls, MN, 39
Temperature. *See* Weather and climate
Tenney, MN, 192
Thief River Falls, MN, 208
Thoen, Bud, 212
Thomson West, 265
Tobkin, Ron, 159–60
Topography and vegetation
  biomes of Minnesota, ***44***
  glacial deposits and surface features, ***27***
  topographic relief, ***24***
  vegetation at time of Public Land Survey, ***47***
Tourism
  Duluth, MN, 237
  "Fall color" as attraction, 19
  forest management for, 102–03
  Iron Range Resources (IRR) and, 232
  seasonal and climate impact on, 16, 21
  See also Lake and resort country
Tower, Charlemagne, 221
Tower, MN, 219, 221, 299 (n1:1)
Traverse des Sioux, MN, 68
Trees. *See* Big Woods; Eastern deciduous forest; Northern coniferous forest
Trommald, MN, 233
Turkey production, 181–83
Twin Cities
  defined, 239–41
  history and heritage, 251–57
  industry and economic development, 260–61
  Interstate Highway system, ***263***
  labor force, 208
  lumber and flour milling (1880–1940), ***260***
  Mall of America, ***264–65***
  Metropolitan area, ***240, 255, 262, 267***
  Minneapolis downtown, ***269***
  minority groups (2000), ***259***
  population and cultural diversity, 192, 204–05, 257–59
  town sites, ***252–53***
  urban and suburban development, 261–73
  See also Minneapolis, MN; St. Paul, MN
Two Harbors, MN, 272

United States Steel Corporation, 222–25, 227
University of Minnesota, 227, 237, 254, 271
U.S. Public Land Office, 46

Vadnais Lake, 31
Vegetation at Time of Public Land Survey, 46, ***47***
Vermilion Range. *See* Iron Range
Virginia, MN, 97, 219, 225, 232
Volcanoes and geologic activity, 23, 25–26
Voyageurs, fur trade industry, 58–67

Wabasha County, 54
Wadena, MN, 35
Wahpeton, ND, 163, 165
Walker, MN, 277
Walker, Thomas B., 270
Warroad, MN, 99, 208
Washburn, Cadwallader, 255

Washington County, 212, 239–40
Weather and climate
  evapotranspiration, 48
  fronts, 18
  global warming, 21
  growing season, length of, ***10–11***
  impact on biological diversity, 43–46, 54–55
  jet stream, 18, 305
  precipitation and humidity, 7, 12, ***13***, 17, ***49***
  seasonal variability, 6
  snowfall and snow cover, ***14***, ***17***
  temperature, 5–7, ***8–11***
  thunderstorms, 19
  tornadoes, 18, ***19***
  winter, 12, 17, 19, 21
Wheat production, 121–29, 161
  cropland as % harvested, ***135–36***
  Duluth harbor shipping and, 234
  flour milling industry, 255, 260
  gristmills, ***129***
  peak year, ***124***
  Red River Valley, 122–25, 135
  total planted, ***123***
  when wheat was "King," 121–29, 161
White Bear Lake, 257, 277
White Earth Reservation, 70, 72
Wildflowers, 51–***52***
Willmar, MN, 181, 206
Winchell, Newton H., 40–41
Wind energy, 36–37, ***38***, 299 (n2:5)
Winona, MN, 92, 108, 126, 192, 206
Winston, MN, 221
Wirth, Theodore, 257
Wisconsin
  dairy industry, 153
  determining MN boundaries, 78
  fur trade routes, 60
  glacial advance and retreat, 36–37
  logging industry, 92
  population migration from, 114
  turkey production, 183
Wright County, 240

Young America, MN, 192

Zumbro River, 105

CREDITS

The front cover aerial photograph is courtesy Bob Firth; the photographs on page ii and 289 are © Explore Minnesota Tourism; the photographs by David Brewster on page vi, 19 (top), and 38 are from the *Minneapolis Star Tribune,* May 31, 1998, and December 27, 2005, by permission of the *Star Tribune;* the photographs on pages viii, 229, and 231 (bottom) are courtesy Cleveland-Cliffs; the map on page 27 was generalized from electronic data of the *Geologic Map of Minnesota: Quarternary Geology,* Minnesota Geological Survey and the Land Management Information Center; the diagrams on page 30 are reproduced from Vernor C. Finch and Glenn T. Trewartha, *Physical Elements of Geography*, 3rd ed. (New York: McGraw-Hill, 1949), p. 301, by permission of McGraw-Hill Book Company; the maps on pages 36 and 37 are reproduced from P. K. Sims and G. B. Morey, eds., *Geology of Minnesota: A Centennial Volume* (St. Paul: Minnesota Geological Survey, 1972), p. 520–22 and 551, by permission of the Minnesota Geological Survey; the diagram on page 41 is based on a drawing in R. K. Hogberg, *Environmental Geology of the Twin Cities Metropolitan Area*, Educational Series 3 (St. Paul: Minnesota Geological Survey, 1971), p. 50, by permission of the Minnesota Geological Survey; the map on page 44 is modified from a map in *Managing Landscapes in the Big Woods Ecosystem* (St. Paul: Minnesota Department of Natural Resources, n.d.), by permission of the Department of Natural Resources; the map on page 47 is based on a map in Keith M. Wendt and Barbara A. Coffin, *Natural Vegetation of Minnesota at the Time of the Public Land Survey, 1847–1907,* Biological Report No. 1 (St. Paul: Minnesota Department of Natural Resources, 1938), by permission of the Department of Natural Resources; the drawing on page 61 (bottom) is reproduced and the plan on page 62 is redrawn by permission of the Superintendent of the Grand Portage National Monument; the map on page 64 is based on the map in Grace Lee Nute, "Posts in the Minnesota Fur-trading Area, 1660-1855," *Minnesota History* 11 (1930): 355; the map on page 67 is compiled from maps in Rhoda R. Gilman, Carolyn Gilman, and Deborah M. Stulz, *The Red River Trails: Oxcart Routes between St. Paul and the Selkirk Settlement, 1820–1870* (St. Paul: Minnesota Historical Society, 1979); the map on page 69 is reproduced from June Drenning Holmquist, Ed., *They Chose Minnesota: A Survey of the State's Ethnic Groups* (St. Paul: Minnesota Historical Society Press, 1981), p. 19; the photograph on page 74 is reproduced by permission of Jackpot Junction Casino Hotel; the diagram on page 81 and the map on page 85 are from Hildegard Binder Johnson, *Order Upon the Land: The U. S. Rectangular Survey and the Upper Mississippi Country* (New York: Oxford University Press, 1976), p. 58 and 120, by permission of Oxford University Press; the map on page 82 is from *Plat Book of*

Rice County: Compiled from County Records and Actual Surveys (Philadelphia: Northwest Publishing Co., 1900), p. 12; the map on page 94 is based on a map in Edward Van Dyke Robinson, *Early Economic Conditions and the Development of Agriculture in Minnesota*, Studies in the Social Sciences, No. 3, *Bulletin of the University of Minnesota* (Minneapolis: University of Minnesota, 1915), p. 112; the photograph on page 99 is reproduced by permission of UPM-Blandin; the photograph on page 101 is reproduced by permission of Boise Paper, International Falls, MN; the drawing on page 102 is reproduced by the permission of its artist, James A. Mattson, North Central Forest Experiment Station, U.S. Department of Agriculture, Forest Service, Houghton, Michigan; the map on page 109 is modified from maps in Harold Austin Meeks, *The Growth of Minnesota Railroads 1857-1957*, unpublished Plan B Master of Arts paper, Department of Geography, University of Minnesota, 1957; the map on page 118 is redrawn from an unpublished map by Sandra Haas, John Rice, and Jon Walstrom, *Culture Groups in Rural Minnesota at the Turn of the Century* (Minneapolis: University of Minnesota, 1979); the map on page 129 is based on a map in Edward Van Dyke Robinson, *Early Economic Conditions and the Development of Agriculture in Minnesota*, Studies in the Social Sciences, No. 3, *Bulletin of the University of Minnesota* (Minneapolis: University of Minnesota, 1915), p. 112; the photograph on page 156 (bottom) is reproduced with the permission of Ron Durst; the map on page 191 is based on William Casey, *Trade Centers of the Upper Midwest: 1999 Update* (Minneapolis: Center for Urban and Regional Affairs, University of Minnesota, 1999), p. 13; the map on page 195 is from the *CURA Reporter*, October 1989, p. 3, by permission of the Center for Urban and Regional Affairs, University of Minnesota; the diagram on page 196 is reproduced from John C. Hudson, "Towns of the Western Railroads" *Great Plains Quarterly*, Winter 1982, 2(1): 47, by permission; the graphs on page 198, 199, 201, 205, and 207 and the map on page 242 are from the *CURA Reporter*, Summer 2002, p. 9, 10, 11, and 19, by permission of CURA; the map on page 220 is based on a map in Richard W. Ojakangas and Charles L. Matsch , *Minnesota's Geology* (Minneapolis: University of Minnesota Press, 1982), p. 36, by permission of the University of Minnesota Press; the map on page 231 (top) is modified from a map prepared by Iron Range Resources, State of Minnesota, by permission; the map on page 252 is from Nevin M. Fenneman, *Physiography of Eastern United States* (New York: McGraw-Hill, 1938), p. 585, by permission of McGraw-Hill Book Company; the graph on page 260 is reproduced from Calvin F. Schmid, *Social Saga of Two Cities: An Ecological and Statistical Study of Social Trends in Minneapolis and St. Paul* (Minneapolis: Bureau of Social Research, Minneapolis Council of Social Agencies, 1937); the map on page 262 is reproduced from Ronald Abler, John S. Adams, and John R. Borchert, *The Twin Cities of St. Paul and Minneapolis* (Cambridge, Massachusetts: Ballinger Publishing Company, 1976), p. 32, by permission; the photograph on page 291 is reproduced with the permission of Madden's Resort; the photograph on page 297 is by Chris Gregerson/cgstock.

The images on page 65 (David Geisler), 66 (bottom) (Steve Woit), 91, 92, 93, 95, 96 (Dave Bjork), and 108 are courtesy of the Minnesota Historical Society.

All the other photographs are courtesy of the authors. Maps and graphs not otherwise attributed were prepared under the direction of Mark B. Lindberg, University of Minnesota Cartography Laboratory.